NEUROHUMORAL CODING OF BRAIN FUNCTION

ADVANCES IN BEHAVIORAL BIOLOGY

Volume 1 • BRAIN CHEMISTRY AND MENTAL DISEASE
Edited by Beng T. Ho and William M. McIsaac • 1971

Volume 2 • NEUROBIOLOGY OF THE AMYGDALA
Edited by Basil E. Eleftheriou • 1972

Volume 3 • AGING AND THE BRAIN
Edited by Charles M. Gaitz • 1972

Volume 4 • THE CHEMISTRY OF MOOD, MOTIVATION, AND MEMORY
Edited by James L. McGaugh • 1972

Volume 5 • INTERDISCIPLINARY INVESTIGATION OF THE BRAIN
Edited by J. P. Nicholson • 1972

Volume 6 • PSYCHOPHARMACOLOGY AND AGING
Edited by Carl Eisdorfer and William E. Fann • 1973

Volume 7 • CONTROL OF POSTURE AND LOCOMOTION
Edited by R. B. Stein, K. G. Pearson, R. S. Smith, and J. B. Redford • 1973

Volume 8 • DRUGS AND THE DEVELOPING BRAIN
Edited by Antonia Vernadakis and Norman Weiner • 1974

Volume 9 • PERSPECTIVES IN PRIMATE BIOLOGY
Edited by A. B. Chiarelli • 1974

Volume 10 • NEUROHUMORAL CODING OF BRAIN FUNCTION
Edited by R. D. Myers and René Raúl Drucker-Colín • 1974

A Continuation Order Plan is available for this series. A continuation order will bring delivery of each new volume immediately upon publication. Volumes are billed only upon actual shipment. For further information please contact the publisher.

NEUROHUMORAL CODING OF BRAIN FUNCTION

Edited by
R. D. Myers
Laboratory of Neuropsychology
Purdue University
West Lafayette, Indiana

and
René Raúl Drucker-Colín
Universidad Nacional
Autónoma de México
México D.F. México

PLENUM PRESS • NEW YORK AND LONDON

Library of Congress Cataloging in Publication Data

International Symposium on the Neurohumoral Coding of Brain Function, Mexico, 1973.
Neurohumoral coding of brain function.

(Advances in behavioral biology, v. 10)
Includes bibliographies.
1. Brain chemistry—Congresses. 2. Neural transmission—Congresses. I. Myers, Robert Durant, ed. II. Drucker-Colín, René Raúl, ed. III. Title. [DNLM: 1. Brain—Physiology—Congresses. 2. Neurohumors—Physiology—Congresses. W2 AD215 v. 10 1973/WL300 I6125n 1973.]
QP356.3.I57 1973 612'.822 74-7408
ISBN 0-306-37910-4

Proceedings of the International Symposium on the Neurohumoral Coding of Brain Function held in Mexico City, June 26-27, 1973

This symposium was co-sponsored by
THE AMERICAN ASSOCIATION FOR THE ADVANCEMENT OF SCIENCE,
CONSEJO NACIONAL DE CIENCIA Y TECNOLOGIA (CONACYT)
and the MEXICAN PHARMACOLOGICAL SOCIETY

Additional financial support for this symposium was kindly provided by Miles Laboratories, Inc. (Elkhart, Indiana), Smith, Kline and French Laboratories (Philadelphia, Pennsylvania), and the Upjohn Company (Kalamazoo, Michigan) to each of whom we are all deeply indebted.

A Division of Plenum Publishing Corporation
227 West 17th Street, New York, N.Y. 10011

United Kingdom edition published by Plenum Press, London
A Division of Plenum Publishing Company, Ltd.
4a Lower John Street, London W1R 3PD, England

Printed in the United States of America

Introduction

It is indeed a pleasure to welcome all of you to this International Symposium on the Neurohumoral coding of Brain Function. Many of you have undertaken a very long trip in order to cross swords with some of the most fascinating issues in all of the neurosciences. Of particular satisfaction in this instance is the geographical representation of the sciences with individuals here from Europe and the Americas - South, Central and North.

As we do battle, so to speak, with each of the questions raised during the next several days, we should remember that the problems faced by each of our fields are exceptionally difficult. In a way, this difficulty stems from two related facts: (1) we are trying to construct a conceptual bridgework between one discipline and another; and (2) the distance between the research fabric of any two of these disciplines is vast.

It would probably not be unfair to say that a large number of scientists feel relatively contented in remaining within the boundaries of their own area of specialization. In a sense, that is certainly justifiable today primarily because of the intensity of such specialization. However, the participants of this symposium, who reflect some of the major thrusts in biochemistry, physiology, pharmacology and psychology, have in the main chosen to explore the pathways that cross the interface between mind and body - between behavior and brain.

To answer the question pertaining to the meaning of the title of this Symposium will require a reading of the papers that follow. The reason is simple: each one of us envisages the coding or rather a coding process in somewhat of a different way. Nevertheless, I think that you would all agree that at the basis of the continuum between each functional process and the molecular events that take place within the brain is the neurohumoral or neurochemical activity that persists as long as life is sustained. Less than fifty years ago, the span of this continuum was huge; in fact, how a tiny portion of behavior could be coded by a chemical mechanism was not even in the realm of conceptualization. Now, the goal of this Symposium is to highlight some of the points of resolution along the continuum and to show how research in all spheres is yielding an ever finer-grained analysis. As we learn about the remarkable findings that are before us today, we shall see that the degree of resolution between the molecular and the functional has certainly moved a quantum to the right - a motion toward the end-point of ultimate explanation.

We should like to express our very warm thanks to our kind hosts, CONACYT and the AAAS, for providing the opportunity for us to come together. We are equally indebted to the management and scientists of Miles Laboratories Inc. (Elkhart, Indiana), Smith, Kline and French Laboratories (Philadelphia, Pennsylvania) and the Upjohn Company (Kalamazoo, Michigan) without whose timely and generous financial support this Symposium would not have been possible.

June, 1973

R. D. Myers
Lafayette, Indiana

En los últimos años se han visto enormes avances en el estudio de la bioquímica de las funciones cerebrales. Por esta razon pensamos que sería oportuno organizar un Simposio el cual tratara de sintetizar algunos aspectos del papel que juegan los diversos neurotransmisores sobre lo fisiología y la conducta tanto normal como anormal. Para esto hemos invitado a un selecto numero de científicos tanto nacionales como extranjeros quienes no solamente han pasado la mayor parte de su vida científica estudiando problemáticas en este campo, sino que ademas han logrado hacer contribuciones de gran importancia.

Este Simposio se realizo gracias al patrocinio del Consejo Nacional de Ciencia y Tecnología de México (CONACYT) y al American Association for the Advancement of Science (AAAS) la cual cumplio su 125° aniversario. Queremos asimismo expresar nuestro agradecimiento a los Laboratorios Miles, Smith, Kline & French y Upjohn quienes gracias a su visión futurista tambien patrocinaron en forma importante este Simposio.

Finalmente quisiera expresar que es un gran placer para nosotros que este evento se lleve a cabo en nuestro país, pues fue precisamente un científico mexicano el Dr. Raúl Hernández-Peón uno de los pioneros en este campo de la ciencia. Las contribuciones del Dr. Raul Hernández-Peón no solo tuvieron impacto dentro de nuestras fronteras sino también fuera de ellas, y habiendo tenido la fortuna de haber sido uno de sus alumnos me es muy grato poder dedicar esto Simposio a su memoria, a escasos seis años de su muerte.

René Raúl Drucker-Colín
México City

June, 1973

Contents

PART IV: NEUROHUMORAL FACTORS AND MENTAL DISORDERS

PART I
MOLECULAR CORRELATES OF BRAIN FUNCTION

THE ROLE OF γ-AMINOBUTYRIC ACID METABOLISM IN THE REGULATION OF CEREBRAL EXCITABILITY

Ricardo Tapia

Instituto de Biología, Universidad Nacional Autónoma de México, México 20, D. F., México

INTRODUCTION

The chemical transmission of nerve impulses across the synaptic gap is probably the most important event in the function of the nervous system, since this phenomenon is a determinant factor for the existence of unidirectional, specific, regulated and plastic communication between neurons. Thus, it can be said that the chemical synaptic transmission is the clue event for the function of the different neuronal circuits underlying the diverse physiological roles of the nervous system.

The recognition that the communication between neurons can be positive (excitation of the postsynaptic neuron) or negative (inhibition of the postsynaptic neuron) has brought to the attention of many investigators the possibility that the activity of the central nervous system is controlled not only by an active excitation of neurons but also, and perhaps more effectively, by inhibition of neurons (Roberts *et al.*, 1960; Florey, 1961; von Euler *et al.*, 1968). From a biochemical point of view, excitatory or inhibitory communication means the release of an excitatory or an inhibitory chemical transmitter, respectively, and its combination with a specific receptor present in the postsynaptic membrane. According to the present concepts, when an excitatory neuron is depolarized, a specific excitatory transmitter is liberated from its axon terminals, and by combining with the receptor in the postsynaptic neuron, the transmitter will produce a depolarization (excitatory postsynaptic potential) that, if it reaches the threshold for the firing of that cell, will trigger a propagated potential. Repolarization of the presynaptic neuron by stopping the stimulus will stop the release of the transmitter and, since there is a mechanism for the elimination of the transmitter (enzymatic breakdown, uptake or diffusion), the postsynaptic neuron will be repolarized. The same sequence of presynaptic events seems to occur in the case of an inhibitory neuron: when depolarized, it will liberate the inhibitory transmitter, which will interact with the specific receptor of the postsynaptic neuron. However, in contrast to the excitatory transmitter, it will hyperpolarize the postsynaptic membrane (inhibitory postsynaptic potential), thus increasing the depolarization necessary for attaining a value sufficient for the production of the propagated potential in the postsynaptic neuron. When the presynaptic inhibitory neuron is repolarized, the inhibitory transmitter will not be released any more, and the postsynaptic neuron will return to the normal state of polarization after the elimination of the remaining inhibitory transmitter (Eccles, 1964; Auerbach, 1972; Hall, 1972).

These generally accepted events imply that, unless the presynaptic neuron is excited, that is, the nerve ending membrane is depolarized, the synaptic transmitter is not released into the synaptic cleft. This means that both the excitation and the inhibition are phasic, transient phenomena, dependent upon the depolarization of the presynaptic neuron. Furthermore, at least for acetylcholine and probably for catecholamines, the release seems to occur in packets (or "quanta") (Hall, 1972; Auerbach, 1972).

In the present paper, some evidence will be presented which suggests that, in addition to the mechanism described above for the release of the synaptic transmitters, in the central nervous system of mammals there seems to exist a continuous release (not in packets), independent of the state of depolarization of the presynaptic membrane, at least for one type of inhibitory neuron, namely the γ-aminobutyric acid (GABA) neurons. In this way, a continuous, tonic inhibition would be exerted by these neurons.

SYNAPTIC TRANSMISSION AND STORAGE OF THE TRANSMITTER

According to the foregoing discussion, and assuming that when a transmitter is liberated it necessarily interacts with the receptor, the central core of the whole process of synaptic transmission is the release of the synaptic transmitter into the synaptic cleft. The pool from which the transmitter is released can be a storage site, such as the synaptic vesicles, or a metabolic pool not necessarily limited by a structural membrane. An example of the first type of transmitter would be acetylcholine, which is most probably stored in synaptic vesicles and released in packets (Auerbach, 1972), whereas norepinephrine is an example of a transmitter released from a newly synthesized pool, which is nevertheless considered as a storage form (Thierry *et al.*, 1971; Kety, 1970; Glowinski *et al.*, 1972). In any case, the depolarization of the presynaptic membrane appears to be an absolute requirement for the release of the synaptic transmitter.

From this model we can conclude that under "resting" conditions of a particular neuronal system using a specific synaptic transmitter, we could block the synthesis of the transmitter and no immediate effect would be noticed on the function of such neuronal system. Only after depletion of the available transmitter in the storage site, *or* when the system is subjected to stimulation (that is, when it is in active state as opposed to the "resting" conditions) the physiological manifestation of the impairment of the normal functioning of the neuronal system involved would be apparent. This is the case with some transmitters as catecholamines or serotonin, since we can inhibit their synthesis with drugs like α-methyl-*p*-tyrosine, *p*-chlorophyenylalanine or α-methyl-DOPA, and no *immediate* physiological alteration produced by the drugs is noticed in the absence of stimulation (Bloom and Giarman, 1968; Kety, 1970; Tapia and Pasantes, 1971; Aprison and Hingtgen, 1972). In other words, in order to observe the physiological alteration produced by the drug it is necessary to study a parameter that requires some kind of stimulation, even for newly synthesized functional pools of the transmitter (Glowinski *et al.*, 1972).

In the following part of the paper, I will give evidence for the hypothesis that in the case of the neuronal system involved in the motor activity of mammals, the motoneurons are held in check by a continuous release of GABA under normal, "resting" conditions. From this evidence I will present a model in which the release of at least a particular functional pool of GABA is not dependent upon the stimulation of the GABA-neuron, but it is continuously released immediately after being synthesized. Blocking the synthesis of GABA, without any kind of stimulation, results in the manifestation of a physiological impairment, which in this case is convulsive activity, immediately after reaching what appears to be a critical level in the rate of GABA synthesis.

GABA, GLUTAMATE DECARBOXYLASE (GAD) AND CONVULSIONS

If we directly stimulate the cortical motoneurons by applying an electroshock to experimental animals, we will observe convulsions. Such an effect can be explained because of a depolarization of some excitatory neurons produced by the electric current. We therefore can assume that under normal conditions the motoneurons are not firing maximally, which can be due to any of these three reasons: they lack the excitatory impulses from other excitatory neurons, they are being continuously inhibited by some inhibitory neurons, or a combination of these two factors. We can assume that the first possibility manifests itself when we give an electrical stimulus, or when a muscle is contracted as a consequence of a stimulus applied to a receptor, as in the case of the reflex arc, or even perhaps when a muscle is voluntarily contracted. To demonstrate the existence of the second possibility, we should block the action of the inhibitory neurons, without any other stimulus, and observe the presence of convulsions due to the released activity of the motoneurons. The impairment of the action of the involved inhibitory transmitter can be carried out through two different mechanisms, which would be expected to produce convulsions: a) blocking of the postsynaptic receptor, or b) blocking the release of the transmitter. In the case of GABA, which is now considered to be an inhibitory transmitter in the central nervous system of mammals, including the cerebral cortex (Krnjević and Schwartz, 1967, 1968; Obata _et al._, 1967; Obata and Takeda, 1969; Curtis _et al._, 1970; Otsuka _et al._, 1971; Iversen _et al._, 1971), it is now possible to block its receptor with bicuculline (Curtis _et al._, 1970a, 1970b; Curtis _et al._, 1971), and convulsions occur (Curtis _et al._, 1970a; Meldrum and Horton, 1971; Pérez de la Mora and Tapia, 1973). There is presently no available method for blocking directly the release of GABA.

Table 1. Changes of GABA levels and of the activity of GAD and GABA aminotransferase (GABA-AT) in brain of animals treated with various convulsant drugs and sacrificed at the moment of convulsions

	Per cent of control				
	GAD	GABA-AT	GABA	Species	Reference
Anthranilic hydroxamic acid (see Fig. 1)	57	-	56	Rat	Utley (1963)
High oxygen pressure (10 min)	75	100	81	Rat	Wood *et al.* (1966)
Thiosemicarbazide (45 min)	74	109	87	Rat	Wood *et al.* (1966)
Thiosemicarbazide (15 min)	80	98	77	Chick	Wood and Abrahams (1971)
Isonicotinyl hydrazide (60 min)	63	86	75	Chick	Wood and Peesker (1972)
Pyridoxal phosphate-γ-glutamyl hydrazone (PLPGH) (40 min)	58	100	66	Mouse	Tapia *et al.* (1967a), Tapia and Awapara (1969)
Allylglycine (120 min)	75	92	40	Rat	Alberici *et al.* (1969)
1,1-dimethyl-hydrazine (90 min)	25	80	63	Rat	Medina (1963), Minard (1967)
Mercaptopropionic acid (15 min)	79[†]	148[†]	57[†]	Rat	Rodríguez de L. A. *et al.* (1972)
Aminooxyacetic acid (15 min)	37	0	141	Mouse	Tapia *et al.* (1967b)
L-glutamic acid-γ-hydrazide (GAH) (180 min)	18	49	575	Mouse	Tapia *et al.* (1967b)

* The time between the administration of the drug and the occurrence of convulsions is indicated in parentheses. The doses varied from 25 mg/kg (thiosemicarbazide) to 2 g/kg (GAH).

† Data obtained in cerebellum.

However, when the rate of GABA synthesis is decreased by inhibition of glutamate decarboxylase (GAD, EC 4.1.1.15) activity (the one-step pathway of GABA synthesis), convulsions occur. The experimental evidence leading to this conclusion will be examined next.

Inhibition of GAD Activity and Convulsions

When the activity of brain GAD is inhibited in vivo by the administration of several substances or by high oxygen pressure, the animals show convulsions (Table 1). Although thiosemicarbazide may be an exception because it possibly acts also through a different mechanism (Baxter, 1970; Sze et al., 1971), the extent of inhibition of GAD activity necessary to produce convulsions varies from 21% to 82%. However, it should be noted that only when GABA aminotransferase (EC 2.6.1.19) (the enzyme that metabolyzes GABA) was also inhibited to a certain degree (aminooxyacetic acid, L-glutamic-γ-hydrazide (GAH), 1,1-dimethylhydrazine), GAD activity could be blocked more than 45%. This can be interpreted to mean that when the only effect of a drug on GABA metabolism is a decrease of its synthesis, there seems to be a threshold value of GAD activity (about 60% of the normal value), below which seizures will occur, whereas this value must be lower when the catabolism of GABA is also inhibited. It is interesting in this respect that in a study of the action of six different derivatives of pyridoxal phosphate (PLP), with similar chemical structures, three of them produced convulsions and inhibited GAD activity, while the other three did not show any of these effects (Table 2).

These results are consistent with the existence of a causal relationship between GAD inhibition and convulsions, but they do not prove it. That GAD inhibition is probably related to the appearance of convulsions is more readily evidenced in the following data.

Changes of GABA Synthesis with Time after the Administration of Convulsant Drugs

When we analyse in more detail the effect of some of the convulsant substances shown in Table 1 (Fig. 1), it will be noticed that convulsions occur only when GAD activity reaches a certain

Table 2. Changes of pyridoxal kinase activity, PLP levels and GAD activity in brains of mice treated with several derivatives of PLP and sacrificed at the moment of convulsions or at the corresponding time *

Drug†	Per cent decrease				
			GAD		
	Pyridoxal kinase	PLP	No PLP	PLP‡	Convulsions
PLPGH	36	56	42	4	Yes (30-45 min)
PLP-unsubstituted hydrazone	32	60	48	7	Yes (30-45 min)
PLP-monomethyl hydrazone	19	55	40	0	Yes (30-45 min)
PLP-isonicotinyl hydrazone	0	7	12	4	No
PLP-thiosemi-carbazone	0	7	1	0	No
PLP-oxime-O-acetic acid	-	-	6	0	No

* Data from Tapia and Awapara (1969) and Tapia *et al*. (1969).

† All derivatives were injected at a dose of 220 μmoles/kg.

‡ PLP added *in vitro* at a 0.92×10^{-4} M final concentration.

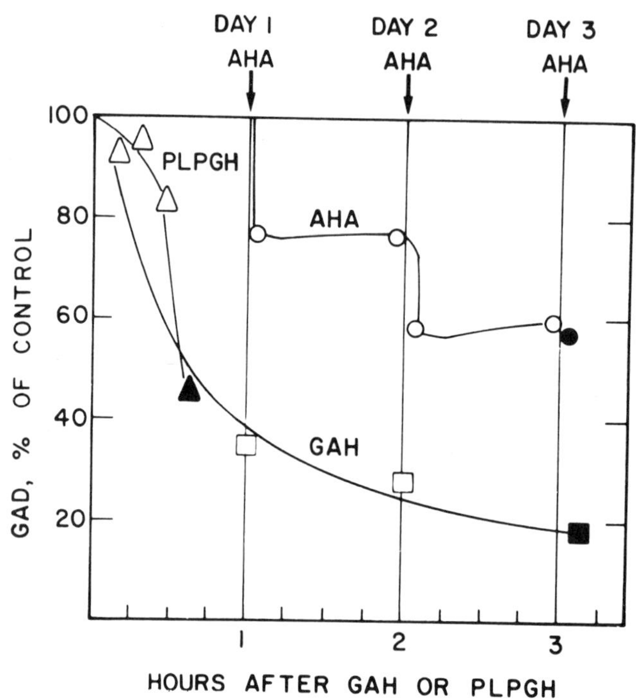

Fig. 1. Changes of GAD activity in brain after administration of anthranilic hydroxamic acid (AHA; Utley, 1963), L-glutamic acid-γ-hydrazide (GAH; Tapia et al., 1967b) and pyridoxal phosphate-γ-glutamyl hydrazone (PLPGH; Tapia and Sandoval, unpublished). AHA (500 mg/kg) was injected once a day (arrows) and GAD activity was measured one hr before and after the injections, as indicated at the top of the figure. A single dose of GAH (2 g/kg) and of PLPGH (80 mg/kg) was injected at zero time and GAD activity was measured at the times indicated at the bottom of the figure. The black symbols indicate that convulsions occurred at the corresponding times.

value, which is of about 60% of the control for anthranilic hydroxamic acid and pyridoxal phosphate-γ-glutamyl hydrazone (PLPGH) (which do not inhibit GABA aminotransferase), and about 20% of the control for L-glutamic acid-γ-hydrazide (GAH) (which considerably inhibits GABA aminotransferase, Table 1) (Fig. 1). In the case of GAH and PLPGH, it has been shown that the decrease of GAD activity observed in these conditions is reflected in a decrease in the rate of synthesis of GABA from labeled glutamic acid, *in vivo* (Fig. 2); this effect was independent of the levels of GABA, which were increased by GAH and decreased by PLPGH. Further evidence of this point is presented in Figure 3, which shows some preliminary results from our laboratory: when the GAH is injected at zero time and pyridoxal phosphate (PLP) is administered 90 min later, convulsions occur approximately 30 min after the injection of PLP, and GAD activity is considerably inhibited precisely at that time, when total brain GABA concentration is greatly increased (Fig. 3).

From the foregoing results, we can conclude that the rate of synthesis of GABA, and not the level of this amino acid, is the determinant for the control of neuronal excitability, at least in the motor system. From part of this evidence and on the basis of the fact that GAD is concentrated in the soluble part of the synaptosomes, whereas GABA aminotransferase is predominantly a mitochondrial enzyme (Salganicoff and De Robertis, 1965; van Kempen *et al.*, 1965; Fonnum, 1968), it has been previously suggested (Wood *et al.*, 1966; Tapia and Awapara, 1967; Tapia *et al.*, 1969) that it is the concentration of GABA *in the synaptic cleft* that is important for the modulation of neuronal excitability. To test this hypothesis, we recently studied the effect of the administration of PLPGH, a drug previously demonstrated to be specific with regard to the inhibition of GAD activity as related to the production of convulsions (Tapia and Pasantes, 1971), on the activity of GAD in nerve endings (synaptosomes) isolated from the brains of the treated animals sacrificed at the moment of convulsions (Pérez de la Mora *et al.*, 1973). The results of these experiments (Table 3) show that GAD activity was notably decreased in the soluble component of the synaptosomes (synaptoplasm), thus indicating that a decrease in the rate of synthesis of GABA induced by PLPGH and presumably by other convulsant drugs with similar metabolic effects, occur in the nerve endings. Thus, these findings strongly support the hypothesis that the synaptic cleft is the critical site for the action of GABA.

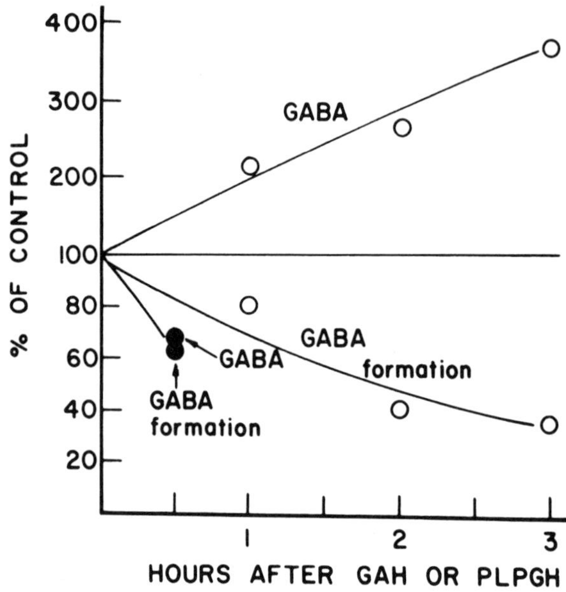

Fig. 2. Changes of GABA levels and GABA formation in brain of mice after the administration of L-glutamic acid-γ-hydrazide (o—— o) and pyridoxal phosphate-γ-glutamyl hydrazone (●——●) (Tapia and Awapara, 1967). GABA formation was determined by measuring the radioactivity in GABA relative to that in glutamic acid, 2 min after the intracranial injection of 3,4 $[^{14}C]$ glutamic acid. GABA aminotransferase was inhibited by the hydrazide and unaffected by the hydrazone (see Table 1).

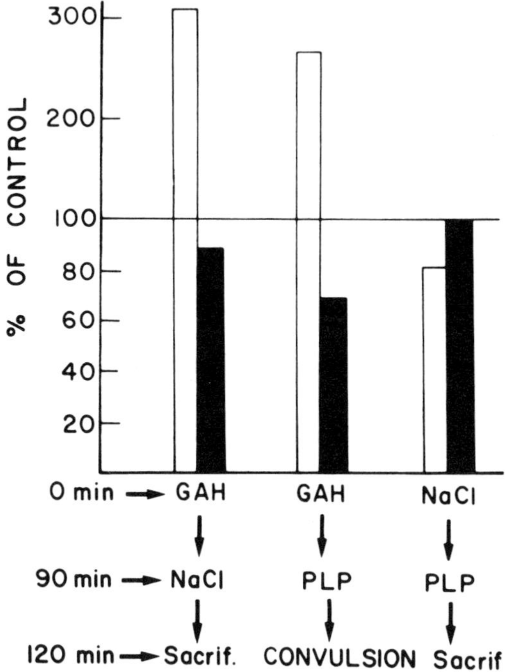

Fig. 3. Changes of GABA levels and GAD activity in brain of mice injected at the times indicated with 0.9% NaCl and GAH (1 g/kg), GAH and PLP (50 mg/kg), and NaCl and PLP. Only the mice injected with GAH and PLP showed fatal tonic convulsions. Control animals were injected twice with 0.9% NaCl. Empty bars indicate GABA levels and black bars refer to GAD activity (Tapia, Sandoval and Contreras, unpublished).

GAD Activity, GABA Release and Convulsions

When the foregoing results and conclusions are considered altogether, they constitute evidence that the release of GABA into the synaptic cleft is directly controlled by the activity of GAD in the axon terminal, and that no storage site seems to exist for this continuous, GAD-dependent, release of GABA. In fact, convulsions occur immediately after the rate of synthesis of GABA is inhibited to a certain value (variable according to the presence or absence of inhibition of GABA aminotransferase), independently of the time necessary to reach this critical value (Fig. 1). If a storage site, replenished by the synthesis of the transmitter (Roberts and Matthysse, 1970) were physiologically present, it would be necessary to deplete it in order to observe the manifestation of the lack of GABA at the synapses. The results are contrary to this expectation, since a four-fold increase of GABA levels, which presumably will cause the filling of the storage sites, does not prevent the convulsions produced immediately after the inhibition of its synthesis (Figs. 2 and 3). Hence, I postulate that under normal conditions the continuous release of GABA will keep the motoneurons in check (tonic inhibition), and, therefore, they will fire uncontrolled (convulsive state) when the synthesis of GABA, and consequently its release, is blocked.

The above described model is presented in Figure 4, which schematizes the two possible ways of functioning of a presynaptic terminal: the phasic release of the transmitter, dependent upon stimulation, which is postulated for both the excitatory and the inhibitory transmitters, and the tonic release of the transmitter, dependent only upon its synthesis, which is postulated for an inhibitory transmitter on the basis of the above discussion. Thus, it is implicit in the model of continuous release that the process of synthesis is coupled to the release of the transmitter, directly from the synaptoplasm. No storage site is required, and the GABA newly synthesized would be "dropped" directly from the GAD into the synaptic cleft.

The requirement of Ca^{++} for the release of neurotransmitters upon electrically induced or K^{+} -induced depolarization is well established (see Eccles, 1964; Auerbach, 1972). In the case of GABA, Ca^{++} appears to be also required when its release is induced by stimulation, but not for its spontaneous release, in

Table 3. Changes of PLP levels and GAD activity in the synaptosomal fraction of brains of mice treated with pyridoxal phosphate-γ-glutamyl hydrazone (PLPGH) and sacrificed at the moment of convulsions *

	Per cent decrease		
		GAD	
Subfraction	PLP	No PLP	PLP†
Whole synaptosomes	23	62	26
Synaptoplasm	47	48	19
Intraterminal mitochondria	8	6	6

* Data from Pérez de la Mora, Feria-Velasco and Tapia (1973).

† PLP added *in vitro* at a 1.7×10^{-4} M final concentration.

cerebral cortex slices (Srinivasan *et al.*, 1969; Iversen *et al.*, 1971; Cutler *et al.*, 1971; Hammerstad and Cutler, 1972). Hence, Ca^{++} is probably not required for the continuous release of GABA postulated here. It is interesting in this regard that the GAD binds to membranes in the presence of Ca^{++} (Fonnum, 1968). Although this binding is too firm to be considered as existing *in vivo*, because the GAD is found to be soluble after the relatively mild conditions used for the studies of subcellular distribution (Fonnum, 1968), its physiological role cannot be discarded yet (see Fig. 5).

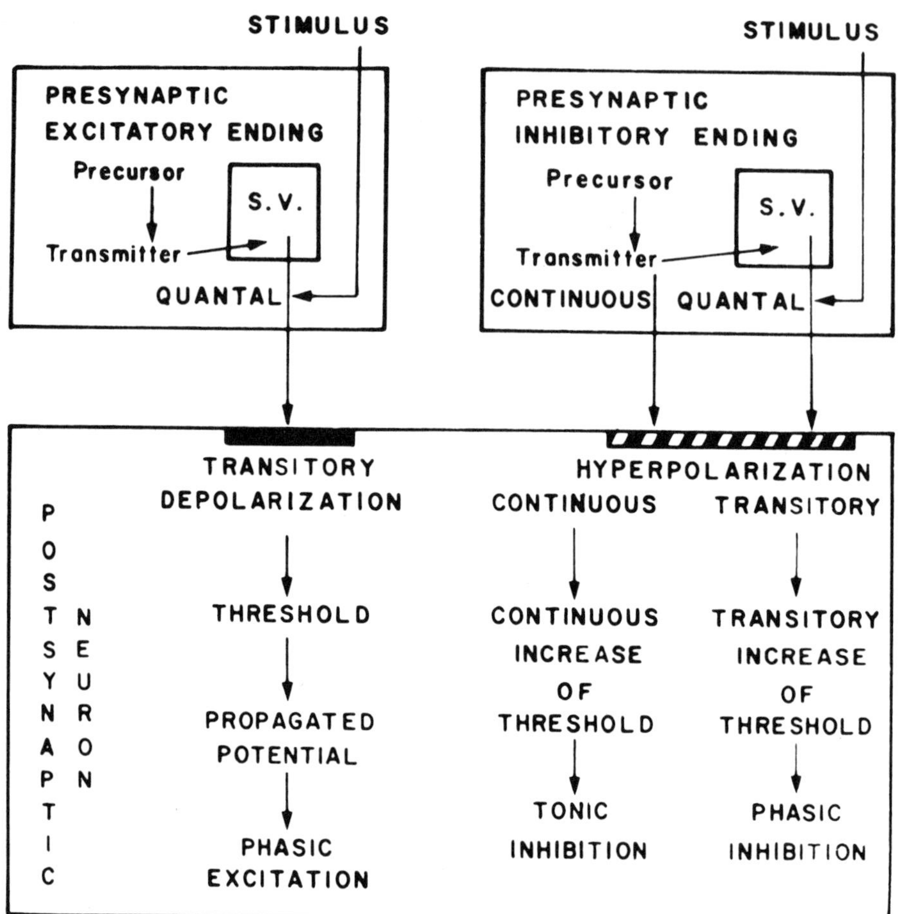

Fig. 4. Postulated scheme for the continuous release of inhibitory synaptic transmitters, besides the phasic release produced by the stimulus, as described in the text. The synaptic vesicles (s.v.) are assumed to be the storage site from where the transmitter is released phasically upon the arrival of the stimulus in both the excitatory and the inhibitory nerve endings. The continuous release of the inhibitory transmitter is postulated to depend only upon the rate of its synthesis from a metabolic precursor, which occurs in the synaptoplasm. The specific receptors for the excitatory and the inhibitory transmitter are indicated by the black and striped bands, respectively.

Another point implicit in the model is that the postsynaptic neuron, in this case the motoneuron, would be continuously firing at a maximum rate, were it not for the fact that the inhibitory GABA neurons are modulating its activity. In other words, both the motoneurons and the inhibitory GABA-neurons probably have an endogenous rhythm of firing in the intact animal. This general hypothesis of the appearance of seizures by disinhibition of neurons (Pérez de la Mora and Tapia, 1973) is in agreement with neurophysiological evidence which also suggests this type of mechanism (Spencer and Kandel, 1969; Meldrum and Horton, 1971).

GAD, PLP AND THE CONTROL OF NEURONAL EXCITABILITY

A major implication of the model proposed is that the controlling, inhibitory GABA-neurons are releasing GABA through a metabolic mechanism, represented by the activity of GAD. Accordingly, the regulation of the activity of GAD, in turn, may be the primary mechanism responsible for the control of neuronal excitability. The following evidence indicates that the concentration of PLP, which depends mainly upon the activity of pyridoxal kinase (EC 2.7.1.35), constitutes such a regulatory control of GAD activity: 1) Pyridoxal kinase, free pyridoxal phosphate, GAD and GABA are localized in the synaptoplasm (Mangan and Whittaker, 1966; Loo and Whittaker, 1967; Fonnum, 1968; Pérez de la Mora *et al.*, 1973). 2) An inhibition of pyridoxal kinase activity is well correlated with a decrease of PLP in the whole brain (Tapia and Awapara, 1969; Tapia *et al.*, 1969; Table 2 of the present paper). 3) A decrease of PLP levels is linearly correlated with an inhibition of GAD activity in the whole brain (Minard, 1967; Tapia and Awapara, 1969; Tapia *et al.*, 1969), and the latter correlation is also observed in the synaptoplasm after the administration of PLPGH (Pérez de la Mora *et al.*, 1973; Table 3). 4) Both in the whole brain (Table 2) and in synaptosomes (Table 3) the inhibition of GAD *in vivo* by drugs that act primarily by decreasing the activity of pyridoxal kinase is reversed by the addition of PLP *in vitro* to the incubation medium (Tables 2 and 3).

The above results suggest, therefore, that the metabolic system pyridoxal kinase-PLP-GAD-GABA is soluble in the synaptoplasm, and that it is continuously functioning (Fig. 5). The activity of the motoneurons, tonically controlled by this system, could be increased under normal conditions by any of the

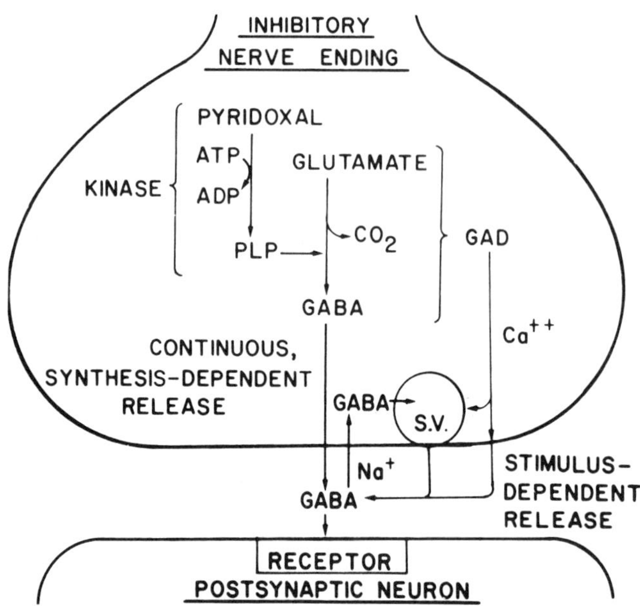

Fig. 5. Diagrammatic summary of the proposed role of GABA metabolism in the regulation of cerebral excitability. There is a continuous release of newly synthesized GABA, coupled to the activity of GAD. This continuously released GABA from the synaptoplasm inhibits tonically the postsynaptic neuron by maintaining its membrane hyperpolarized. The concentration of PLP in the synaptoplasm, which depends upon the activity of pyridoxal kinase, acts as a control mechanism, by regulating the activity of GAD. In addition, as a response to a stimulus, the GAD could be bound to the presynaptic membrane or to the synaptic vesicle (s.v.) membrane in the presence of Ca^{++} (Fonnum, 1968), and GABA could be released also from these structures when the presynaptic inhibitory neuron is depolarized. The storage sites could be also replenished by GABA taken up from the extracellular space in the presence of Na^+ (Iversen and Neal, 1968). Further explanation in the text.

following mechanisms, or by a combination of them: 1) increasing the depolarization of the motoneuron by a stimulus reaching it from afferent excitatory neurons; 2) decreasing the activity of pyridoxal kinase, and thus decreasing the activity of GAD by lowering the concentration of free PLP in the synaptosome; 3) since it seems plausible that a phasic release of GABA from synaptic vesicles, induced by depolarization of the presynaptic membrane, is also occurring in the inhibitory synapses (Srinivasan et al., 1969; Cutler et al., 1971; Hall, 1972) (see Figs. 4 and 5), an inhibition of the inhibitory GABA neuron would also result in a decreased rate of release of GABA, with the consequent activation of the postsynaptic neuron.

Epilepsy and Vitamin B_6

Besides the experimental studies described above, it seems appropriate to mention the possible relationship of vitamin B_6 with epileptic seizures in man. The best example of such relationship is provided by the cases of pyridoxine-dependent children, who showed seizures unless treated with pyridoxine, although their vitamin B_6 intake was in the normal range (see the review by Tower, 1969). The model presented in the present paper could account for these observations, assuming that because of a genetic defect, the cerebral GAD of these children was much more dependent upon PLP than the normal enzyme. A result of this defect would be a decreased activity of GAD, with the consequent decrease of GABA release and the lack of adequate tonic inhibition of the motoneurons. The well known role of vitamin B_6 in the function of the central nervous system (Tower, 1969) can be explained at the molecular level, at least in part, by the model proposed.

SUMMARY

The quantal release of acetylcholine from the presynaptic endings, and the consequent depolarization of the postsynaptic membrane, is a well established phenomenon, at least for neuromuscular junctions. Catecholamines, both in the peripheral and in the central nervous system, appear to be released by electrical stimulation preferentially from a newly synthesized pool. From this evidence, it is generally accepted that the synaptic transmitters are released from one particular metabolic pool, which may be either a storage site such as the synaptic vesicles or a recently synthesized pool, following the depolarization of the nerve ending membrane. Thus, the release of the transmitter is considered to be a phasic phenomenon, ocurring only when a propagated potential reaches the axon terminal. In this paper a hypothesis is presented, which postulates that, in addition to this phasic release of both the excitatory and the inhibitory transmitters, a continuous, tonic release, dependent only upon the activity of the enzyme responsible of the synthesis of the transmitter, and independent of the depolarization of the presynaptic membrane, may occur in certain inhibitory neurons. In this way, these neurons would act tonically by constantly maintaining an elevated threshold in the excitatory neurons, so that the latter would start firing when a decrease in the continuous release of the inhibitory transmitter acting upon them occurs.

The model is presented for the γ-aminobutyric acid (GABA), on the grounds of the following evidence: 1) Both glutamate decarboxylase (GAD) and GABA are located in the soluble component of the nerve endings (synaptoplasm). In contrast to the excitatory transmitter acetylcholine, or to catecholamines, no GABA has been found in synaptic vesicles of the central nervous system. 2) It has been observed in many different experimental conditions that a decrease of GABA synthesis, through the inhibition of GAD produced by a variety of substances, induces convulsions whatever the concentration of GABA in the whole brain may be. 3) There is a good correlation between the inhibition of GAD in the synaptoplasm and the appearance of convulsions produced by some drugs.

According to this model, the continuous release of GABA is coupled to its synthesis, and hence the concentration of GABA in the synaptic cleft is dependent upon the activity of GAD. Thus, the regulation of GAD activity would simultaneously be responsible for the control of the inhibitory postsynaptic effect of GABA. Since there is experimental evidence that this regulation of GAD activity is exerted by the concentration of pyridoxal phosphate in the synaptoplasm, the role of this B_6 vitamin in nervous tissue is postulated to be of paramount importance for the regulation of cerebral excitability.

REFERENCES

Alberici, M., Rodríguez de Lores Arnaiz, G., and De Robertis, E., 1969, Glutamic acid decarboxylase inhibition and ultrastructural changes by the convulsant drug allylglycine, Biochem. Pharmac. 18:137.

Aprison, M. H., and Hingtgen, J. N., 1972, Serotonin and behavior: a brief summary, Fed. Proc. 31:121.

Auerbach, A. A., 1972, Transmitter release at chemical synapses, in "Structure and Function of Synapses", (G. D. Pappas and D. P. Purpura, eds.), pp. 137-159, Raven Press, New York.

Baxter, C. F., 1970, The nature of γ-aminobutyric acid, in "Handbook of Neurochemistry" (A. Lajtha, ed.), Vol. 3, pp. 289-353, Plenum Press, New York.

Bloom, F. E., and Giarman, N. J., 1968, Physiologic and pharmacologic considerations of biogenic amines in the nervous system, Ann. Rev. Pharmac. 8:229.

Curtis, D. R., Duggan, A. W., Felix, D., and Johnston, G. A. R., 1970a, GABA, bicuculline and central inhibition, Nature (Lond.) 226:1222.

Curtis, D. R., Duggan, A. W., Felix, D., and Johnston, G. A. R., 1970b, Bicuculline and central GABA receptors, Nature (Lond.) 228:676.

Curtis, D. R., Duggan, A. W., Felix, D., Johnston, G. A. R., and McLennan, H., 1971, Antagonism between bicuculline and GABA in the cat brain, Brain Res. 33:57.

Cutler, R. W. P., Hammerstad, J. P., Cornick, L. R., and Murray, J. E., 1971. Efflux of amino acid neurotransmitters from rat spinal cord slices. I. Factors influencing the spontaneous efflux of [^{14}C] glycine and ^{3}H-GABA, Brain Res. 35:337.

Eccles, J. C., 1964, "The Physiology of Synapses", Springer-Verlag, Berlin.

Florey, E. (Ed.), 1961, "Nervous Inhibition", Pergamon Press, New York.

Fonnum, F., 1968, The distribution of glutamate decarboxylase and aspartate transaminase in subcellular fractions of rat and guinea pig brain, Biochem. J. 106:401.

Glowinski, J., Besson, M. J., Cheramy, A., and Thierry, A. M., 1972, Disposition and role of newly synthesized amines in central catecholaminergic neurons, in "Studies of Neurotransmitters at the Synaptic Level", (E. Costa, L. L. Iversen and R. Paoletti, eds.), pp. 93-109, Raven Press, New York.

Hall, Z. W., 1972, The storage, synthesis and inactivation of the transmitters acetylcholine, norepinephrine and gamma-aminobutyric acid, in "Structure and Function of Synapses", (G. D. Pappas and D. P. Purpura, eds.), pp. 161-171, Raven Press, New York.

Hammerstad, J. P., and Cutler, R. W. P., 1972, Sodium ion movements and the spontaneous and electrically stimulated release of $[^3H]$ GABA and $[^{14}C]$ glutamic acid from rat cortical slices, Brain Res. 47:401.

Iversen, L. L., and Neal, M. J., 1968, The uptake of ^3H-GABA by slices of rat cerebral cortex, J. Neurochem. 15:1141.

Iversen, L. L., Mitchell, J. F., and Srinivasan, V., 1971, The release of γ-aminobutyric acid during inhibition in the cat visual cortex, J. Physiol. (Lond.) 212:519.

Kety, S. S., 1970, The biogenic amines in the central nervous system: their possible roles in arousal, emotion, and learning, in "The Neurosciences: Second Study Program", (F. O. Schmitt, ed.), pp. 324-336, Rockefeller University Press, New York.

Krnjević, K., and Schwartz, S., 1967, The action of γ-amino-butyric acid on cortical neurons, Exp. Brain Res. 3:320.

Krnjević, K., and Schwartz, S., 1968, The inhibitory transmitter in the cerebral cortex, in "Structure and Function of Inhibitory Neuronal Mechanisms", (C. von Euler, S. Skoglund and U. Söderberg, eds.), pp. 419-427, Pergamon Press, Oxford.

Loo, Y. H., and Whittaker, V. P., 1967, Pyridoxal kinase in brain and its inhibition by pyridoxilidene-β-phenylethylamine, J. Neurochem. 14:997.

Mangan, J. L., and Whittaker, V. P., 1966, The distribution of free amino acids in subcellular fractions of guinea pig brain, Biochem. J. 98:128.

Medina, M. A., 1963, The *in vivo* effects of hydrazines and vitamin B_6 on the metabolism of gamma-aminobutyric acid, J. Pharmac. Exp. Ther. 140:133.

Meldrum, B. S., and Horton, R. W., 1971, Convulsive effects of 4-deoxypyridoxine and of bicuculline in photosensitive baboons (*Papio papio*) and in rhesus monkeys (*Macaca mulatta*), Brain Res. 35:419.

Minard, F. N., 1967, Relationships among pyridoxal phosphate, vitamin B_6-deficiency, and convulsions induced by 1,1-dimethylhydrazine, J. Neurochem. 14:681.

Obata, K., and Takeda, K., 1969, Release of γ-aminobutyric acid into the fourth ventricle induced by stimulation of the cat's cerebellum, J. Neurochem. 16:1043.

Obata, K., Ito, M., Ochi, R., and Sato, N., 1967, Pharmacological properties of the postsynaptic inhibition by Purkinje cell axons and the action of γ-aminobutyric acid on Deiters neurons, Exp. Brain Res. 4:43.

Otsuka, M., Obata, K., Miyata, Y., and Tanaka, Y., 1971, Measurement of γ-aminobutyric acid in isolated nerve cells of cat central nervous system, J. Neurochem. 18:287.

Pérez de la Mora, M., and Tapia, R., 1973, Anticonvulsant effect of 5-ethyl, 5-phenyl, 2-pyrrolidinone and its possible relationship to γ-aminobutyric acid-dependent inhibitory mechanisms, Biochem. Pharmac. In press.

Pérez de la Mora, M., Feria-Velasco, A., and Tapia, R., 1973, Pyridoxal phosphate and glutamate decarboxylase in subcellular particles of mouse brain and their relationship to convulsions, J. Neurochem. In press.

Roberts, E., and Matthysse, S., 1970, Neurochemistry: at the crossroads of neurobiology, Ann. Rev. Biochem. 39:777.

Roberts, E., Baxter, C. F., van Harreveld, A., Wiersma, C. A. G., Adey, W. R., and Killam, K. F. (Eds.), 1960, "Inhibition in the Nervous System and Gamma-Aminobutyric Acid", Pergamon Press, Oxford.

Rodríguez de Lores Arnaiz, G., Alberici de Canal, M., and De Robertis, E., 1972, Alteration of GABA system and Purkinje cells in rat cerebellum by the convulsant 3-mercaptopropionic acid, J. Neurochem. 19:1379.

Salganicoff, L., and De Robertis, E., 1965, Subcellular distribution of the enzymes of the glutamic acid, glutamine and γ-aminobutyric acid cycles in rat brain, J. Neurochem. 12:287.

Spencer, W. A., and Kandel, E. R., 1969, Synaptic inhibition, in "Basic Mechanisms of the Epilepsies", (H. H. Jasper, A. A. Ward and A. Pope, eds.), pp. 575-603, Little, Brown and Co., Boston.

Srinivasan, V., Neal, M. J., and Mitchell, J. F., 1969, The effect of electrical stimulation and high potassium concentrations on the efflux of [^{3}H] γ-aminobutyric acid from rat brain slices, J. Neurochem. 16:1235.

Sze, P. Y., Kuriyama, K., and Roberts, E., 1971, Thiosemicarbazide and γ-aminobutyric acid metabolism, Brain Res. 25:387.

Tapia, R., and Awapara, J., 1967, Formation of γ-aminobutyric acid (GABA) in brain of mice treated with L-glutamic acid-γ-hydrazide and pyridoxal phosphate-γ-glutamyl hydrazone, Proc. Soc. Exp. Biol. Med. 126:218.

Tapia, R., and Awapara, J., 1969, Effects of various substituted hydrazones and hydrazines of pyridoxal-5'-phosphate on brain glutamate decarboxylase, Biochem. Pharmac. 18:145.

Tapia, R., and Pasantes, H., 1971, Relationships between pyridoxal phosphate availability, activity of vitamin B_6-dependent enzymes and convulsions, Brain Res. 29:111.

Tapia, R., Pérez de la Mora, M., and Massieu, G. H., 1967a, Modifications of brain glutamate decarboxylase activity by pyridoxal phosphate-γ-glutamyl hydrazone. Biochem. Pharmac. 16:1211.

Tapia, R., Pasantes, H., Pérez de la Mora, M., Ortega, B. G., and Massieu, G. H., 1967b, Free amino acids and glutamate decarboxylase activity in brain of mice during drug-induced convulsions, Biochem. Pharmac. 16:483.

Tapia, R., Pérez de la Mora, M., and Massieu, G. H., 1969, Correlative changes of pyridoxal kinase, pyridoxal-5'-phosphate and glutamate decarboxylase in brain, during drug-induced convulsions, Ann. N. Y. Acad. Sci. 166:257.

Thierry, A. M., Blanc, G., and Glowinski, J., 1971, Effect of stress on the disposition of catecholamines localized in various intraneural storage forms in the brain stem of the rat, J. Neurochem. 18:449.

Tower, D. B., 1969, Neurochemical mechanisms, in "Basic Mechanisms of the Epilepsies", (H. H. Jasper, A. A. Ward and A. Pope, eds.), pp. 611-638, Little, Brown and Co., Boston.

Utley, J. D., 1963, The effects of anthranilic hydroxamic acid on rat behaviour and rat brain γ-aminobutyric acid, norepinephrine and 5-hydroxytryptamine concentrations, J. Neurochem. 10:423.

van Kempen, G. M. J., van den Berg, C. J., van der Helm, H. J., and Veldstra, H., 1965, Intracellular localization of glutamate decarboxylase, γ-aminobutyrate transaminase and some other enzymes in brain tissue, J. Neurochem. 12:581.

von Euler, C., Skoglund, S., and Söderberg, U. (Eds.), 1968, "Structure and Function of Inhibitory Neuronal Mechanisms", Pergamon Press, Oxford.

Wood, J. D., and Abrahams, D. E., 1971, The comparative effects of various hydrazides on γ-aminobutyric acid and its metabolism, J. Neurochem. 18:1017.

Wood, J. D., and Peesker, S. J., 1972, A correlation between changes in GABA metabolism and isonicotinic acid hydrazide-induced seizures, Brain Res. 45:489.

Wood, J. D., Watson, W. J., and Stacey, N. E., 1966, A comparative study of hyperbaric oxygen-induced and drug-induced convulsions with particular reference to γ-aminobutyric acid metabolism, J. Neurochem. 13:361.

BIOCHEMICAL NEUROANATOMY OF THE BASAL GANGLIA

P. L. McGeer, H. C. Fibiger, T. Hattori, V. K. Singh,
E. G. McGeer and L. Maler

Kinsmen Laboratory of Neurological Research
Department of Psychiatry
University of British Columbia, Vancouver, Canada

It is evident that reliable information on the neurohumoral coding of brain function must have as its basis a sound knowledge of biochemical neuroanatomy. It is not enough merely to establish the details of neuronal connections between various areas of brain. The biochemical modus operandi of the cells making up the pathways must also be known. The great majority of the mammalian nervous system operates via chemical synapses where specific neurotransmitters are responsible for modulating the excitability of post-synaptic elements. Biochemical neuroanatomy is concerned with defining the neurotransmitters associated with particular cell groups.

The field is in its infancy. Some assignments have been made for dopamine, noradrenaline and serotonin (Dahlstrom and Fuxe, 1964c; Anden et al., 1966; Ungerstedt, 1971) which have depended largely on the excellent fluorescence histochemical method of Falck et al. (1962). Some postulates have been made for acetylcholine (Fonnum, 1970; McGeer et al., 1971b), GABA (Fonnum, 1970; Hattori et al., 1973; Graham, 1972; Storm-Mathisen and Fonnum, 1971), and glycine (Aprison et al., 1970) based on physiological and biochemical measurements. But the neurotransmitters associated with the overwhelming majority of synapses in brain are still a mystery.

The extrapyramidal system is an excellent model for studying biochemical neuroanatomy and the specific ways that amine pathways are interconnected. The highest levels in brain of dopamine and acetylcholine occur in the caudate and putamen, while the highest levels of GABA occur in the globus pallidus and substantia nigra.

Two important human diseases of the extrapyramidal system, Parkinson's disease and Huntington's chorea, involve disturbances in these amines and their synthesizing enzymes. Important classes of behavioural drugs such as the rauwolfia alkaloids, the phenothiazines, and the butyrophenones, which are administered to modify behaviour, interact with these amines and produce extrapyramidal reactions as a side effect.

We have studied the basal ganglia using lesioning techniques and axoplasmic flow in an attempt to unravel the anatomical and biochemical relationships between these nuclei. Neurotransmitters and their synthetic enzymes are localized primarily to nerve endings. If their axon is sectioned, complete degeneration, distal to the lesion, takes place. Unless there is a sustaining axon collateral, degeneration of the cell body and proximal axonal stump also takes place. Thus, rapid disappearance of neurotransmitter synthetic enzymes in an area remote from a brain lesion usually signifies that axons have been cut. The nerve endings themselves are nourished by the process of axoplasmic flow. Proteins, synthesized in the rough endoplasmic reticulum of the cell body, are transported along with other material down the axon to the terminals. A radiolabelled amino acid such as leucine, injected into a localized area of brain, can therefore be picked up by cell bodies, synthesized into protein and the radiolabelled protein transported along axons to distant brain areas.

LESION STUDIES

If the rat brain is hemitransected between the entopeduncular nucleus and the substantia nigra (posterior hemitransection), all known descending pathways from the basal ganglia to the substantia nigra will be transected. Similarly, all ascending pathways from the substantia nigra will also be transected. Such a lesion produced a decrease in tyrosine hydroxylase, the rate-controlling enzyme for dopamine synthesis, in both the corpus striatum and the substantia nigra (Table I). Glutamic acid decarboxylase (GAD), the synthetic enzyme for GABA, decreased to 17.8% in the substantia nigra but was unchanged in the caudate-putamen and globus pallidus. Choline acetylase, the synthetic enzyme for acetylcholine, was unaffected by the lesion. Details of the methods have been provided elsewhere (McGeer et al., 1971c; Hattori et al., 1973; McGeer et al., 1973a).

When the rat brain was hemitransected between the caudate-putamen and globus pallidus (anterior hemitransection) similar results were obtained for tyrosine hydroxylase but quite different results for GAD. There was no change in GAD in the substantia nigra, but the caudate-putamen dropped to 86% and the globus pallidus to 77% of controls (Table I). As with posterior hemitransections, choline acetylase was unaffected. Finally, discrete electrolytic lesions of the globus pallidus resulted in large

TABLE I

Effect of lesions or hemitransections on GAD and tyrosine hydroxylase activities in various brain regions (% control values)

	Caudate-Putamen	Globus Pallidus	Substantia Nigra
		Posterior Hemitransection	
GAD	107.6 ± 6.9	-	17.8 ± 3.5
Tyrosine Hydroxylase	15.5 ± 8.2	-	55.8 ± 8.7
		Anterior Hemitransection	
GAD	86.0 ± 7.7	76.8 ± 9.1	100.0 ± 9.2
Tyrosine Hydroxylase	19.7 ± 4.9	53.1 ± 10.1	65.9 ± 9.6

V_{max} of GAD (in μM/G/h) in the control samples was: caudate-putamen 15.8; globus pallidus 24.9; substantia nigra 29.0; V_{max} of tyrosine hydroxylase (in μM/g/h) was: caudate-putamen 102.7; globus pallidus 53.2; substantia nigra 64.6. 10-11 animals were used in each experiment.

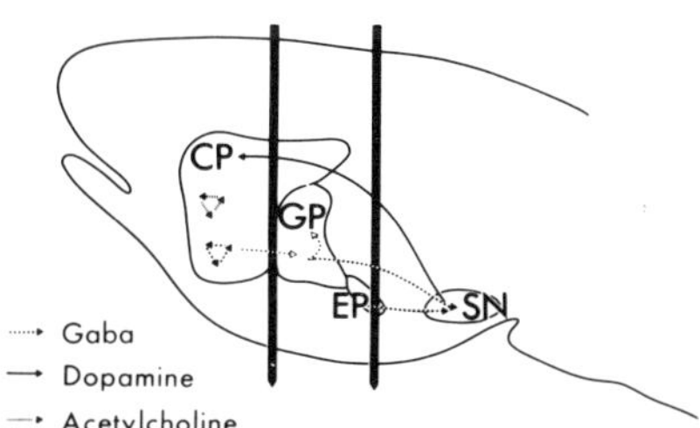

FIGURE 1. Schematic drawing of amine pathways in the extrapyramidal system as indicated by axoplasmic flow and lesion data.

decreases in GAD activity in the substantia nigra (McGeer et al., 1971c; McGeer et al., 1973a). A summary of the pathways suggested by lesion studies between extrapyramidal structures is shown in Figure 1.

The well-known dopaminergic pathway ascending from the substantia nigra to the caudate-putamen was confirmed. Descending GABA-containing pathways from the globus pallidus-entopeduncular nucleus to the substantia nigra were indicated as well as internally arranged acetylcholine-containing pathways within the caudate-putamen. There was some suggestive evidence of a descending GABA-containing pathway from the caudate-putamen to the globus pallidus. Since the drop in GAD in the caudate-putamen following anterior hemitransections was only a few percent, the pathway would either have to have sustaining collaterals or there would need to be a separate internally arranged GABA-containing system in the caudate-putamen. Similarly, the failure of GAD to be reduced in the globus pallidus following posterior hemitransections would suggest either that there were sustaining axon collaterals or a high proportion of intrinsic GABA neurons in the globus pallidus. Stripping of the cerebral cortex failed to affect striatal choline acetylase (McGeer et al., 1971b) or GAD (McGeer and McGeer, unpublished observations). These data indicate that there are not descending acetylcholine or GABA-containing pathways from the cortex to the striatum. The neurotransmitter of this projection remains unknown therefore.

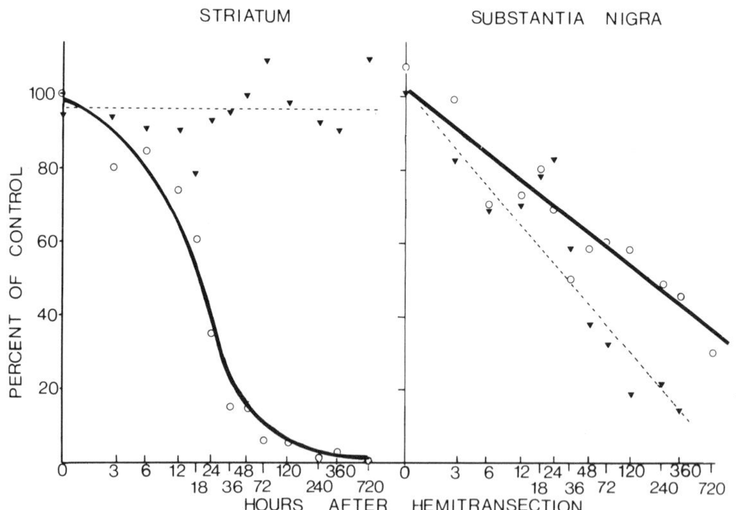

FIGURE 2. The effect of hemitransection on amine synthesizing activities in the corpus striatum and the substantia nigra. Hours after hemitransection plotted on a log scale. O———, tyrosine hydroxylase; ▼- - - , glutamic acid decarboxylase. Each point represents the average value for 3-7 rats.

Figure 2 shows the rate of decline of tyrosine hydroxylase and GAD in the corpus striatum and substantia nigra following posterior hemitransections. It can be seen from the figure that there is a rapid decline of tyrosine hydroxylase in the corpus striatum and of GAD in the substantia nigra. Tyrosine hydroxylase, on the other hand, declines more slowly in the substantia nigra while GAD shows no significant change in the caudate-putamen. Since terminal degeneration is more rapid than that of cell bodies, these data are consistent with sectioning of descending GABA-containing pathways to the substantia nigra and ascending dopamine-containing pathways to the caudate-putamen (McGeer et al., 1973a).

AXOPLASMIC FLOW STUDIES

We have used the nigro-striatal pathway as a model for learning some of the characteristics of axoplasmic flow within the central nervous system. If radioactive leucine is injected into the substantia nigra of a rat, then radioactive protein can be detected in the caudate-putamen (Fibiger et al., 1972). In a typical experiment 10 μC of L-leucine-4,5-^{3}H (specific activity 37 Ci/mmol) in 1 μl of solution is injected stereotaxically into the zona compacta of the substantia nigra, over a period of 5 minutes. The rat is sacrificed at an appropriate time interval after the injection, and the brain quickly dissected on ice. The radioactivity is measured in both the TCA-insoluble and TCA-soluble fractions of tissue homogenates in 10% TCA. If the radioactive protein is to be fractionated, the tissue is homogenized in 0.5% (v/v) Triton X-100 and an aliquot of the homogenate counted. The remaining homogenate is centrifuged and the supernatant subjected to DEAE-cellulose chromatography. Elution of radioactive protein from the column is accomplished with phosphate buffers of concentrations increasing in the following steps: 0.01M, 0.1M, 0.2M and 0.5M. The profile of non-radioactive proteins is monitored by measuring the absorbence at 280 nm using a Beckman model DU spectrophotometer, and the tyrosine hydroxylase and choline acetylase activities are assayed by previously described methods (McGeer et al., 1971a; McGeer et al., 1971b).

Figure 3 illustrates the time course observed for accumulation of transported protein in the caudate-putamen. As early as 6 hours after the injection, ^{3}H-proteins were detected. The maximum accumulation was after about 5 days and there was still residual protein radioactivity after 30 days. Several methods were used for establishing that the protein in the caudate-putamen arrived by genuine axoplasmic flow in the nigro-striatal tract. One of these was to examine the general distribution of protein in all areas of the brain. Such a distribution is shown in Table 2. As can be seen from the table, radioactivity in the corpus striatum on the

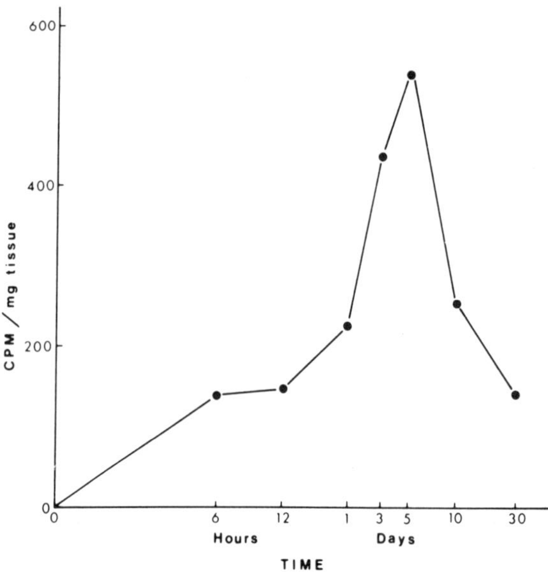

FIGURE 3. Accumulation of total radioactive protein in the rat caudate-putamen at various time intervals after the injection of 10 μCi of L-leucine-4,5-^{3}H (specific activity 37 Ci/mmole) into the substantia nigra.

injected side was over 70 times as great as on the contralateral control side. The areas from the rostral control side of the brain were all low in radioactivity, indicating that the labelled leucine had not been taken up by the blood stream or CSF and distributed in a nonspecific way.

Significant levels of labelled protein were, however, found in the ipsilateral thalamus and hypothalamus. The thalamus receives projections from the substantia nigra (Moore et al., 1971; Carpenter and Strominger, 1967) just as does the caudate-putamen. Although the hypothalamus is not thought to contain efferents from the substantia nigra, nigro-striatal axons pass through the lateral hypothalamus (Moore et al., 1971; Ungerstedt, 1971). The radioactivity in the hypothalamus may represent, therefore, labelled protein within nigro-striatal axons. Considerable radioactivity was retained near the site of injection and some activity, possibly resulting from axonal transport, appeared bilaterally in the pons-medulla. At none of the times (6 hours to 30 days) was radioactivity detected in low molecular weight fractions except at the site of injection up to 12 hours.

Other methods for establishing that the protein was arriving by axoplasmic flow included administration of ^{14}C-leucine 1.5 mm

TABLE II

Regional distribution in rat brain of ^{14}C-protein at 4 days after administration of ^{14}C-leucine to the left substantia nigra.

Brain Area	Left	Right
	(d.p.m./mg of tissue)	
Pons/medulla	6.0 ± 1.3	6.0 ± 1.2
Hippocampus	1.1 ± 0.1	0.4 ± 0.04
Midbrain	236.3 ± 26.4	7.1 ± 1.2
Hypothalamus	19.1 ± 2.8	2.7 ± 0.5
Thalamus	28.4 ± 5.4	2.9 ± 0.9
Corpus striatum	87.6 ± 13.1	1.2 ± 0.8
Cortex	1.8 ± 0.4	0.5 ± 0.2

The data represent the means (± S.D.) of four animals. Rats were injected into the left substantia nigra with 0.3 μCi of L-leucine-U-^{14}C (262 μCi/μmol) in 3 μl of acetate buffer pH 6.5. The TCA insoluble fraction of brain tissue homogenate was counted to give ^{14}C-protein activity.

dorsal to the substantia nigra; and administration of ^{14}C-leucine into the substantia nigra of animals that had been lesioned pharmacologically with 6-hydroxydopamine or electrolytically in the region of the lateral hypothalamus. When ^{14}C-leucine was injected 1.5 mm dorsal to the zona compacta of the substantia nigra, the amount of ^{14}C-protein recovered in the ipsilateral striatum was only 3.6% as high as when the injection was into the substantia nigra itself. In animals where the dopaminergic neurons had been lesioned by administering 200 μg of 6-hydroxydcpamine intraventricularly 30 minutes after the intraperitoneal injection of 5 mg/kg of tranylcypromine sulfate, the amount of protein arriving in the substantia nigra was reduced to 31% of the untreated value. The striatal tyrosine hydroxylase activity was reduced to an even greater extent, being only 7.8% of the control value. When the nigro-striatal pathway had been damaged by an electrolytic lesion, the amount of protein arriving in the caudate-putamen was reduced to 25% of the untreated value whereas the striatal tyrosine hydroxylase activity was reduced to 19% (Fibiger et al., 1972).

It has been repeatedly noted in our hands that there is a good correlation between the extent of reduction of axonal transport and striatal tyrosine hydroxylase activity when the nigro-striatal pathway is physically damaged but that tyrosine hydroxylase activity is always reduced to a substantially greater degree than axonal transport when a pharmacological lesion is induced with 6-hydroxydopamine. This discrepancy suggests the possibility of a non-dopaminergic as well as a dopaminergic pathway from the substantia nigra to the caudate putamen. The data suggest that about 20% of the protein transported may be in non-dopaminergic neurons.

When the proteins extracted from the caudate-putamen were fractionated by DEAE-cellulose chromatography, 4 peaks (A, B, C, and D) appeared (Singh et al., 1973). A typical profile 5 days after the injection of ^{3}H-leucine into the substantia nigra is shown in Figure 4. As can be seen from the figure, peak B was the most prominent radioactive peak at the time of maximum labelling.

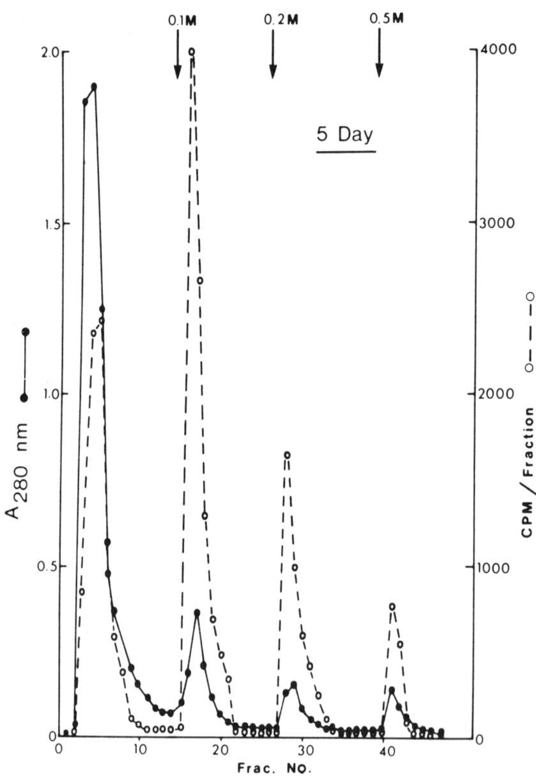

FIGURE 4. DEAE-cellulose chromatographic profile of ^{3}H-protein of rat caudate-putamen 5 days after the injection of ^{3}H-leucine into the substantia nigra.

The peaks were differently labelled, however, as a function of time. Only peak A had significant radioactivity 6 hours after the administration of ^{3}H-leucine, and this peak appeared to reach a maximum after about 24 hours. Radioactive labelling of peak B was just detectable after 6 hours and reached a maximum at about 5 days. Peaks C and D were smaller and were substantially labelled only during the period 3-5 days after the injection.

The relative specific activities (dpm per minute / 280 nm) for peaks A, B, C and D as a function of time are shown in Figure 5. During the period 6-24 hours (fast axonal transport) the relative specific activity of peak A was high compared to that of the other peaks. This ratio had, however, declined to the lowest of all peaks by 5 days. At this time, peak B had the highest relative specific activity. Radioactivity of all peaks decayed in a similar fashion with only a small amount of activity remaining after 30 days.

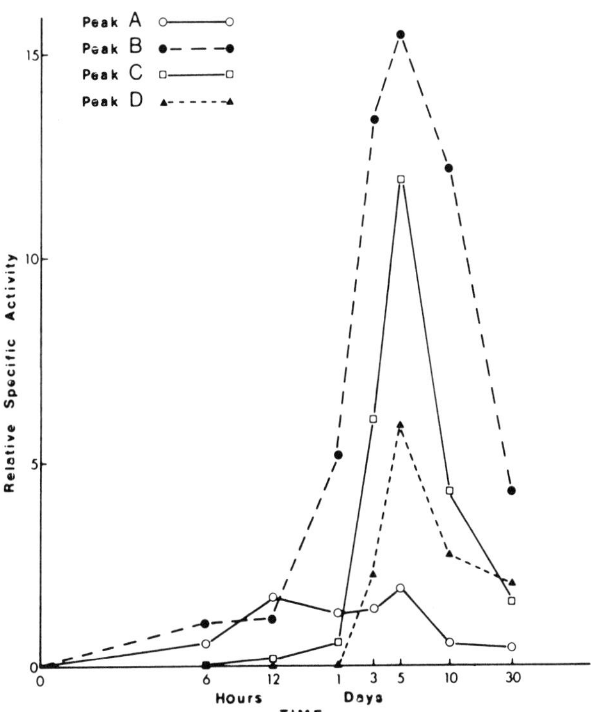

FIGURE 5. Relative specific activities of different protein peaks from a DEAE-cellulose column as a function of time.

The radioactive protein in peaks A and B obtained after 5 days was subjected to Sephadex G-200 column chromatography. By comparison with the behaviour of lysosyme and catalase on this column, the molecular weight of the proteins was estimated to be between 13,000 and 300,000 for each peak.

Tyrosine hydroxylase and choline acetylase activities were also determined in all the fractions. Tyrosine hydroxylase activity was localized exclusively to peak B which was the major radioactive peak. Choline acetylase, on the other hand, was found primarily in peak A which contained the lowest specific activity. Peak A was further resolved into a fast and a slow moving peak indicating that it was made up of more than one component. The protein transported in 6-hydroxydopamine-treated animals was also subjected to DEAE-cellulose fractionation. It was found that there was almost a total elimination of radioactivity in peak B which is in accord with the disappearance of tyrosine hydroxylase activity following such treatment.

To determine whether nigro-striatal dopaminergic neurons would transport their own transmitter, experiments were performed where labelled dopa was injected into the substantia nigra of rats pretreated with tranylcypromine sulfate (5 mg/kg). Significant quantities of labelled material were recovered in the injected substantia nigra and in the ipsilateral globus pallidus and caudate-putamen of the rats. The activity in the substantia nigra consisted of dopa, dopamine, methoxytyramine and other unidentified metabolites. In the caudate-putamen, however, 85% of the activity was recovered as dopamine. No dopa was recovered from the caudate-putamen. A significant amount of radioactivity was found 2 hours post injection in the globus pallidus. The peak was at 4 hours with a gradual decline thereafter. Significant radioactivity was not observed in the caudate-putamen until 4 hours after the injection. This activity peaked at 8 hours and then gradually declined. There was an exponential decline of radioactivity in the substantia nigra from the time of injection.

Electrolytic lesions of the nigro-striatal bundle at the level of the lateral hypothalamus, pretreatment with 6-hydroxydopamine, or injections of ^{14}C-dopa dorsal to the substantia nigra each produced profound reductions in the amount of radioactivity subsequently recovered in the caudate-putamen. Pretreatment with the dopa decarboxylase inhibitor Ro 4-4602 or reserpine also produced significant reductions (Fibiger et al., 1973). These various experiments suggest that the activity recovered from the caudate-putamen reflects axonal transport of dopamine rather than such processes as diffusion or transport via the circulation. The reduction in transport brought about by treatment with the dopa decarboxylase inhibitor indicates that the decarboxylation of dopa to dopamine is a prerequisite for transport. Furthermore, the reduction following reserpine pretreatment indicates that the binding of dopamine to amine storage granules may be essential to the transport.

Although there is a well known caudato-nigral pathway (Voneida, 1960) and injections of radioactive leucine into the caudate lead to the appearance of labelled protein in the substantia nigra, there was no evidence of dopamine transport to the substantia nigra when radioactive L-dopa was injected into the caudate-putamen.

Thus, the nigro-striatal pathway transports proteins including such enzymes as tyrosine hydroxylase necessary for amine production. It is also capable of transporting dopamine itself, possibly by binding the dopamine to storage granules which are also being transported. But dopamine will not be transported by non-dopaminergic neurons such as in the caudato-nigral tract.

ELECTRON MICROSCOPIC RADIOAUTOGRAPHY

The fine-structure of the nigro-striatal projection was traced by injection of 10 μC of L-leucine-4,5-H^3 (specific activity 38.6 Ci/mmol) into the substantia nigra followed by electron microscopic radioautographic examination of the caudate-putamen. At 1 and 4-day intervals after the injection, the animals were sacrificed by perfusion with 4% paraformaldehyde, 0.5% gluteraldehyde, and 0.54% glucose in 0.1 M phosphate buffer (pH 7.4). After dissection into small cubes, the partially fixed corpus striatum was fixed for a further two hours _in vitro_, and then post-fixed in 1% buffered osmium tetroxide. The tissue was embedded in epon-araldite and gold sections cut on an LKB microtome. These sections were placed on carbon-coated grids and Ilford L4 emulsion applied by the standard loop technique. After 4-6 weeks exposure, the sections were developed in microdol X and fixed in Kodak Rapidfix. The subcellular localization of radioactive proteins was then revealed by the location of silver grains from the photographic emulsion.

One day after the injection, synaptic terminals in the striatum were preferentially labelled to a high degree. Figure 6A shows a silver grain appearing over an asymmetrical contact between a small nerve ending and a dendritic spine. The vesicles are moderately pleomorphic. Figure 6B shows a silver grain over a second type of synapse. This one is symmetrical with highly pleomorphic vesicles. Figure 6C shows a silver grain appearing over a dendritic spine which is in contact with an asymmetrical synapse with moderately pleomorphic vesicles.

By examining many electron microscopic radioautograms, it has

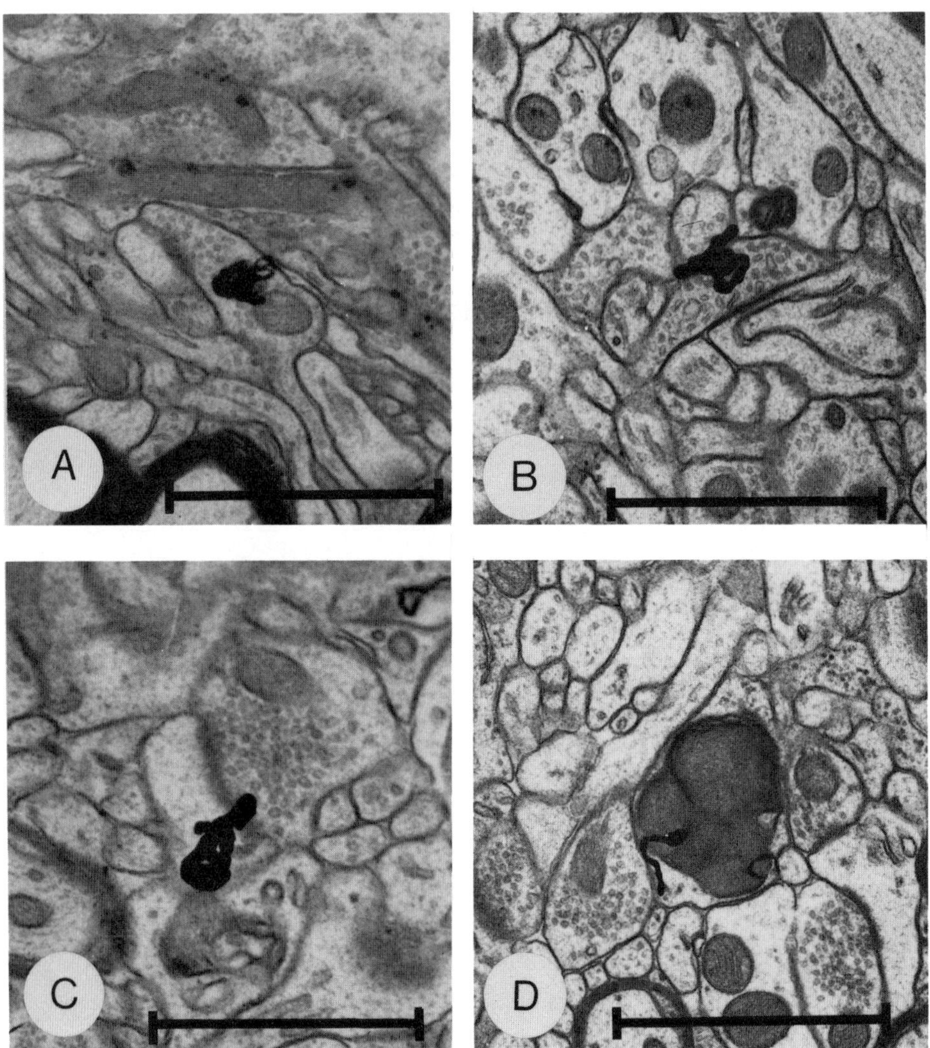

FIGURE 6. Electron microscopic autoradiographs of rat caudate-putamen following injection of ^{3}H-leucine into the substantia nigra. Electron dense silver grains appear over structures containing labelled protein. Calibration bars represent 1 μ. A - labelled asymmetrical synapse; B - labelled symmetrical synapse. Note highly pleomorphic vesicles; C - labelled dendritic spine; D - labelled degenerating profile containing two silver grains engulfed by astrocyte. Degeneration caused by intraventricular 6-hydroxydopamine. See text for experimental details.

been possible to put together a picture of what is taking place. Table III summarizes the information.

TABLE III

	24 Hours		96 Hours	
	Number of grains	Percent	Number of grains	Percent
Synapse:				
Asymmetrical Symmetrical	81 32 } 113	56	62 12 } 74	37
Axon	42	21	59	29
Dendrite or dendritic spine	31	15	46	23
Soma	9	4.5	16	8
Glia and others	7	3.5	6	3
	202		201	

As can be seen from the table, 24 hours after the injection of leucine into the substantia nigra (which is the time of the rapid flow peak), 77% of the grains were over synapses or axons. There were two kinds of synapse that were labelled in a ratio of about 3:1. The asymmetrical type (Figure 6A) was the most common. There was some labelling over dendritic spines and dendrites (15%) and even a few grains over cell somata and glial elements.

After 96 hours when protein transported by slow flow was arriving, there was a lower accumulation of grains in the nerve endings themselves but a higher accumulation in axons. There was a significantly greater number of grains appearing over dendrites with a few grains again appearing in glial cells and neuronal cell bodies.

The accumulation of grains over dendrites and their increase with time suggest the possibility of trans-synaptic movement of protein. There is other evidence in the literature to support this concept (Globus et al., 1971; Grafstein, 1969; Grafstein, 1971) but clearly different kinds of experiments will be needed to prove the point and to demonstrate, for example, that the data does not reflect local protein breakdown and resynthesis.

Information regarding which of the two types of labelled

nerve endings might be dopaminergic was obtained by administering 6-hydroxydopamine intraventricularly following the intra-nigral injection of ^{3}H-leucine. Three days after the 6-hydroxydopamine administration significant numbers of degenerating profiles were observed in the caudate and putamen. Some of the profiles contained silver grains (Figure 6D). Without exception, the profiles degenerating following 6-hydroxydopamine treatment were of the asymmetrical type, which by axoplasmic tracing were found to make up 75% of the labelled terminals. It can be postulated that the symmetric terminals are associated with the second nigro-striatal pathway which was inferred from the discrepancy between axoplasmic flow and tyrosine hydroxylase levels following 6-OHDA treatment (Fibiger et al., 1972). Such a second pathway has also been predicted from physiological studies (Feltz and De Champlain, 1972; Frigyesi and Purpura, 1967; York, 1970).

Axoplasmic flow was also used to gain information concerning the descending pathways making contact with the dendrites of dopaminergic cell bodies in the substantia nigra. Figure 7A shows a light microscopic radioautograph of the substantia nigra of a rat injected with 10 μC of ^{3}H-leucine in the globus pallidus 24 hours previously. The injection was reasonably well confined to the globus pallidus proper with no grains being detected in the caudate-putamen. There was heavy labelling in the subthalamic nucleus which is the principal known efferent pathway of the external globus pallidus. However, there was also significant labelling in the rostral substantia nigra confirming the existence of a pallido-nigral pathway (McGeer et al., 1973b). The pars compacta, which is shown in Figure 7A, is more heavily labelled than the pars reticulata.

Figure 7B shows an electron microscopic radioautograph from a comparable experiment. The silver grain is over a symmetrical type of nerve ending making contact with a main stem dendrite in the substantia nigra.

Figure 7C shows the same type of main stem dendrite with symmetrically arranged nerve endings around it. In this case the dendrite is degenerating because of the prior administration of intraventricular 6-OHDA, establishing that it is dopaminergic. The labelled protein from a ^{3}H-leucine injection into the globus pallidus here appears to be in the degenerating dendrite, implying trans-synaptic movement into the dopaminergic cell.

When labelled GABA was injected into the caudate no radioactivity could be detected in the substantia nigra. However, when it was injected into the globus pallidus, GABA was picked up in the substantia nigra, both chemically, and as illustrated by Figure 7D, by electron microscopic radioautography. Figure 7D shows

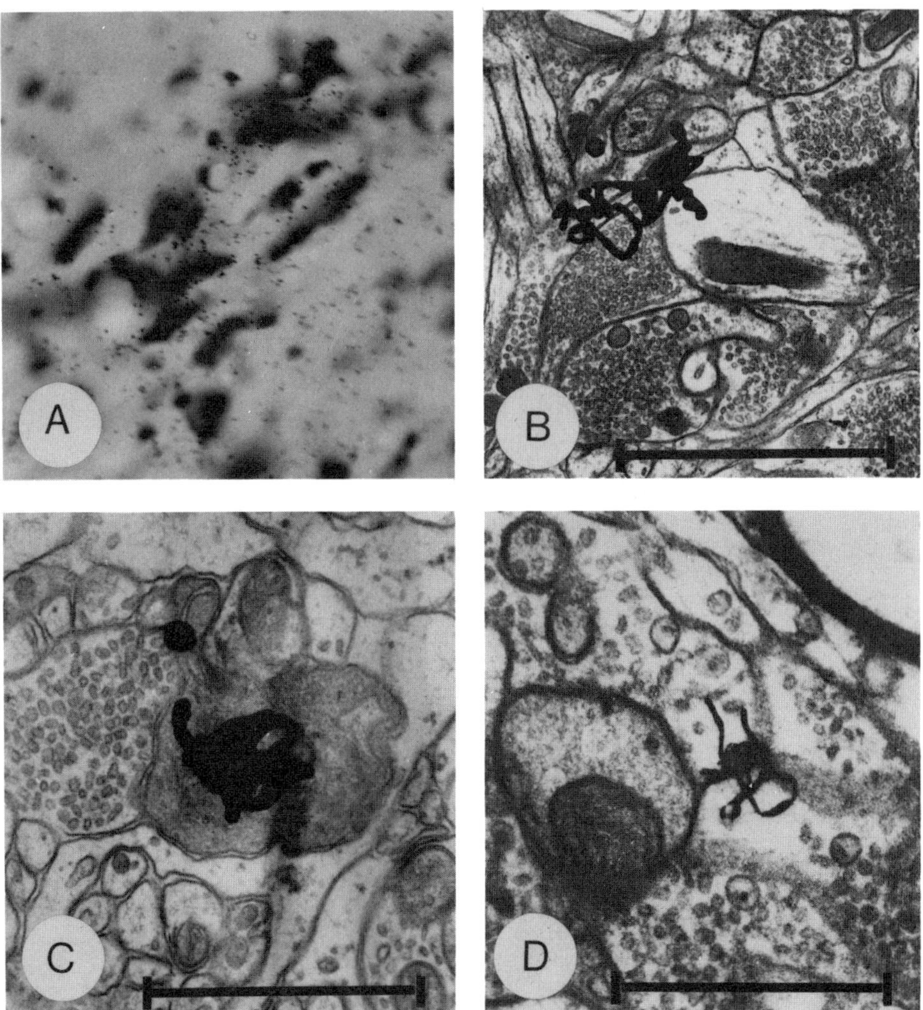

FIGURE 7. Light (A) and electron microscopic autoradiographs (B, C, D) of rat substantia nigra following injection of ^{3}H-leucine or ^{3}H-GABA into the globus pallidus. A - pars compacta × 650. Small dots due to dense silver grains appearing over labelled protein. B - protein labelled terminal making symmetrical contact with healthy main stem dendrite. C - protein labelled main stem dendrite degenerating after intraventricular 6-hydroxydopamine. Note healthy nerve terminals from which protein may have originated. D - GABA labelled nerve terminal making symmetrical contact with main stem dendrite degenerating after intraventricular 6-hydroxydopamine. Calibration bars indicate 1μ. See text for details.

a silver grain over a nerve ending in the substantia nigra in contact with a degenerating dendrite. The degeneration was due to the intraventricular administration of 6-hydroxydopamine 2 days prior to the injection of ^{3}H-GABA into the globus pallidus. The animal was sacrificed 4 hours after the GABA injection and the electron microscopic radioautogram is of substantia nigra tissue. Thus, there is evidence that descending fibers from the globus pallidus make symmetrical contacts with main stem dopaminergic dendrites of the substantia nigra pars compacta. At least some of these fibers would appear to be GABA-containing since GABA itself is transported along these pathways to make similar contacts.

In summary, the lesion and axoplasmic flow studies done to date suggest the following:

(1) There is an ascending dopaminergic pathway from the substantia nigra to the caudate-putamen which has fine terminals containing slightly pleomorphic vesicles. The contacts are principally axo-dendritic of an asymmetrical type.

(2) Protein is transported along this pathway by slow and fast axonal flow. The major peak (B) travels mainly by slow flow and tyrosine hydroxylase activity coincides with this peak.

(3) Dopamine is also transported along this pathway by rapid flow, but does not move in the reverse direction.

(4) A second, non-dopaminergic type of neuron accounts for about 20% of the protein transport from the region of the substantia nigra which has symmetrical nerve endings containing highly pleomorphic vesicles. This type of neuron is not attacked by 6-hydroxydopamine under the conditions we have used and does not transport dopamine.

(5) The numerous cholinergic nerve endings in the striatum are probably almost entirely associated with intrinsic neurons.

(6) A GAD-containing path descends from the globus pallidus and/or entopeduncular nucleus to the substantia nigra where it makes symmetrical synapses on main stem dendrites of dopaminergic neurons. This pallido-nigral tract will transport both labelled protein and radioactive GABA but will not transport dopamine.

(7) The possibility of trans-synaptic movement of protein is suggested by the increase with time of the labelling of dendrites in the caudate-putamen following axoplasmic transport of labelled protein from the substantia nigra; similar labelling of dendrites in the substantia nigra has been found following injections of ^{3}H-leucine into the globus pallidus.

APPLICATIONS TO HUMAN DISEASE

The relationships between the amine pathways in the basal ganglia may have interesting implications for at least two diseases of the extrapyramidal system, Parkinson's disease and Huntington's chorea. These diseases are characterized by opposite difficulties in movement. In Parkinson's disease there is rigidity and akinesia, while in Huntington's chorea there is a dancing sort of gait. The principal pathological finding in Parkinson's disease is a loss of cells in the substantia nigra pars compacta, while in Huntington's chorea the loss is in the caudate and putamen. Parkinson's disease involves a loss of dopamine (Ehringer and Hornykiewicz, 1960) and its synthesizing enzyme, tyrosine hydroxylase (McGeer and McGeer, 1973a) in the caudate and putamen but choline acetylase is unaffected (McGeer and McGeer, 1971b). Huntington's chorea involves a patchy loss of choline acetylase in the caudate and putamen (McGeer and McGeer, 1973b; McGeer et al., 1973c; Bird et al., 1973) but tyrosine hydroxylase is unaffected (McGeer and McGeer, 1973a). Drugs which deplete dopamine or block its receptor sites produce a Parkinsonian-like rigidity, while agents which block central acetylcholine receptor sites produce a chorea-like effect (McGeer, 1963). Thus, the opposite actions of dopamine and acetylcholine containing cells on extrapyramidal movement seem to be established.

The role of GABA containing cells in extrapyramidal function is less certain. Agents which deplete or block GABA tend to produce convulsions while those that raise its levels tend to produce sedation. There are no reports on a more restricted effect on motor function.

We have reported reduced GAD in the globus pallidus and substantia nigra of Parkinsonian patients (McGeer et al., 1971c; McGeer et al., 1971d) while Hornykiewicz and colleagues (Bernheimer and Hornykiewicz, 1962; Lloyd and Hornykiewicz, 1973) have also reported reduced GAD in the caudate and putamen. However, Lloyd and Hornykiewicz report that GAD returns to normal following prolonged L-dopa therapy and suggest that the GAD reduction is a secondary effect of the disease.

Perry et al. (1973) reported low levels of GABA in the caudate-putamen, globus pallidus and substantia nigra of Huntington's chorea cases. We have also found reduced GAD in these nuclei (McGeer and McGeer, 1973b; McGeer et al., 1973c), and Bird et al. (1973) have reported even sharper losses in the caudate, putamen and globus pallidus. Whether the decrease implies death of GABA-containing cells or a secondary depression of function is not known. However, the therapeutic possibilities of changing GABA levels in the basal ganglia of both Parkinsonians and Huntington's choreics needs to be explored.

SUMMARY

The biochemical neuroanatomy of the rat basal ganglia was investigated using enzymatic, axoplasmic flow and electron microscopic techniques.

Tyrosine hydroxylase in both striatum and substantia nigra decreased following mid-hypothalamic or anterior commissural brain hemitransections. Glutamic acid decarboxylase (GAD) in the nigra decreased sharply after the mid-hypothalamic, but not anterior hemitransections. GAD in both the globus pallidus and striatum decreased slightly after anterior hemitransections. Choline acetylase in the striatum was unaffected by transections or by extensive thalamic or cortical lesions. These data are consistent with an ascending nigro-striatal dopaminergic tract, a descending pallido-nigral gabaminergic tract and internal striatal cholinergic and gabaminergic neurons.

After the injection of ^{3}H-leucine or ^{3}H-dopa into the substantia nigra, labelled protein or dopamine appeared in the striatum. Electron microscopic autoradiography showed the protein to be accumulated preferentially in striatal nerve endings with slightly pleomorphic vesicles making asymmetrical axo-dendritic contacts. Fractionation of the labelled protein on DEAE-cellulose columns indicated four radioactive protein peaks differently labelled with time. The peak containing tyrosine hydroxylase activity coincided with slow flow.

6-hydroxydopamine treatment caused losses in nigro-striatal tyrosine hydroxylase activity correlatable with decreased nigro-striatal dopamine transport and specific loss of the labelled protein peak corresponding to tyrosine hydroxylase activity. Electron microscopic autoradiography showed degeneration of asymmetrical nerve endings in the corpus striatum containing labelled protein.

When ^{3}H-leucine or ^{3}H-GABA was injected into the globus pallidus, labelled protein or GABA was found in the substantia nigra and appeared, on electron microscopic autoradiography, in nerve endings making symmetrical contacts with main stem dendrites. Moreover, in 6-OHDA-treated animals, labelled, healthy nerve endings were seen in contact with degenerating main stem dendrites. These results indicate the descending GABA-containing fibers make contact with dopaminergic cells in the substantia nigra.

ACKNOWLEDGEMENTS

The authors are grateful for support from the Medical Research Council of Canada, the Sloan Foundation, the Muscular Dystrophy Association of America and the Province of British Columbia.

REFERENCES

Anden, N. E., Dahlstrom, A., Fuxe, K., Larsson, K., Olson, L. and Ungerstedt, U., 1966, Ascending monoamine neurons to the telencephalon and diencephalon, Acta Physiol. Scand. 67:313.

Aprison, M. H., Davidoff, R. A. and Werman, R., 1970, Glycine: its metabolic and possible roles in nervous tissue, in "Handbook of Neurochemistry, Volume 3" (A. Lajtha, ed.), pp. 381-397, Plenum Press, New York, London.

Bernheimer, H. and Hornykiewicz, O., 1962, Das verhalten einiger enzyme im gehirn normaler und Parkinson-kranker menschen, Naunyn-Schmiedeberg Arch. Exp. Path. Pharmak. 243:295.

Bird, E. D., Mackay, A. V. P., Rayner, C. N. and Iversen, L. L., 1973, Reduced glutamic acid decarboxylase activity of post mortem brain in Huntington's chorea, Lancet i:1090.

Carpenter, M. B. and Strominger, N. L., 1967, Efferent fibers of the subthalamic nucleus in the monkey. A comparison of the efferent projections of the subthalamic nucleus, substantia nigra and globus pallidus, Am. J. Anat. 121:41.

Dählstrom, A. and Fuxe, K., 1964, Evidence for the existence of monoamine-containing neurons in the central nervous system, Acta Physiol. Scand. 62, Suppl. 232:55.

Ehringer, H. and Hornykiewicz, O., 1960, Verteilung von noradrenalin und dopamin (3-hydroxytyramin) im gehirn des menschen und ihr verhalten bei erkrankungen des extrapyramidal systems, Klin. Wschr. 38:1236.

Falck, B., Hillarp, N. A., Thieme, G. and Torp, A., 1962, Fluorescence of catecholamines and related compounds condensed with formaldehyde, J. Histochem. Cytochem. 10:348.

Feltz, P. and De Champlain, J., 1972, Persistence of caudate unitary responses to nigral stimulation after destruction and functional impairment of the striatal dopaminergic terminals, Brain Res. 43:595.

Fibiger, H. C., Pudritz, R. E., McGeer, P. L. and McGeer, E. G., 1972, Axonal transport in nigro-striatal and nigro-thalamic neurons: Effects of medial forebrain bundle lesions and 6-hydroxydopamine, J. Neurochem. 19:1697.

Fibiger, H. C., McGeer, E. G. and Atmadja, S., 1973, Axoplasmic transport of dopamine in nigro-striatal neurons, J. Neurochem., in press.

Fonnum, F., 1970, Topographical and subcellular localization of ChAc, J. Neurochem. 17:1029.

Frigyesi, T. L. and Purpura, D. P., 1967, Electrophysiological analysis of reciprocal caudato-nigral relations, Brain Res. 6:440.

Globus, A., Lux, H. D., Schubert, P. and Kaups, P., 1971, Labelling of nearby neurons following the intracellular iontophoresis of H^3-glycine, Anat. Rec. 169:325.

Grafstein, B., 1969, Axonal transport: Communication between soma and synapse, in "Advances in biochemical psychopharmacology", (E. Costa and P. Greengard, eds.), pp. 11-25, Raven Press, New York, Volume 1.

Grafstein, B., 1971, Transneuronal transfer of radioactivity in the central nervous system, Science 172:177.

Graham, L. T. Jr., 1972, Intraretinal distribution of GABA content of GAD activity, Brain Res. 36(2):476.

Hattori, T., McGeer, P. L., Fibiger, H. C. and McGeer, E. G., 1973, On the source of GABA-containing terminals in the substantia nigra. Electron microscopic autoradiographic and biochemical studies, Brain Res. 54:103.

Lloyd, K. G. and Hornykiewicz, O., 1973, L-glutamic acid decarboxylase in Parkinson's disease: effect of L-dopa therapy, Nature, in press.

McGeer, E. G. and McGeer, P. L., 1973a, New concepts in neurotransmitter regulation, (A. J. Mandell, ed.), pp. 69-89, Plenum Press, New York, London.

McGeer, E. G., McGeer, P. L. and Wada, J. A., 1971, Distribution of tyrosine hydroxylase in human and animal brain. J. Neurochem. 18:1647.

McGeer, E. G., Fibiger, H. C., McGeer, P. L. and Brooke, S., 1973a, Temporal changes in amine synthesizing enzymes of rat extrapyramidal structures after hemitransections or 6-hydroxydopamine administration, Brain Res. 52:289.

McGeer, P. L., 1963, Central amines and extrapyramidal function, J. Neuropsych. 4:247.

McGeer, P. L. and McGeer, E. G., 1971b, Cholinergic enzyme systems in Parkinson's disease, Arch. Neurol. 25:265.

McGeer, P. L. and McGeer, E. G., 1973b, Neurotransmitter synthetic enzymes, Progress in Neurobiology 2:67.

McGeer, P. L., McGeer, E. G., Fibiger, H. C. and Wickson, V., 1971b, Neostriatal choline acetylase and cholinesterase following selective brain lesions, Brain Res. 35:308.

McGeer, P. L., McGeer, E. G., Wada, J. A. and Jung, E., 1971c, Effect of globus pallidus lesions and Parkinson's disease on brain glutamic acid decarboxylase, Brain Res. 32:425.

McGeer, P. L., McGeer, E. G. and Wada, J. A., 1971d, Glutamic acid decarboxylase in Parkinson's disease and epilepsy, Neurology 21:1000.

McGeer, P. L., Fibiger, H. C., Maler, L., Hattori, T. and McGeer, E. G., 1973b, Evidence for descending pallido-nigral GABA-containing neurons, in "Advances in neurology, Volume 5" (F. McDowell and A. Barbeau, eds.), Parkinson's Disease - Proceedings of the 2nd Canadian-American Conference, Raven Press.

McGeer, P. L., McGeer, E. G. and Fibiger, H. C., 1973c, Choline acetylase and glutamic acid decarboxylase in Huntington's chorea, Neurology, in press.

Moore, R. Y., Bhatnagar, R. K. and Heller, A., 1971, Anatomical and chemical studies of a nigro-neostriatal projection in the cat, Brain Res. 30:119.

Perry, T. L., Hansen, S. and Kloster, M., 1973, Huntington's chorea, deficiency of γ-aminobutyric acid in brain, New Eng. J. Med. 288:337.

Singh, V. K., Fibiger, H. C., McGeer, E. G. and McGeer, P. L., 1973, Biochemical studies on axonal transport of proteins in nigro-striatal system, Abstracts of Canadian Federation of Biological Societies 20:105.

Storm-Mathisen, J. and Fonnum, F., 1971, Quantitative histochemistry of glutamate decarboxylase in the rat hippocampal region, J. Neurochem. 18:1105.

Ungerstedt, U., 1971, Stereotaxic mapping of monoamine pathways, Acta Physiol. Scand. Suppl. 367.

Voneida, T. J., 1960, An experimental study of the course and destination of fibers arising in the head of caudate nucleus in the cat and monkey, J. Comp. Neurol. 115:75.

York, D. H., 1970, Possible dopaminergic pathway from substantia nigra to putamen, Brain Res. 20:233.

CYCLIC AMP AND THE INHIBITION OF CEREBELLAR PURKINJE CELLS BY NORADRENERGIC SYNAPSES

Floyd E. Bloom

Laboratory of Neuropharmacology, DSMHR, NIMH

St. Elizabeths Hospital, Washington, D. C. 20032

INTRODUCTION

When considering the many important concepts implicit in the title of this Symposium, there are three major questions which quickly emerge. What is the nature of the chemical which can transmit information in cellular terms between nerve cells, how can these substances be identified as to site and mechanism of action, and how is the information coded within simple molecules such as synaptic transmitters? In our laboratory, we have gone to considerable lengths to establish the identification of norepinephrine (NE) as one of the inhibitory transmitters which function to inhibit cerebellar Purkinje cells. This system of chemical inhibition would appear to offer a unique insight into the conceptual problems underlying our present converence, and this paper will explore the extent to which this insight can presently be helpful.

The function of any central noradrenergic synapses requires the identification and selective activation of isolatable discrete noradrenergic pathways and the detection of the activity patterns of these synapses in the unrestrained experimental animal. In the past, emphasis has been placed mainly upon data which demonstrate that iontophoretically administered NE can act with sufficient potency and regularity to merit consideration as a synaptic transmitter. At present, many differences in interpretation are still associated with data obtained by applying the iontophoretic method to CNS areas in which the precise anatomy of the NE-containing terminals has yet to be related to the cells undergoing testing.

Beyond these interpretive problems, however, it has been possible to determine the actions of NE fiber systems arising from the pontine nucleus, locus coeruleus (LC), and projecting to the cerebellar Purkinje cells (Hoffer _et al._, 1973) and to the hippocampal pyramidal cells (Segal and Bloom, 1973) and to record from the LC neurons themselves during unrestrained observations of sleeping and waking behavior (Chu and Bloom, 1973). Let us first concentrate upon two questions: what are the principal effects of iontophoretically applied NE; what is the nature of the physiological central NE receptor and the molecular mechanisms by which NE synapses can exert these effects?

Principal Actions of Iontophoretically Applied NE

The extensive work of the past 5 years has amply supported the view that NE is able to affect the discharge of neurons in all portions of the neuraxis (see Table I of Hoffer and Bloom, 1973). Present work is more realistically directed at determining the functional significance of the presence or absence of responses to NE and the qualitative nature of the response, i.e., excitatory or inhibitory. Retrospective analysis of experimental inconsistencies suggests that major variables were previously uncontrolled. Thus, excitatory responses to NE in the cerebral cortex were found to be more frequent when the animal was unanesthetized or under the influence of halothane anesthesia while inhibitory cortical responses were more frequent with barbiturate anesthesia (Johnson _et al._, 1969). Similarly, the pH of the NE solution to be iontophoretically applied also takes on critical importance, as pH values less than 4 are associated with major increases in the frequency of excitatory NE responses (Frederickson _et al._, 1972). An extreme view of the anesthesia-pH controversy is that all excitatory responses to NE are the result of vasoconstriction artefacts (Stone, 1971), a suggestion which was quickly rejected (Boakes _et al._, 1972).

The eclectic view of these controversies must await additional observations on several other factors such as the cytological or functional heterogeneity of the cell population being tested: it is known that when defined populations of neurons are tested, the responses are far more reproducible than when all randomly encountered neurons in a given CNS region are lumped together (Salmoiraghi and Stefanis, 1967). An additional cytological index which must be considered is whether or not the population of cells to be tested receives a demonstrable synaptic input of NE-containing terminals.

When we only consider results obtained from those populations of defined post-synaptic cells for which NE-containing nerve terminals have been demonstrated by light or electron microscopic techniques, NE almost always inhibits (see Bloom and Hoffer, 1973).

The most serious obstacle to interpretation of these results has been that the NE-containing pathway has not been amendable to selective electrical activation for comparisons of qualitative and pharmacological results on the specified post-synaptic population of test neurons. Conversely, synaptic effects which were subjectable to selective activation were neither exclusively NE-containing nor sensitive to NE antagonists. Thus, in neither the olfactory mitral cells (Salmoiraghi _et al._, 1964) nor the hypothalamic neurosecretory cells (Nicoll and Barker, 1971) could the recurrent antidromic synaptic inhibition be removed by the results of acute or chronic NE depletion, or NE receptor blockade.

Over the past several years we have pursued the mechanisms by which NE slows the discharge of cerebellar Purkinje cells (see Hoffer _et al._, 1971a, b). We have used light and electron microscopy to establish that these cells receive NE-containing synapses onto their dendrites (Bloom _et al._, 1971) and that these NE fibers arise from the LC (Olson and Fuxe, 1971; Bloom _et al._, 1972a). By electrophysiological methods, we have analyzed the pharmacological receptors of the Purkinje cells (Siggins _et al._, 1971) and the effects of electrical activation of the pathway (Siggins _et al._, 1971; Hoffer _et al._, 1973). Briefly, these experiments indicate that NE slows Purkinje cells by interaction with a beta receptor, and that NE prolongs the pauses between bursts of single spikes without effect on climbing fiber responses. By intracellular recordings, NE hyperpolarizes the membrane of Purkinje cells and this hyperpolarization is generally accompanied by increased membrane resistance (but never by increased membrane conductance). The actions of norepinephrine on the Purkinje cell are blocked by iontophoretic application of either a beta antagonist (MJ-1999), prostaglandins of the E series or nicotinate; the latter two both inhibit adenylate cylcase in some autonomic tissues. The effects of norepinephrine on discharge rate and membrane parameters are precisely emulated by iontophoretic application of cyclic AMP, and the effects of both the applied cyclic AMP and of NE are potentiated by any of several phosphodiesterase inhibitors. On the basis of these data, we proposed (Hoffer _et al._, 1971b) that the synaptic action of NE was mediated by an interaction via the adenyl cyclase of the cerebellar cortex, known to be highly responsive to NE.

With the anatomical information that the cerebellar NE fibers arose from the LC (see above), it was possible to test this proposal by activating and analyzing the effects of the pathway on Purkinje cell properties. These experiments disclosed that stimulation of the pathway inhibited Purkinje cell discharge, especially single spike bursts, that the inhibitory effects of stimulating LC required active synthesis of NE, and that no effects on cerebellar neuronal discharge were observed when the area of the locus was stimulated in animals pretreated with 6-hydroxydopamine to eradicate the adrenergic projection to the cerebellum (Hoffer *et al.*, 1973). By intracellular recording during the activation of the LC, Purkinje cells were found to be hyperpolarized and this hyperpolarization was usually accompanied by a definitive increase in the resistance of the membrane (Siggins *et al.*, 1971). Similar effects of NE have been observed on motoneurons (Engborg and Marshall, 1971). Pharmacologically, activation of the LC led to an inhibition of spontaneous discharge which could be potentiated by local iontophoresis of phosphodiesterase inhibitors onto the Purkinje cell and could be blocked by local iontophoretic administration of prostaglandins of the E series (see Hoffer *et al.*, 1973). All these results supported the concept that this adrenergic projection could be operating by the trans-synaptic elevation of cyclic AMP in Purkinje cells. The latter observation has now been documented by application to tissue sections of an immunocytochemical method for cyclic AMP (Bloom *et al.*, 1972b). Using this method we have observed that topical application of NE or electrical activation of the LC will elevate the number of Purkinje cells showing positive immunocytological staining for cyclic AMP from resting frequencies of 5-15% to levels greater than 75% (Siggins *et al.*, 1973). Neither topical application of GABA, glycine, histamine, or acetylcholine, or electrical activation of other cerebellar pathways has this effect on Purkinje cell cyclic AMP (Siggins *et al.*, 1973).

Recently we have applied the same techniques to determine that some cells of the locus coeruleus also project to hippocampal pyramidal cells and their inhibitory action, like that of iontophoretically applied NE, appears also to be potentiated by phosphodiesterase inhibition and antagonized by prostaglandins of the E series (Segal and Bloom, 1973). Furthermore, the inhibitory actions of dopamine in the caudate nucleus (Connor, 1970) also may be mediated by activation of adenylate cyclase (in preparation) as they are in the bovine sympathetic ganglion in vitro (Kebabian *et al.*, 1972). However, we still do not know whether the ability to activate the adenylate cyclase of the Purkinje cell or that of other specific post-synaptic cells reflects special properties of: 1) the receptive cell; 2) the nucleus whose axons make the synapses onto the cell or 3) the transmitter molecule. Both dopamine and 5-HT can also activate the adenylate cyclase of aplysia ganglia (Cedar and Schwartz, 1972), but here the effects of 5-HT on cyclic AMP synthesis appear to be independent of transmitter release or of changes in synaptic potentials.

Thus, nature may have selected various monoamines to mediate a function between nerve cells by activation of adenylate cyclase. However these transmitters produce their effects on membrane potentials and ionic fluxes, the same transmitter molecules may still be called on to transmit more "classically" defined synaptic messages at other junctional sites.

Neurohumoral Coding

The co-existence of so many transmitters of varying effectiveness and variable mechanisms demands that some theoretical assessment be attempted (see Bloom, 1973). Clearly, nature could have devised a sophisticated and sensitive nervous system without this chemical redundancy. For example, interneuronal operations could be designated only by terms of circuitry (i.e., cell A transmits to B and C, but not to D...N) and a binary state of activity/inactivity determined by the actions of one simple excitant and one simple inhibitor agent. However, it is equally clear that this was not the way in which presently available nervous systems evolved. This fact, therefore, implies that some additional purposes must be served by the non-redundant properties of each transmitter substance.

One index against which the actions of the various transmitter substances could be analyzed is time. As revealed by cellular electrophysiological testing and behavioral observations, amino acids, certain nicotinic cholinergic actions and perhaps substance P produce effects with rapid onset. Monoamines and muscarinic cholinergic actions proceed more slowly and for longer duration and in part via cyclic nucleotides, while hypothalamic releasing factors and peripheral polypeptides with central actions (see Bloom, 1973) act over periods of hours-to-days in duration.

However, it seems likely that when additional experimental results are available, it may be more appropriate to categorize the various chemical modes of transmitter actions along cytological or functional lines (e.g., special environmental or affective state monitoring circuits) or in terms of the succinctness or diffuseness with which the chemical messages must be transmitted. Thus, cyclic nucleotide mediation of a synaptic junctional transmission can also trigger a cascade of neurochemical events affecting protein conformations in areas as separated as the plasma membrane and the nucleus (see Greengard _et al._, 1973). Sequential intracellular mediation of synaptic messages by cyclic nucleotides offers at least the possibility for longer term changes in neuronal membrane and in carbohydrate and protein metabolism by the intermediary role of cyclic nucleotide dependent protein kinases (Johnson _et al._, 1971). It should at least be considered that some chemical synapses whose actions are "coded" into intracellular nucleotides or other forms of second messenger molecules may thus carry out trophic effects which far outweigh the more transient electrophysiological actions generated immediately at the cell surface membrane.

REFERENCES

Bloom, F. E., 1973, Dynamic synaptic communications: finding the vocabulary, Brain Res., in press.

Bloom, F. E., Hoffer, B. J., and Siggins, G. R., 1971, Studies on norepinephrine-containing afferents to purkinje cells of rat cerebellum. I. Localization of the fibers and their synapses, Brain Res. 25:501.

Bloom, F. E., Hoffer, B. J. and Siggins, G. R., 1972, Norepinephrine mediated synapses: a model system for neuropsychopharmacology, Biol. Psychiat. 4:157.

Bloom, F. E., Hoffer, B. J., Battenberg, E. F., Siggins, G. R., Steiner, A. L., Parker, C. W. and Wedner, H. J., 1972b, Adenosine 3',5'-monophosphate is localized in cerebellar neurons: immunofluorescence evidence, Science 177:436.

Boakes, R., Bradley, P., Candy, J. and Dray, A., 1972, Noradrenaline artefacts?, Nature (Lond.) 239:151.

Cedar, H. and Schwartz, J. H., 1972, Cyclic adenosine monophosphate in the nervous system of Aplysia californica. II. Effect of serotonin and dopamine, J. Gen. Physiol. 60:570.

Chu, N-s. and Bloom, F. E., 1973, Norepinephrine-containing neurons: changes in spontaneous discharge patterns during unrestrained sleeping and waking, Science 179:908.

Connor, J. D., 1970, Caudate nucleus neurones; correlation of the effects of substantia nigra stimulation with iontophoretic dopamine, J. Physiol. (Lond.) 208:691.

Engborg, I. and Marshall, K. C., 1971, Mechanism of noradrenaline hyperpolarization in spinal cord motorneurones of the cat, Acta Physiol. Scand. 83:142.

Frederickson, R., Jordan, L. and Phillis, J., 1972, The action of noradrenalin on cortical neurons, Brain Res. 35:556.

Greengard, P., Kebabian, J. W. and McAffee, D., 1973, Studies on the role of cyclic AMP in neural function, in "Pharmacology and the Future of Man," (R. A. Maxwell and G. H. Acheson, eds.), vol. V, p. 207, S. Karger, Basel.

Hoffer, B. J., Siggins, G. R., Oliver, A. P. and Bloom, F. E., 1973, Activation of the pathway from locus coeruleus to rat cerebellar Purkinje neurons: pharmacologic evidence of noradrenergic central inhibition, J. Pharmacol. Exp. Ther. 184:553.

Hoffer, B. J., Siggins, G. R. and Bloom, F. E., 1971a, Studies on norepinephrine-containing afferents to Purkinje cells of rat cerebellum. II. Sensitivity of Purkinje cells to norepinephrine and related substances administered by microiontophoresis, Brain Res. 25:522.

Hoffer, B. J., Siggins, G. R., Oliver, A. P. and Bloom, F. E., 1971b, Cyclic AMP mediation of norepinephrine inhibition in rat cerebellar cortex: a unique class of synaptic responses, Ann. N. Y. Acad. Sci. 185:513.

Hoffer, B. J. and Bloom, F. E., 1973, Effects of norepinephrine on central neurons, in "Actions on Neurotransmitters in the CNS," (C. Hockman, ed.), Charles Thomas Co., in press.

Johnson, E., Roberts, M. and Straughan, D., 1969, The responses of cortical neurons to monoamines under differing conditions, J. Physiol. (Lond.) 203:261.

Johnson, E. M., Maeno, H. and Greengard, P., 1971, Phosphorylation of endogenous protein of rat brain by cyclic adenosine 3',5'-monophosphate-dependent protein kinase, J. Biol. Chem. 246:7731.

Kebabian, J. W., Petzold, G. L. and Greengard, P., 1972, Dopamine sensitive adenylate cyclase in caudate nucleus of rat brain and its similarities to the dopamine receptor, Proc. Nat. Acad. Sci. (U.S.A.) 69:2145.

Nicoll, R. A. and Barker, J. L., 1971, The pharmacology of recurrent inhibition in the supraoptic neurosecretory system, Brain Res. 35:501.

Olson, L. and Fuxe, K., 1971, On the projections from the locus coeruleus noradrenaline neurons, Brain Res. 28:165.

Salmoiraghi, G. C. and Stefanis, C., 1967, Central synapses and suspected transmitters, Int. Rev. Neurobiol. 10:1.

Salmoiraghi, G. C., Bloom, F. E. and Costa, E., 1964, Adrenergic mechanisms in rabbit olfactory bulb, Amer. J. Physiol. 207:1417.

Segal, M. and Bloom, F. E., 1973, A projection of the nucleus locus coeruleus to the hippocampus of the rat, Abst., 3rd Ann. Meeting Soc. Neurosci., in press.

Siggins, G. R., Hoffer, B. J. and Bloom, F. E., 1971, Studies on norepinephrine-containing afferents to Purkinje cells of rat cerebellum. III. Evidence for mediation of norepinephrine effects by cyclic 3,5'-adenosine monophosphate, Brain Res. 25:535.

Siggins, G. R., Hoffer, B. J., Oliver, A. P. and Bloom, F. E., 1971, Activation of a central noradrenergic projection to cerebellum, Nature 233:481.

Siggins, G. R., Battenberg, E. F., Hoffer, B. J., Bloom, F. E. and Steiner, A. L., 1973, Noradrenergic stimulation of cyclic adenosine monophosphate in rat Purkinje neurons: an immunocytochemical study, Science 179:585.

Stone, T., 1971, Are noradrenaline excitations artefact?, Nature (Lond.) 234:145.

ABSTRACT:

Interdisciplinary cellular level studies indicate that cerebellar Purkinje neurons receive norepinephrine-containing synapses. On these neurons, iontophoretic application of norepinephrine produces slowing of discharge by evoking a hyperpolarizing membrane response accompanied by increased membrane resistance. These unique effects are reproduced by iontophoresis of cyclic AMP. Additional pharmacologic evidence has suggested that cyclic AMP can mediate this inhibitory noradrenergic response. Extensions of this work have shown that electrical stimulation of the nucleus locus coeruleus will reproduce the inhibitory effects of iontophoretically applied norepinephrine and of cyclic AMP. Most recently, by the use of immunocytochemical staining for cyclic AMP, it has been possible to localize the nucleotide within individual Purkinje cells and to compare the intensity of this reactivity after application of pharmacologic and physiologic tests.

EVIDENCE FOR CHOLINERGIC TRANSMISSION IN THE CEREBRAL CORTEX

John W. Phillis

Department of Physiology, College of Medicine,

University of Saskatchewan, Saskatoon, Canada.

The postulate that acetylcholine (ACh) is a synaptic transmitter in the cerebral cortex has received support from several lines of investigation. All the components of the cholinergic metabolic system are found in the cortex. ACh is present (MacIntosh, 1941; Elliott et al., 1950), being especially concentrated in the nerve ending and synaptic vesicle fractions of subcellular preparations. The highest levels of ACh in the feline cortex are found in layers II, III and IV (Sastry, 1956). Choline acetyltransferase, the enzyme which synthesizes ACh from choline and acetylcoenzyme A, and acetylcholinesterase (AChE), the enzyme which hydrolyzes ACh, are both present in the cortex (Hebb and Silver, 1956; Burgen and Chipman, 1951).

Acetylcholinesterase-containing fibres have been observed histochemically ramifying throughout the various cortical layers of the cat although the majority are present in the deeper layers of the cortex (Krnjević and Silver, 1965). Some of the spindle or polymorph cells in layer VI stain for AChE and in these the enzyme appears to be located intracellularly. These neurones may give rise to some of the enzyme-containing nerve fibres which can be visualized in the other cortical layers. Many large pyramidal cells in layer V are lightly, but definitely, stained and here the enzyme is probably associated with the cell membrane.

The majority of AChE-containing nerve fibres in the feline cortex may originate from a system of fibres running beneath the cortex, which can be traced back to the corpus striatum and septal area (Krnjević and Silver, 1965). This projection may be comparable to the ascending system of fibres, originating in the ventral tegmental area of the brain stem, which has been described by Shute and Lewis (1967). Septal lesions in the rat reduce the ACh content of the cerebral cortex by about 40% (Pepeu et al., 1971). Studies on the development of AChE staining in embryos have shown that the primary cortical elements of the cat are devoid of the enzyme (Krnjević and Silver, 1966). In the undeveloped forebrain, AChE is found only in the lenticular nucleus and septum. Fibres spread out from these areas to innervate the rest of the hemisphere.

Histochemical localization of choline acetyltransferase in the cerebral cortex has been studied by Burt (1971) using a recently developed technique. Neurones of the neocortex were nearly devoid of enzymic activity, although the large pyramidal cells of the primary motor and sensory areas stained lightly. Densely staining granular structures resembling boutons terminaux were present in the cortical neuropil. These boutons were most abundant in the large pyramidal cell layers, both in the neuropil and on the surface of pyramidal neurones.

After isolation, the choline acetyltransferase and cholinesterase contents of cat cortical slabs fall rapidly for three days and then reach fairly stable levels (choline acetyltransferase at 35% of control value; cholinesterase at 43% of control) which are sustained for at least a further 14 days (Green et al., 1970). Histochemical studies of large isolated slabs have revealed that there is substantial retention of the normal AChE staining pattern with preservation of a system of fibres running beneath and ramifying into the cortex. In small slabs, where there was damage to the deeper layers of the cortex, there was an almost complete disappearance of cholinesterase-staining nerve fibres (Krnjević et al., 1970). These findings on isolated slabs suggest the presence of at least two systems of cholinergic fibres in the cerebral cortex. An extrinsic system, arising in subcortical structures including the septum and corpus striatum, supplying the majority of cholinergic nerve terminals in the cerebral cortex, supplemented by an intrinsic system, which may have its origins in the polymorph cells of cortical layer VI.

ACH RELEASE STUDIES

This concept of an extrinsic and intrinsic cholinergic innervation of cortical neurones derives further support from studies on ACh release from the cerebral cortex. In the presence of an inhibitor of cholinesterase, ACh is released from the cerebral cortex and can be collected in cups resting on the cortical surface or by push-pull cannulae inserted into the cortical gray matter. A considerable volume of literature on ACh release from the cerebral cortex has accumulated since Elliott et al., (1950) and

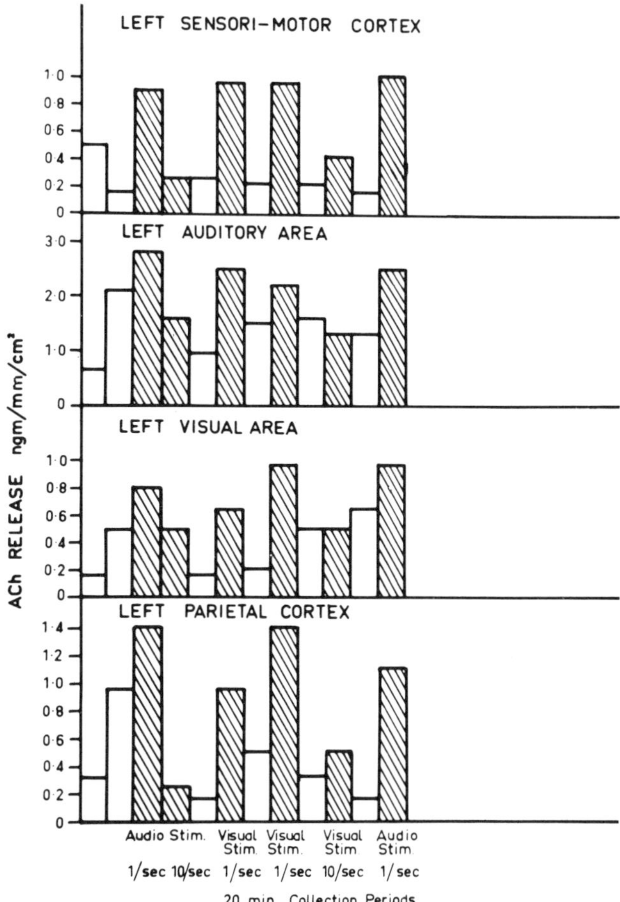

Fig. 1. Rates of release of ACh from different cortical areas of a diethyl ether anaesthetised cat. Cups were placed unilaterally on the left sensori-motor, auditory, visual and parietal cortices. The unhatched columns represent unstimulated release and cross-hatched columns the release during stimulation. The modality and frequency of stimulation is indicated at the base of each histogram. A 10 min gap was interposed between each stimulated collection and the subsequent unstimulated period. (From Phillis, 1968).

MacIntosh and Oborin (1953) published the original reports on the utilization of this method for studying cholinergic mechanisms. Release occurs spontaneously from the surface of the cortex of both anaesthetized and unanaesthetized animals and the rate of release from both hemispheres can be enhanced by stimulation of a variety of inputs, including somatic sensory, auditory and visual stimulation (Fig. 1) as well as stimulation of such subcortical structures as the brain stem reticular formation, the geniculate nuclei, the hypothalamus, medial thalamus and septum (Mitchell, 1963; Phillis and Chong, 1965; Szerb, 1967; Phillis, 1968; Collier and Mitchell, 1966; Hemsworth and Mitchell, 1969). ACh is released from chronically isolated cortical slabs, although the rate of release is only about one-fifth of that from the intact contralateral cortex (Collier and Mitchell, 1967). Direct stimulation of isolated slabs increases the rate of ACh release.

The ACh released from the cerebral cortex originates from synaptic terminals since it can be influenced by agents which affect neuronal activity or transmitter release from nerve terminals. The presence of calcium is known to be necessary for the release of ACh at peripheral cholinergic synapses and removal of this ion from the fluid bathing the cortex reduces the rate of ACh release (Randić and Padjen, 1967; Hemsworth and Mitchell, 1969). Addition of the ACh depletor hemicholinium-3 to the bathing fluid decreases both the spontaneous and evoked release of ACh (Hemsworth and Mitchell, 1969; Szerb *et al.*, 1970). Tetrodotoxin, which blocks conduction in nerve fibres also reduces ACh release from the cortex (Dudar and Szerb, 1969; Bjegović *et al.*, 1969).

The rate of release of ACh from the cerebral cortex is influenced by the state of arousal of the animal. Jasper and Tessier (1971) have shown that release is highest during waking and paradoxical (REM) sleep and lower during slow wave sleep. It appears therefore that desynchronization of the EEG is associated with an increase in ACh release and that it is the mechanism controlling the state of the EEG rather than the level of behavioral arousal of the animal which determines the rate of ACh release.

Release studies have yielded a considerable amount of information about the organization of the cholinergic projections to the cerebral cortex. The widespread nature of the enhanced release following stimulation of a specific sensory modality suggests that the ascending cholinergic system is a diffusely projecting pathway reaching most, if not all, areas of the cerebral cortex. The pathway is almost certainly involved in the control of EEG and behavioral states, although it may control these functions in conjunction with other aminergic projections. It is equally clear that the main afferent pathways to the specific sensory areas are non-cholinergic, although Mitchell and his colleagues (Collier and Mitchell 1966, 1967; Hemsworth and Mitchell, 1969) have presented some evidence to support the existence of a group of cholinergic thalamo-cortical and geniculo-cortical fibres associated with the augmenting and repetitive after-discharge responses that are evoked in the cortex by thalamic or geniculate stimulation.

The experiments on isolated cortical slabs are consistent with the presence of intracortical cholinergic neurones, previously inferred from studies on the ACh and enzymes contents of such slabs. Further evidence for the existence of these neurones will be cited in the section on cholinergic inhibition in the cerebral cortex.

EFFECTS OF ACH ON CORTICAL NEURONES:

MICROIONTOPHORETIC STUDIES

Iontophoretically applied acetylcholine depresses many cortical neurones, especially those in the more superficial layers of the cortex and excites others in the deeper layers.

ACh Excitation

ACh-excited neurones have been observed in all the regions of the neocortex in which they have been sought in rats, rabbits, cats and monkeys (Krnjević and Phillis, 1963a; Stone, 1972a). The most consistently excited neurones are the deep pyramidal cells, including those of the primary sensory areas, which exhibit irregular spontaneous activity of the projection type (Fig. 2) (Krnjević and Phillis, 1963b; Spehlmann, 1963; Crawford and Curtis, 1966). The excitation induced by ACh is of slow onset and very prolonged, often persisting for 10-60 sec after the termination of the ACh-ejecting current. The receptors are of the muscarinic type; acetyl-β-methylcholine and muscarine are powerful excitants and the effects of cholinomimetic agents are blocked by atropine and hyoscine but not by D-tubocurarine or dihydro-β-erythroidine (Krnjević and Phillis, 1963c). Inhibitors of cholinesterases, including neostigmine and tensilon, potentiate the action of ACh and frequently cause an excitation of cortical neurones. An excitation of superficial cortical neurones in the rat which is mediated by nicotinic receptors has been reported by Stone (1972a). The latency of ACh-excitation of these cells is shorter than that for the deeper neurones; nicotine is an effective agonist (it has little or no effect on the deep pyramidal cells) and D-tubocurarine the most effective antagonist.

Intracellular recordings have revealed that ACh initiates a slow depolarization of cortical neurones, which may or may not cause an outright discharge, but which potentiates the effect of other excitatory inputs (Krnjević *et al.*, 1971). Repetitive firing tends to occur, apparently due to a slowing of the repolarization after each action potential. During the depolarization, the cell membrane

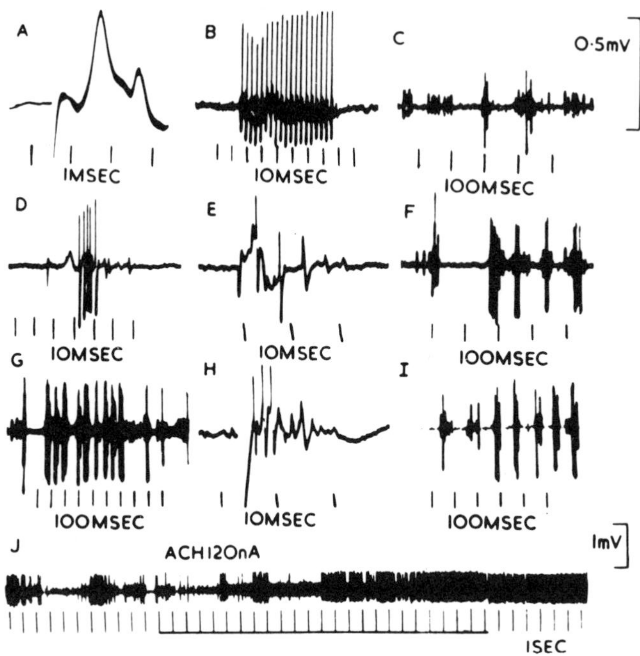

Fig. 2. Two Betz cells found at depth of 1•1 mm in lateral precruciate cortex and identified by antidromic activation from medullary pyramid (A); large and small spikes had latencies of 0•7 and 2•0 msec respectively. Both were driven at 300/sec antidromically (B) and fired spontaneously in the absence of stimulation (C). They gave early responses to peripheral (contralateral forepaw) stimulation (D), and fired after a short latency when stimulating specific afferent nucleus in thalamus (E). There were also marked late repetitive discharges after peripheral and specific thalamic excitation with single shocks (F and G). Similar early firing and later repetitive responses were evoked by transcallosal stimulation in H and I. Both units were strongly excited by ACh applied iontophoretically during period indicated by white line in J. (Note lower amplification). Cat under Dial. (Krnjević and Phillis, 1963b).

resistance is increased and studies on the voltage-current relationship indicate a decrease in membrane conductance for an ion with a highly negative equilibrium potential. Krnjević *et al.* (1971) suggest that ACh depolarizes by decreasing both the resting potassium conductance of the membrane and the delayed potassium currents associated with the repolarizing phase of the action potential. A similar mechanism has been postulated to account for the generation of slow cholinergic EPSPS in frog sympathetic ganglion cells (Weight and Votava, 1970).

Metabolic inhibitors, such as sodium azide and 2,4-dinitrophenol, have a selective inhibitory action on the slow EPSP of ganglion cells (Kobayashi and Libet, 1968), suggesting that activation of metabolic pathways in the postsynaptic membrane may be involved in the synaptic inactivation of resting potassium conductance. 2, 4-Dinitrophenol also antagonizes the excitant action of ACh on cerebral cortical neurones (Krnjević *et al.*, 1971).

The identity of the postulated excitatory cholinergic afferent pathway to the cerebral cortex has yet to be established. Systemically administered atropine depresses the spontaneous activity of ACh-excited neurones and reduces the repetitive firing (after-discharge) of cortical neurones evoked by stimulation of peripheral nerves or thalamic relay nuclei (Krnjević and Phillis, 1963b; Stone, 1972a). The short latency specific afferent projections from the thalamus and geniculate nuclei to the primary sensory cortices are unlikely to be cholinergic since many of the synaptically-evoked responses are not affected by atropine. Likewise, the majority of the transcallosal fibres are unlikely to be cholinergic. Spehlmann (1971) has reported that iontophoretically applied atropine depresses the repetitive firing of neurones in the visual cortex evoked by stimulation of the mesencephalic reticular formation, a finding that would be consistent with the results obtained from ACh release experiments.

The evidence therefore supports the postulate that an excitatory cholinergic projection forms the final link in an ascending pathway from the mesencephalic reticular formation and striatum to the cerebral cortex. This pathway, by controlling the level of excitability of deep pyramidal cells, would determine the degree of responsiveness of cortical efferent pathways. The relatively slow excitatory action of ACh would make it especially suitable for such a modulatory role, probably performed in conjunction with overlapping aminergic projections from the brain stem.

Cholinergic Inhibition

The existence of an inhibitory cholinergic system in the cerebral cortex has been suggested by various investigators. The con-

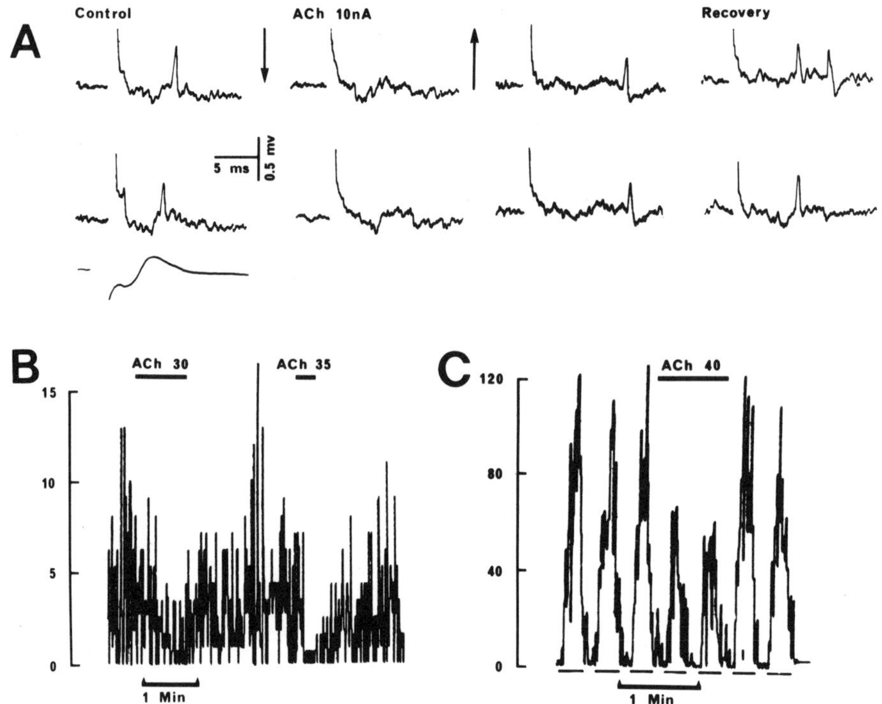

Fig. 3. ACh inhibition of synaptically-evoked, spontaneous and glutamate-induced firing in a single cell. Blockage of synaptic firing during two separate applications of ACh (10 nA) is shown in A. The third trace in the control column is the N-wave direct cortical response recorded from the surface of the cortex. B and C show ACh inhibition of spontaneous and glutamate-induced firing, respectively. The ordinates represent firing rate in spikes per second, and the periods during which drugs were applied are indicated above the traces. Glutamate pulses (40 nA) were applied during the periods indicated by horizontal bars below the trace in C. (Jordan & Phillis, 1972).

cept was initiated by Chatfield and his colleagues to account for the augmentation of primary cortical responses by topically applied atropine (Chatfield and Purpura, 1954; Chatfield and Lord, 1955). Iontophoretic studies have demonstrated that ACh can depress the firing of neurones in the cerebral cortex of cats (Fig. 3) and rats, especially if the animals are not anaesthetized with a barbiturate preparation (Randic et al., 1964; Phillis and York, 1967a, 1968a, b; Jordan and Phillis, 1972; Stone, 1972b). Neurones that are

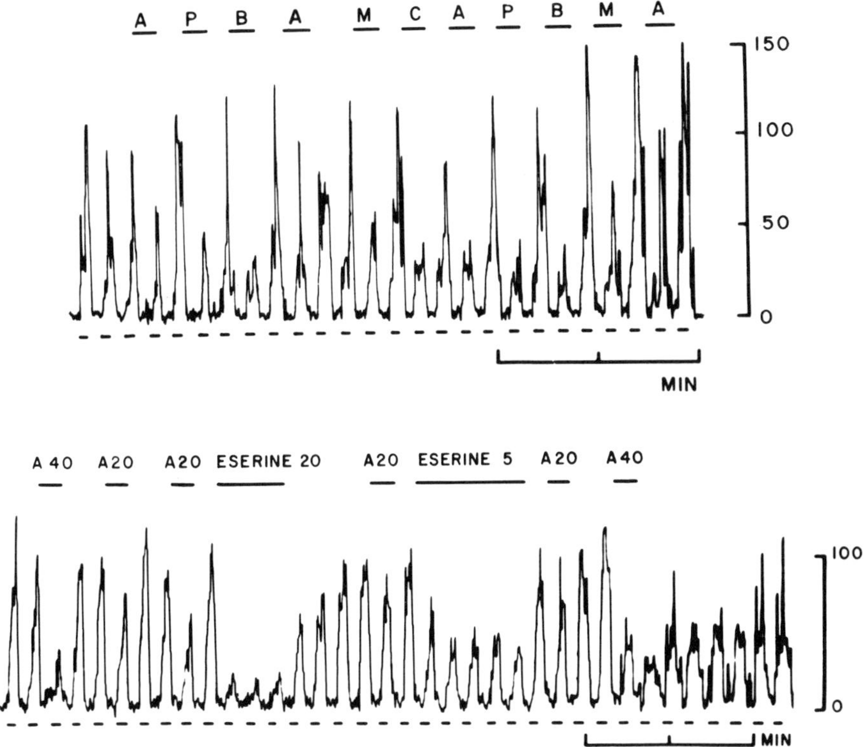

Fig. 4A. Above. Responses of a precruciate cortical cell to constant duration (7 sec) pulses of L-glutamic acid (50 nA) applied iontophoretically every 15 sec (indicated by horizontal bars below the trace in all figures). These responses were used as a measure of cell excitability. The iontophoretic application of the various cholinomimetic drugs, acetylcholine (A), propionylcholine (P), butyrylcholine (B), acetyl-β-methylcholine (M) and carbachol (C) (all 40 nA) is indicated by horizontal bars above the trace. The scale on the right is cell firing frequency in spikes per second. B. Below. Acetylcholine (A) caused a marked depression of this cortical neurone at 40 nA whereas at 20 nA a smaller depression was evident. Iontophoretically applied eserine (20nA) had a potent depressant action followed by a slow recovery. A second application at 1/4 the previous dose (5 nA) still had strong depressant action. After eserine the depressant response of this cell to ACh (40 nA) was considerably prolonged in its timecourse. (Phillis and York, 1968b).

depressed by ACh have been located in all primary cortical areas as well as in associational areas and tend to occur predominantly in the more superficial layers of the cerebral cortex.

A range of cholinomimetic drugs, including acetyl-β-methylcholine, carbamylcholine, nicotine, propionylcholine, butyrylcholine (see Fig. 4A) also depressed cells that were inhibited by ACh. Pilocarpine and tetramethylammonium ion were relatively ineffective. The anticholinesterase agents eserine, neostigmine and tensilon were all effective in potentiating the depressant actions of ACh and had potent depressant actions of their own (Fig. 4B). Atropine and hyoscine were the most effective ACh antagonists; strychnine was also on ACh antagonist (Fig. 5). The nicotinic antagonists, D-tubocurarine, dihydro-β-erythroidine and gallamine triethiodide also exhibited some ACh-antagonist properties.

Repetitive stimulation of the adjacent cortical surface, the mesencephalic reticular formation and the lateral hypothalamus inhibits the responses of many of these ACh-depressed neurones, and this synaptically evoked inhibition is also abolished by atropine and strychnine (Fig. 6) (Phillis and York, 1967a; 1968b; Jordan and Phillis, 1972). Inhibitions of this type may have a duration of over 1 min, and can be potentiated by anticholinesterases. Comparable inhibitions can be evoked by direct stimulation of the surface of both acutely and chronically isolated cortical slabs, indicating that cholinergic inhibitory interneurones may be present in

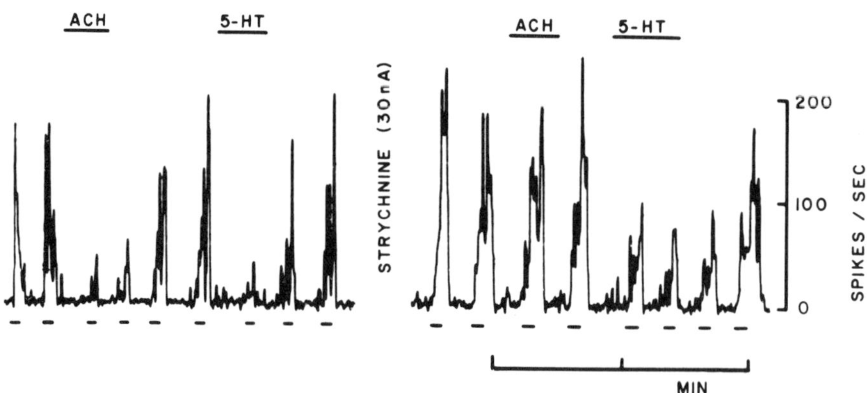

Fig. 5. Effect of strychnine (30 nA for 2 min) on the depression of L-glutamate firing by ACh (30 nA) and 5-hydroxytryptamine. The recording was interrupted during and for a period of 3 min immediately after the application of strychnine. Strychnine blocked the action of ACh and reduced that of 5-hydroxytryptamine (Phillis and York, 1967b).

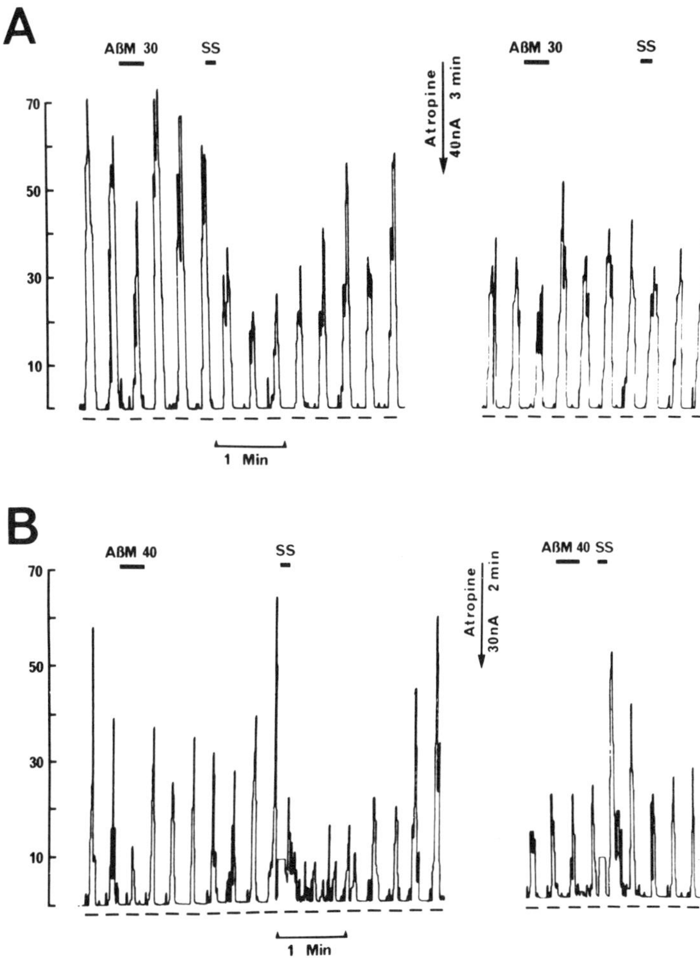

Fig. 6. Cholinergic and long duration inhibition in chronically isolated cortical slabs in the suprasylvian gyrus. Glutamate firing of a cell (A) 570 microns below the surface of a slab isolated for 14 days was inhibited both by acetyl-β-methylcholine (AβM) (30 nA) and by surface stimulation. A cell (B) located 777 microns below the surface of a slab isolated for 4 months was also inhibited by AβM (40 nA) and by surface stimulation. The glutamate current was 40 nA in both cases. The inhibition induced by drug application and the long duration inhibition produced by surface stimulation (SS) were both antagonized by atropine in these cells. (Jordan and Phillis, 1972).

the cerebral cortex. Histochemical observations would suggest that these may be the AChE-containing polymorph cells of layer VI.

Although intracellular studies of the mechanism of ACh-inhibition of cortical neurones are not as yet available, there is an indication that calcium ions are involved in the process. It has recently been reported (Phillis et al., 1973) that calcium antagonists such as lanthanum, verapamil and manganese, can abolish the inhibitory effects of monoamines on cortical neurones. Further studies have shown that these substances, as well as nickel and cobalt, will also antagonize the inhibitory action of ACh on neurones whilst not affecting its excitatory action. This effect is shared by ethanol and some of the local anaesthetics, which are also known to affect membrane calcium (Ritchie and Greengard, 1966; Seeman et al., 1971). Of interest in this context are the reports that strychnine, which antagonizes the depression of cortical neurones by both ACh and the monoamines, is also a calcium antagonist (Hauser and Dawson, 1968; Klee et al., 1973).

The basis for the involvement of calcium in ACh inhibition is not readily apparent. It is possible that net fluxes of calcium across the neuronal membrane resulting from amine application (perhaps through increased membrane calcium permeability, or activation of a calcium pump, or a release of calcium from internal stores) cause alterations in membrane properties which decrease neuronal excitability. Alternatively calcium may be a co-factor for an enzymatic process which is important in the regulation of neuronal excitability.

A tentative indication of the mechanism of action of ACh on cortical neurones may be forthcoming from studies on amphibian sympathetic ganglia. The slow inhibitory hyperpolarizing postsynaptic potential generated in sympathetic ganglion cells of the bullfrog by ACh are associated with a marked increase in membrane resistance (Weight and Padjen, 1972). The reversal potential for this IPSP when the membrane was polarized was in the region of the sodium equilibrium potential and it was suggested that ACh generates the potential by inactivating the resting sodium conductance.

Neuropharmacological evidence in support of the concept of an intracortical cholinergic inhibitory synapse has been forthcoming from the studies of other investigators. Ilyutchenok and Gilinsky (1969) have shown that the discharges of some spontaneously active cortical neurones can be facilitated by anticholinergic agents and that the inhibition of such neurones produced by repetitive stimulation of the reticular formation can be abolished by benactyzine or atropine. Vasquez and his associates (Vasquez et al., 1969; Krip and Vasquez, 1971) have confirmed the involvement of cholinergic interneurones in cortical inhibition by demonstrating that cholinergic agents shorten and anticholinergic muscarinic agents lengthen the duration of the epileptiform afterdischarges in chronically isolated cortical slabs.

GUANOSINE 3':5'-CYCLIC MONOPHOSPHATE AND ACH

Recent studies have shown that ACh can cause an accumulation of guanosine 3':5'-cyclic monophosphate (cyclic GMP) in heart, brain and intestinal tissue (Kuo, et al., 1972). Acetylcholine and muscarinic agonists caused an increase in the cyclic GMP levels in rabbit cerebral cortical slices and this action was abolished by atropine, but not by the nicotinic antagonist hexamethonium. Nicotinic agonists did not affect cyclic GMP levels.

These findings gave rise to the postulate (Lee et al., 1972) that the physiological response to activation of muscarinic receptors is mediated by cyclic GMP. Such a biochemical reaction would be compatible with the slowly developing nature of the responses resulting from activation of muscarinic receptors on cerebral neurones.

TABLE I

Response to Acetylcholine	Response to Cyclic GMP[a]		
	Excited	Depressed	Nil
Excited	1/10[b]	1/10	8/10
Inhibited	9/9	0/9	0/9
Nil	7/13	0/13	6/13

[a] The cyclic GMP current was increased to 100 nA before recording a negative result.

[b] The ratios indicate the number of cells in which an effect, or lack thereof, with cyclic GMP was observed over the number of cells responding to ACh with either excitation, inhibition or not responding.

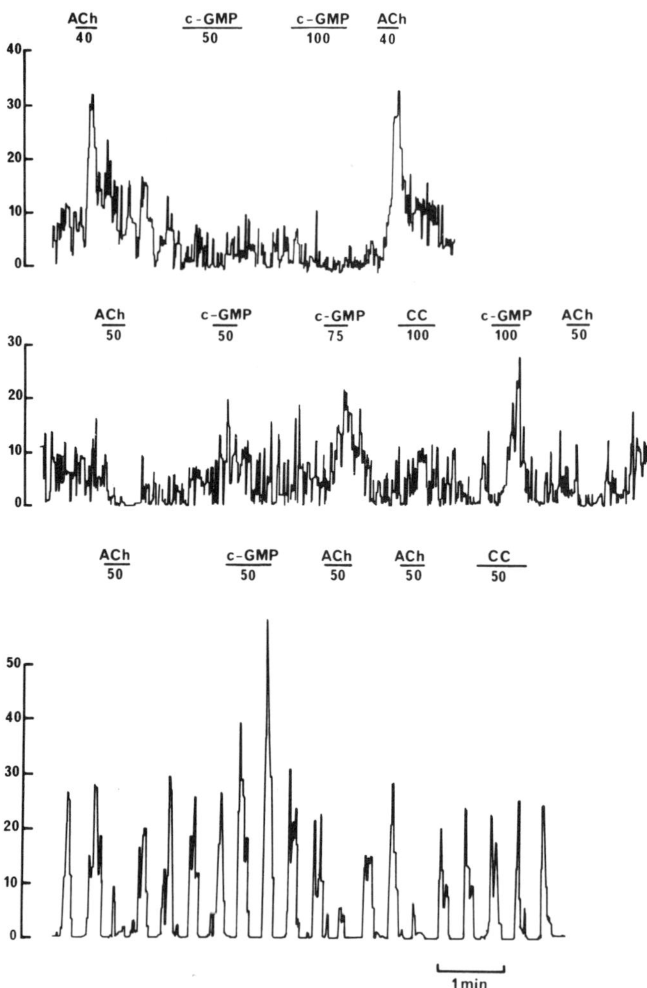

Fig. 7. Three examples of the actions of cyclic GMP on cerebral cortical neurones. The spontaneously firing neurone in the upper record was excited by ACh and depressed by cyclic GMP applied by a current of 100 nA (but not at 50 nA). The middle and lower records show examples of neurones that were inhibited by ACh and excited by cyclic GMP. The neurone in the middle record was firing spontaneously, that in the lower trace was excited by pulses of L-glutamate. The excitation of this neurone is revealed as an enhancement of the response to L-glutamate. c.c. = current control for cyclic GMP.

Cyclic GMP (sodium salt, Calbiochem, 0.2M, pH 6.3) was tested by iontophoretic application on 32 neurones in cat cerebral cortex. The results which are summarized in Table I, show that there was little correlation between the responses to ACh and those to cyclic GMP. The most consistent findings in these studies were that cyclic GMP (in amounts of up to 100 nA) excited neurones which were inhibited by ACh and usually had no effect on those which were excited by ACh. Three examples of the actions of cyclic GMP are presented in Fig. 7.

In studies of this nature it is always difficult to be certain that the exogenously applied cyclic nucleotide is reaching an appropriate locus in or on the cell membrane. The results, however, offer no support for the postulate that the muscarinic actions of ACh on cortical neurones are mediated through cyclic GMP.

SUMMARY

The evidence that acetylcholine (ACh) is a synaptic transmitter in the cerebral cortex has been reviewed. All components of the cholinergic metabolic system are present in the cortex, being particularly concentrated in the synaptosomal fraction of cortical homogenates. Acetylcholinesterase (AChE)-containing nerve fibres are present in all cortical layers and the subcortical connections of this system can be traced to the corpus striatum and septal area. A portion of the cortical ACh content survives even in isolated cortical slabs, suggesting that there may also be intracortical cholinergic neurones.

ACh is released spontaneously from the surface of the cortex, and this release can be enhanced by stimulation of a variety of inputs, including somatic sensory, visual and auditory stimulation, as well as stimulation of the brain stem reticular formation. The increase in release coincides with desynchronization of the electroencephalogram. ACh is also released spontaneously from isolated cortical slabs and its release can be augmented by direct stimulation.

Iontophoretically applied ACh depressed many neurones, especially those in the more superficial layers of the cortex and excites others in the deeper layers. The latter are frequently characterized by their irregular spontaneous firing and include the cells of origin of the corticospinal tract. Both the inhibitory and excitatory effects of ACh are antagonized by atropine and hyoscine and can be potentiated by inhibitors of AChE. Repetitive stimulation of the adjacent cortical surface inhibits many of the neurones which are depressed by ACh, and atropine frequently blocks this synaptically evoked inhibition, demonstrating its cholinergic nature. Similar atropine-sensitive inhibitions can be evoked by surface stimulation of isolated cortical slabs, indicating the presence of an intracortical inhibitory cholinergic system. The excitatory cholinergic pathway is somewhat more obscure. Atropine does depress the spontaneous activity of ACh-excited neurones and reduces the repetitive firing (afterdischarge) of these neurones evoked by stimulation of the somatic afferent pathway.

Cyclic guanosine monophosphate (c-GMP), which has been proposed as a "second messenger" mediating the muscarinic action of ACh, excites many cortical neurones but its actions appear to be unrelated to those of ACh. Thus c-GMP excites cells that are depressed by ACh as well as those which are unaffected by ACh.

REFERENCES

1. Bjegović, M., Geber, J., and Randić, M., 1969, Effect of tetrodotoxin on the spontaneous release of acetylcholine from the cerebral cortex, Jugoslav. Physiol. Pharmacol. Acta 5:345.

2. Burgen, A.S.V. and Chipman, L.M., 1951, Cholinesterase and Succinic dehydrogenase in the central nervous system of the dog, J. Physiol. (Lond.) 114:296.

3. Burt, A.M., 1971, The histochemical localization of choline acetyltransferase, Progr. Brain Res. 34:327.

4. Chatfield, P.O. and Lord, J.T., 1955, Effects of atropine, prostigmine and acetylcholine on evoked cortical potentials, Electroen. Neurophysiol. 7:553.

5. Chatfield, P.O. and Purpura, D.P., 1954, Augmentation of evoked cortical potentials by topical application of prostigmine and acetylcholine after atropinisation of cortex, Electroen. Neurophysiol. 6:287.

6. Collier, B. and Mitchell, J.F., 1966, The central release of acetylcholine during stimulation of the visual pathway, J. Physiol. 184:239.

7. Collier, B. and Mitchell, J.F., 1967, The central release of acetylcholine during consciousness and after brain lesions, J. Physiol. 188:83.

8. Crawford, J.M. and Curtis, D.R., 1966, Pharmacological studies on feline Betz cells, J. Physiol. (Lond.) 186:121.

9. Dudar, J.D. and Szerb, J.C., 1969, The effect of topically applied atropine on resting and evoked cortical acetylcholine release, J. Physiol. 203:741.

10. Elliott, K.A.C., Swank, R.L. and Henderson, N., 1950, Effects of anaesthetics and convulsants on acetylcholine content of brain, Am. J. Physiol. 162:469.

11. Green, J.R., Halpern, L.M. and Van Niels, S., 1970, Choline acetylase and acetylcholine esterase changes in chronic isolated cerebral cortex of cat, Life Sciences 9:481.

12. Hauser, H. and Dawson, R.M.C., 1968, The displacement of calcium ions from phospholipid monolayers by pharmacologically active and other organic bases, Biochem. J. 109:909.

13. Hebb, C.O. and Silver, A., 1956, Choline acetylase in the central nervous system of man and some other mammals, J. Physiol. (Lond.) 134:718.

14. Hemsworth, B.A. and Mitchell, J.F., 1969, The characteristics of acetylcholine release mechanisms in the auditory cortex. Br. J. Pharmac. 36:161.

15. Ilyutchenok, R.Yu. and Gilinsky, M.A., 1969, Anticholinergic drugs and neuronal mechanisms of reticulo-cortical interaction, Pharmac. Res. Comm. 1:242.

16. Jasper, H.H. and Tessier, J., 1971, Acetylcholine liberation from cerebral cortex during paradoxical (REM) sleep, Science 172:601.

17. Jordan, L.M. and Phillis, J.W., 1972, Acetylcholine inhibition in the intact and chronically isolated cerebral cortex. Br. J. Pharmac. 45:584.

18. Klee, M.R., Faber, D.S. and Heiss, W.D., 1973, Strychnine and pentylenetetrazol-induced changes of excitability in Aplysia neurons, Science 179:1133.

19. Kobayashi, H. and Libet, B., 1968, Generation of slow postsynaptic potentials without increases in ionic conductance, Proc. Natn. Acad. Sci. U.S.A. 60:1304.

20. Krip, G. and Vazquez, A.J., 1971, Effects of diphenylhydantoin and cholinergic agents on the neuronally isolated cerebral cortex, Electroenceph. Clin. Neurophysiol. 30:391.

21. Krnjević, K. and Phillis, J.W., 1963a, Iontophoretic studies of neurones in the mammalian cerebral cortex, J. Physiol. (Lond.) 165:274.

22. Krnjević, K. and Phillis, J.W., 1963b, Acetylcholine-sensitive cells in the cerebral cortex, J. Physiol. (Lond.) 166:296.

23. Krnjević, K. and Phillis, J.W., 1963c, Pharmacological properties of acetylcholine-sensitive cells in the cerebral cortex, J. Physiol. (Lond.) 166:328.

24. Krnjević, K., Pumain, R. and Renaud, L., 1971, The mechanism of excitation by acetylcholine in the cerebral cortex, J. Physiol. (Lond.) 215:247.

25. Krnjević, K., Reiffenstein, R.J. and Silver, A., 1970, Chemical sensitivity of neurons in long-isolated slabs of cat cerebral cortex, Electroenceph. Clin. Neurophysiol. 29:269.

26. Krnjević, K. and Silver, A., 1965, A histochemical study of cholinergic fibres in the cerebral cortex, J. Anat. 99:711.

27. Krnjević, K. and Silver, A., 1966, Acetylcholinesterase in the developing forebrain, J. Anat. 100:63.

28. Kuo, J.F., Lee, T.P., Reyes, P.L., Walton, K.G., Donnelly, T.E., and Greengard, P., 1972, Cyclic nucleotide-dependent protein kinases, X. An assay method for the measurement of guanosine 3':5'-monophosphate in various biological materials and a study of agents regulating its levels in heart and brain, J. Biol. Chem. 247:16.

29. Lee, T.P., Kuo, J.F., and Greengard, P., 1972, Role of muscarinic cholinergic receptors in regulation of guanosine 3':5'-cyclic monophosphate content in mammalian brain, heart muscle and intestinal smooth muscle, Proc. Nat. Acad. Sci. U.S.A. 69:3287.

30. MacIntosh, F.C., 1941, The distribution of acetylcholine in the peripheral and the central nervous system, J. Physiol. (Lond.) 99:436.

31. MacIntosh, F.C. and Oborin, P.E., 1953, Release of acetylcholine from intact cerebral cortex, Abstract. XIX International Physiological Congress p. 580.

32. Mitchell, J.F., 1963, The spontaneous and evoked release of acetylcholine from the cerebral cortex, J. Physiol. 165:98.

33. Pepeu, G., Mulas, A., Ruffi, A., and Sotgiu, P., 1971, Brain acetylcholine levels in rats with septal lesions, Life Sciences 10:181.

34. Phillis, J.W., 1968, Acetylcholine release from the cerebral cortex: its role in cortical arousal, Brain Res. 7:378.

35. Phillis, J.W., and Chong, G.C., 1965, Acetylcholine release from the cerebral and cerebellar cortices: its role in cortical arousal, Nature (Lond.) 207:1253.

36. Phillis, J.W., Lake, N. and Yarbrough, G.G., 1973, Calcium mediation of the inhibitory effects of biogenic amines on cerebral cortical neurones, Brain Res. 53:465.

37. Phillis, J.W. and York, D.H., 1967a, Cholinergic inhibition in the cerebral cortex, Brain Res. 5:517.

38. Phillis, J.W. and York, D.H., 1967b, Strychnine block of neural and drug induced inhibition in the cerebral cortex, Nature (Lond.) 216:922.

39. Phillis, J.W. and York, D.H., 1968a, An intracortical cholinergic inhibitory synapse, Life Sci. Oxford 7:65.

40. Phillis, J.W. and York, D.H., 1968b, Pharmacological studies on a cholinergic inhibition in the cerebral cortex, Brain Res. 10:297.

41. Randić, M. and Padjen, A., 1967, Effect of calcium ions on the release of acetylcholine from the cerebral cortex, Nature 215:990.

42. Randić, M., Siminoff, R. and Straughan, D.W., 1964, Acetylcholine depression of cortical neurons, Expl. Neurol. 9:236.

43. Ritchie, J.M. and Greengard, P., 1966, On the mode of action of local anaesthetics, Ann. Rev. Pharmac. 6:405.

44. Sastry, P.B., 1956, The functional significance of acetylcholine in the brain, Doctoral Dissertation, McGill University, Montreal.

45. Seeman, P., Chau, M., Goldberg, M., Sauks, T. and Sax, L., 1971, The binding of Ca^{2+} to the cell membrane increased by volatile anesthetics (alcohols, acetone, ether) which induce sensitization of nerve or muscle, Biochem. Biophys. Acta. 225:185.

46. Shute, C.C.D. and Lewis, P.R., 1967, The ascending cholinergic reticular system: neocortical, olfactory and subcortical projections, Brain 90:497.

47. Spehlmann, R., 1963, Acetylcholine and prostigmine electrophoresis at visual cortex neurons, J. Neurophysiol. 26:127.

48. Spehlmann, R., 1971, Acetylcholine and the synaptic transmission of non-specific impulses to the visual cortex, Brain 94:139.

49. Stone, T.W., 1972a, Cholinergic mechanisms in the rat somatosensory cortex, J. Physiol. (Lond.) 225:485.

50. Stone, T.W., 1972b, Cholinergic mechanisms in the rat cerebral cortex, J. Physiol. (Lond.) 222:155.

51. Szerb, J.C., 1967, Cortical acetylcholine release and electroencephalographic arousal, J. Physiol. (Lond.) 192:329.

52. Szerb, J.C., Malik, H., and Hunter, E.G., 1970, Relationship between acetylcholine content and release in the cat's cerebral cortex, Can. J. Physiol. Pharmacol. 48:780.

53. Vazquez, A.J., Krip, G. and Pinsky, C., 1969, Evidence for a muscarinic inhibitory mechanism in the cerebral cortex, Expl. Neurol. 23:318.

54. Weight, F. F. and Padjen, A., 1972, Slow postsynaptic inhibition and sodium inactivation in frog sympathetic ganglion cells, Abstract. Fifth International Pharmacological Congress 1489.

55. Weight, F.F. and Votava, J., 1970, Slow synaptic excitation in sympathetic ganglion cells: Evidence for synaptic inactivation of potassium conductance, Science, N.Y. 170:755.

PART II
NEUROHUMORAL MECHANISMS IN BASIC PHYSIOLOGIC FUNCTIONS

CODING OF METABOLIC INFORMATION BY HEPATIC GLUCORECEPTORS

Mauricio Russek & Sergio Grinstein

Department of Physiology
Escuela Nacional de Ciencias Biológicas
Instituto Politécnico Nacional México D. F.

Most authors in the field of feeding behavior have adopted the point of view that the control of food intake is "multifactorial" (Adolph, 1943, 1947; Stellar 1954; Brobeck, 1960; Stevenson, 1964). In our opinion this only reflects our lack of knowledge about the main factor controlling hunger and satiety, and a certain misunderstanding about the operation of regulatory systems.

Therefore, we would like to postulate some principles or rules about regulatory systems that I think are in the mind of many people but that I have not seen explicitly stated by any author in our field:

a) Whenever a physiological parameter is regulated by a high priority system, this parameter may be able, under special circumstances, to control another system of lower priority, but cannot be the main controlling factor of this second system. Thus, temperature, osmotic pressure of body fluids or glucose consumption (delta-glucose) cannot be the main parameter controlling food intake as the thermostatic (Brobeck, 1947; Hamilton, 1963), osmostatic (Mc Cleary, 1953; Kakolewsky & Deaux 1970), and classical glucostatic (Mayer, 1955, 1956) hypothesis postulate, because none of these parameters is regulated by food intake. We do not eat in order to maintain our temperature, or our body concentration, or the immediate availability of glucose to the tissues. However, these parameters are regulated by systems of higher priority and are able to modulate food intake under certain circumstances (high and low temperatures, large increases in osmotic pressure, strong hypoglycemia).

b) A parameter that affects a system only when changed in one direction, cannot be the parameter regulated by that system and, therefore, cannot be the main parameter controlling the effectors of that system. Thus, hyperosmolarity and hypoxia produce anorexia, but there is no convincing evidence that hypoosmolarity and hyperoxia induce increased feeding. Hypoglycemia produces hyperphagia, but hyperglycemia has a negligible effect on food intake.

If we establish a sequence of priorities among the regulatory systems mentioned, I think everybody would agree that oxygen and CO_2 concentrations in the blood would have the highest, followed by temperature, osmotic pressure, glucose availability, and whatever is regulated by food intake. Therefore, changes in blood gases can control all the other systems; changes in temperature outside the neutral region would be able to control water intake, glucose availability and food intake; hyperosmolarity and glucose deficiency in the brain, are able to control food intake. But, the regulation of oxygen and carbon dioxide in the blood is attained by the control of respiration and circulation; the main regulation of temperature is achieved by the control of cutaneous blood flow, metabolic rate and sweating; osmotic pressure is regulated by the control of water intake and kidney function; glucose availability is regulated by the hormonal control of metabolic reserves. Is it not logical that the parameter being regulated by food intake control would be something related to the metabolic reserves of the body?

The question is, how and where is the information about metabolic reserves coded and conveyed to the central nervous system in order to control feeding behavior? In this paper we will present evidence that carbohydrate and protein reserves may be sensed and codified into nervous impulses by hepatic glucoreceptors, and that this information may be the basic one operating on the short-term or meal-to-meal control of feeding, that is in the elicitation of hunger and satiety. There is also evidence about a long-term modulation of feeding by the amount of lipid reserves which may be the basis of the regulation of body weight. (Kennedy, 1952). But the possibility exists that this lipostatic control is effected by a modulation of the excitability of the hepatic "glycogenostatic" control, either at the level of the integrative nervous centers or by modifying the sensitivity of the hepatic receptors.

The existence of hepatic glucoreceptors was postulated (Russek, 1963) on the basis of a number of data obtained in our laboratory that could not be explained by the classical glucostatic hypothesis or any of the other current hypotheses (Russek _et al_, 1968; Rodríguez-Zendejas, _et al_, 1968). A few years later their existence was confirmed electrophysiologically (Niijima, 1969). An extensive review of all the data from our own and other laboratories, supporting the hypothesis that these receptors have an important role in the control of food intake has been published (Russek, 1971) and another one is to be published soon (Russek, 1973). Thus, in the present paper I will indulge myself in speculating how these receptors may be coding information about glycogen and protein reserves, and I will present some data that seem to support the proposed mechanism.

The hepatic glucoreceptors might correspond to the nerve fibers running in the Disse spaces (Alvarez-Fuertes, Montemurro, Islas Chaires & Russek, Unpublished) and to the nerve endings described as "intracellular" by some authors (Riegele, 1928, Nicolescu, 1958), and observed with the electronmicroscope to be in direct contact with the hepatocyte membrane and sometimes surrounded by the hepatocyte (Tanikawa, 1968). If we postulate that there are "gap junctions" between the innervated hepatocytes and their nerve fibers, as there are between adjacent hepatocytes (Kreutziger, 1968), the discharge frequency of the fibers would be modulated by the hepatocyte membrane potential. This membrane potential varies with the rate of glucose output, because of the outward current of K ions, which hyperpolarizes the membrane (Haylett & Jekinson, 1969; Daniel _et al_, 1970; Friedman, _et al_, 1971). This outward transport of glucose is dependent mostly on the intracellular concentration of glucose, which, in its turn, depends on the rate of glycogenolysis and gluconeogenesis, and these reflect the amount of glycogen and protein reserves. Therefore, the discharge frequency of these receptors is actually coding the amount of glycogen and protein reserves available during the intermeal post-absorptive periods. Their discharges would only reach the "hunger threshold" when glycogen and protein reserves in the liver attain a certain minimum and less amino acids liberated by muscle are reaching the liver (which means that glycogen and protein reserves in muscle have also reached a certain degree of depletion). In this condition, glucose output from the hepatocytes would reach a minimum, which would cause a decrease in membrane potential (depolarization) of hepatocytes and an increase in the firing rate of the glucoreceptors (innervated hepatocytes). The membrane potential of all hepatocytes is bound to be very similar, due to the gap junctions between them, so a rather small number of fibers could "sample" some hepatocytes and inform about the average metabolic reserves of the whole liver.

Is there any evidence for all this? Recently, we have observed that the average hepatocyte membrane potential measured _in situ_ in rats with _ad libitum_ feeding is significantly higher ($p < 0.01$) than that of rats fasted for 24 h. (Russek and Grinstein, unpublished). Moreover, the hepatocyte membrane potential of fed rats shows a significant inverse correlation with the weight of the animals while the potential of fasted rats does not, varying less among different rats (fig. 1). This suggests that the hyperpolarization produced by feeding ("satiating effect") is smaller the fatter the rat, being perhaps the reason for its obesity.

According to this hypothesis, substances that produce anorexia should hyperpolarize the hepatocytes. We have observed that intraperitoneal injections of glucose and 3-0-methyl-glucose (a non metabolizable monosaccharide), produce an increase in the hepatocyte membrane potential which follows a similar time course as the decrease in food intake that these substances elicit (figs. 2 & 3). Both substances also decrease the discharge frequency of the hepatic glucoreceptors (Niijima, 1969; see fig. _6 B_ for the effect of glucose). Ammonium chloride, which produces anorexia when injected intraperitoneally or intraportally (Russek, 1970a, 1970b, 1971), also hyperpolarizes the hepatocytes (Russek and Grinstein, unpublished). The effect of ammonium chloride might be the result of its glycogenolytic effect (Prior _et al_, 1971). Thus, glucogenic aminoacids would have an immediate satiating effect through the glycogenolytic action of the ammonium liberated in their catabolism, and a more prolonged one through the increase of glycogen by gluconeogenesis. Hence the strong satiating power of proteins could be explained.

These effects of glucose and aminoacids easily explain the postabsorptive satiation. But, is it possible that hepatic glucoreceptors are also involved in the mechanism of "preabsorptive satiation"? It has been observed that adrenaline and noradrenaline hyperpolarize the hepatocytes, partially due to their glyco-

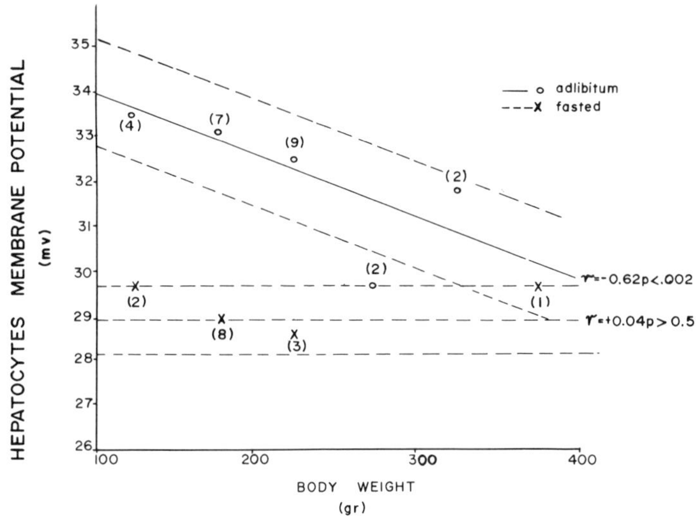

Fig. 1. - Membrane potential of liver cells of rats with food ad libitum (dots) and fasted for 24 hours (crosses). Pentobarbital anesthesia Abscissa: Body weight in grams. Ordinate: Membrane potential in millivolts (obtained with 3M KCl intracellular microelectrodes and recorded on a polygraph). ( ) Number of rats for each 50 gram interval. The value for each rat was the mean of 40-50 cells. (S.E.M. within each rat was 0.1-0.2 mV) Discontinuous lines: ± 1 Sy.x. (Russek & Grinstein, unpublished).

genolytic effect, and partially due to an effect on alpha adrenergic receptors, independent of glycogenolysis (Haylett & Jenkinson, 1969; Daniel et al, 1970).

We have shown that adrenaline produces an anorexia very similar to physiological satiation (Russek, 1965; Russek et al, 1968; Russek & Teitelbaum, 1968). We have also observed that the time courses of the hyperpolarization produced by adrenaline and noradrenaline are similar to the time courses of the anorexia produced by these catecholamines (figs. 4 & 5) which supports the idea that the hyperpolarization of hepatocytes, regardless of its origin, is related to satiation. Besides, we have observed in fasting rats that half an hour of feeding lowers somewhat hepatic noradrenaline and reduces very markedly hepatic adrenaline,

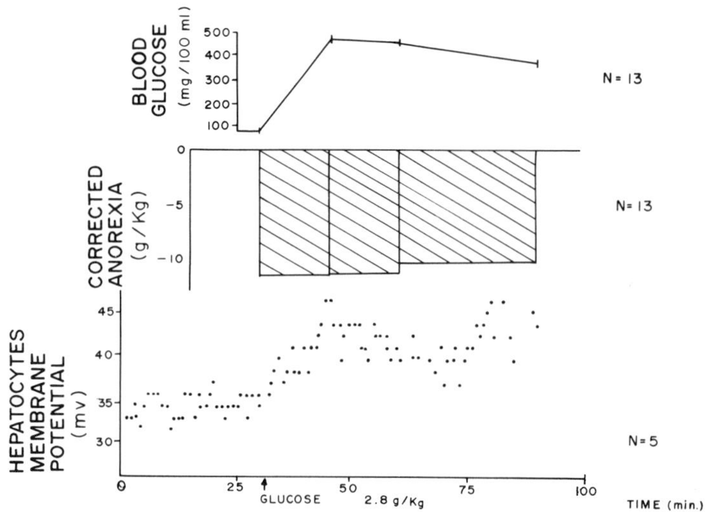

Fig. 2.- Time courses of the hyperglycemia (top), anorexia (middle) and hepatocytes hyperpolarization (bottom) produced by glucose 2.8 g/kg. The hyperglycemia and anorexia were determined, on different days, during thc first 2 hours of ingestion of rats on a 24h-feeding 24h-fasting schedule, after injecting the glucose intraperitoneally. ("Corrected anorexia" was the actual amount eaten in 10 min minus the amount predicted by regression line 6C, for a similar previous ingestion in a control day). Each dot of the bottom graph represents the average of 5-10 electrode impalements performed every minute in the 5 rats. The animals were anesthetized with pentobarbital 35 mg/kg and the glucose was administered intraintestinally to avoid dilution in the Ringer covering the liver. (Russek, Grinstein & Racotta, unpublished).

while blood concentrations of both hormones do not change (Racotta et al, 1972). This suggested that preabsorptive satiation might be the consequence of a reflex secretion of noradrenaline by the hepatic sympathetic fibers, and of adrenaline by hepatic chromaffin cells (Russell, 1965; Martinez, Racotta & Russek, unpublished; fig. 7). The alpha hyperpolarization elicited by these substances would be ideal to produce a quantitative satiation without too much "waste" of the almost depleted glycogen reserves.

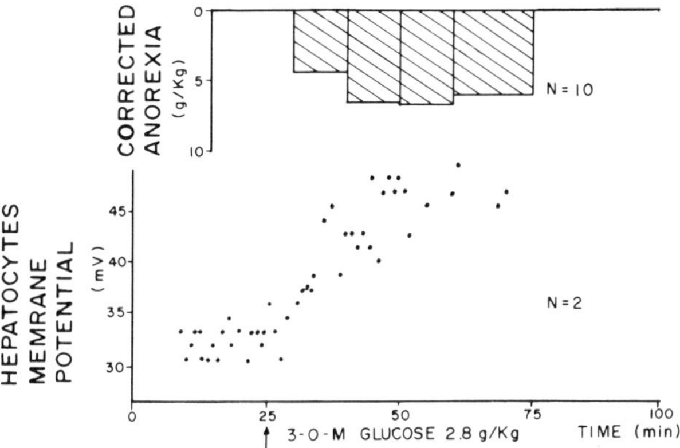

Fig. 3.- Time courses of the anorexia (top) and hepatocytes hyperpolarization (bottom) produced by 3-0-methyl-glucose 2.8 g/kg (a non metabolizable monosaccharide). Details as in fig. 2.

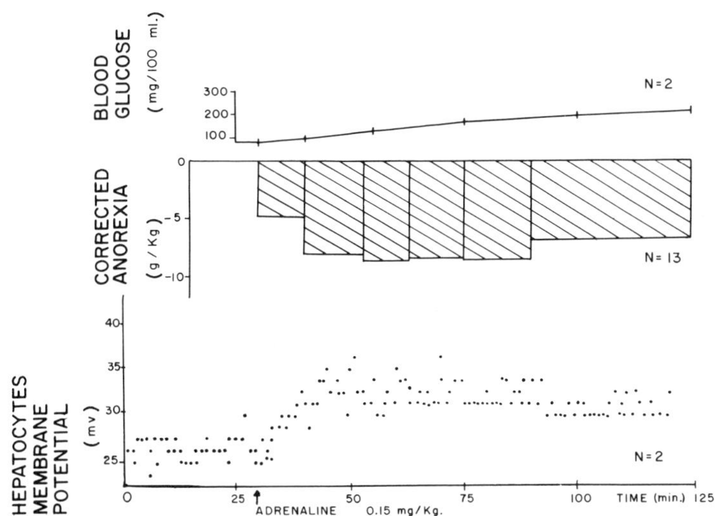

Fig. 4.- Time courses of the hyperglycemia (top), anorexia, (middle) and hepatocytes hyperpolarization (bottom) produced by adrenaline 0.15 mg/kg i.p. Details as in fig. 2, except that the blood glucose was determined in the anesthetized rats simultaneously with the membrane potential measurements.

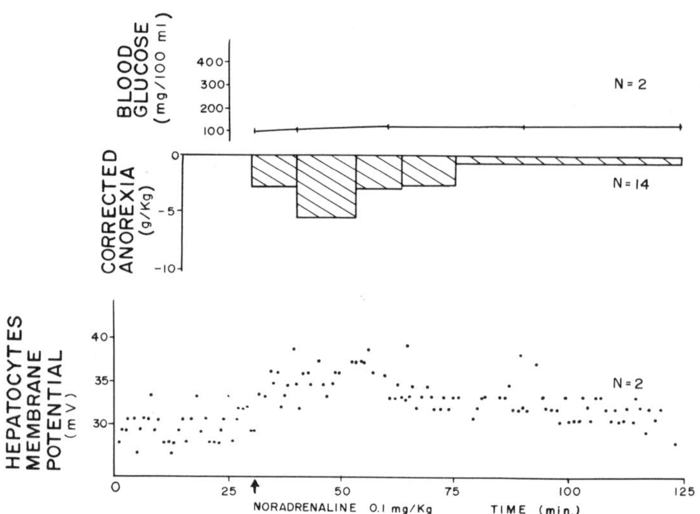

Fig. 5.- Time courses of the hyperglycemia (top), anorexia (middle) and hepatocytes hyperpolarization (bottom) produced by noradrenaline 0.1 mg/kg i.p. Details as in fig. 4.

The source of this "reflex satiation" might be gluco- and amino acid- receptors in the stomach, similar to those described in the intestine (Sharma & Nasset, 1963). The integrating center would be the ventromedial hypothalamus, where gastric stimulation is known to produce electrical changes (Sharma et al, 1961), and whose direct stimulation is known to produce glycogenolysis and hyperglycemia, no doubt effected through the hepatic sympathetic (Shimazu, et al, 1965; Shimazu et al, 1966; Booth et al, 1969). This mechanism will "match" the amount of food that has been ingested with the amount of reserves still present in the liver, because the satiating effect of adrenaline depends partially on the amount of liver glycogen present at that moment (Russek & Stevenson, 1972).

Another fact that supports the idea of a role of catecholamines in preabsorptive satiation is the following: there is a high linea correlation between the amount of food ingested every ten minutes during the first two hours of feeding and the accumulated food ingested up to that moment (fig. 6C). If the amount predicted by this correlation was subtracted from the amount eaten every ten minutes after the injections of noradrenaline, adrenaline and glucose, these "corrected anorexias" showed a high correlation with the hepatocyte membrane potential recorded at the equivalent time (fig. 6d). This suggests that the satiating effects of noradrenaline and adrenaline add freely and linearly, with the early satiating effect of ingested food (preabsorptive satiation) and with the effects of glucose.

Now the question arises; how specific are all these hepatocyte hyperpolarizations? There is evidence that other tissues do not respond in the same way; glucose depolarizes pancreatic islet cells (Dean and Matthews, 1970); catecholamines <u>depolarize</u> vascular smooth muscle, cardiac muscle and adipose cells and produce a small hyperpolarization followed by depolarization in striated muscle (Daniel <u>et al</u>. 1970).

Summing up all that has been said, we can conclude that an hepatic "potentiostatic" hypothesis offers a better probability of explaining feeding control than all the other current hypotheses. This hypothesis postulates that the main factor controlling food intake (and, therefore, regulated by it) is the electric potential of the innervated hepatocytes (glucoreceptors). This potential depends on the amount of glycogen and protein reserves, and when these decrease to a critical point, the glucoreceptor hepatocytes depolarize and "hunger discharges" appear in the afferent glucoreceptor nerve fibers. Preabsorptive satiation would result from reflex secretion of hepatic noradrenaline and adrenaline which would hyperpolarize the glucoreceptor hepatocytes in an amount linearly related to the amount of food in the stomach and partially dependent on the amount of hepatic glycogen present at that moment. The afferents for this reflex satiation would be gluco- and amino acid-gastric-receptors of the type described in the intestine. Oropharyngeal receptors and gastric distention receptors may also have or acquire by conditioning the capacity to mobilize this reflex which would explain the slight satiating effects of sham feeding and mechanical distention of the stomach

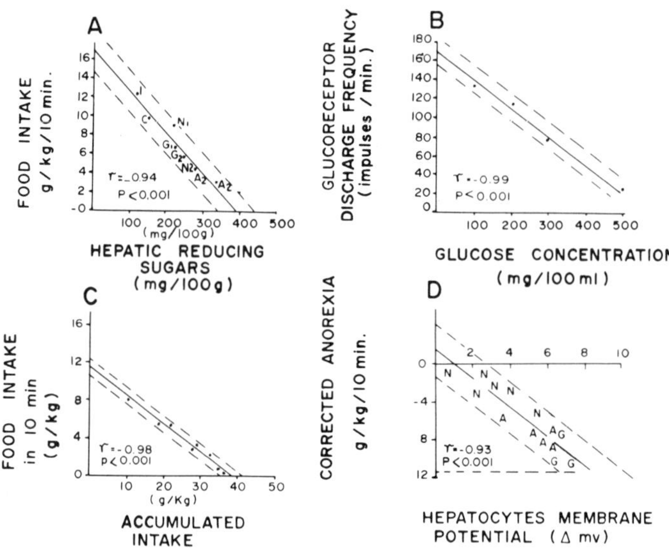

Fig. 6. - A: Correlation between the effects of insulin 1 U/kg (1), isotonic saline (C), noradrenaline 0.1 (N_1) and 0.15 mg/kg (N_2), glucose 1.2 (G_1) and 2.4 g/kg (G_2) and adrenaline 0.1 (A_1) and 0.15 mg/kg (A_2) on the food intake of rats fed for 1 hr daily, and on the hepatic concentration of reducing sugars. Abscissa: hepatic reducing sugars, in milligrams per 100 grams of wet tissue (averages of 6-8 rats for each substance). Ordinate: food intake in 1 hour, in grams per kilogram of body weight per 10 minutes (averages of 12-18 rats). (Recalculated from data published in Russek and Stevenson, 1970). B: Correlation between the discharge frequency of glucoreceptors and the concentration of glucose in the perfusion fluid, in an in vitro hepatic vagus preparation. Abscissa: glucose concentration in milligrams per 100 milliliters of fluid. Ordinate: nerve impulses per minute in a filament of the vagus nerve. (Calculated from data of Niijima, 1969). C: Correlation between the accumulated food intake and the amount eaten in each ten minute period during the first 2 hr of ingestion in rats on a 24h-feeding 24 h-fasting schedule. Abscissa: accumulated food intake in grams, per kilogram of body weight. Ordinate: food ingested each successive 10 minute period, in grams per kilogram body weight per 10 minutes. (Grinstein & Russek, unpublished). D: Correlation between the anorexigenic effects of noradrenaline 0.1 mg/kg (N), adrenaline 0.15 mg/kg (A) and glucose 2.8 g/kg (G) during the first 2 hours of ingestion in rats on a 24 h-feeding 24h-fasting schedule and the effects of

and the hyperglycemia and rise in R.Q. produced by food in the mouth. The magnitude of this reflex may readjust to adapt it against variations of caloric concentration of the food. Postabsorptive satiation would be the natural result of the hyperpolarization of the glucoreceptor hepatocytes, due to replenishment of glycogen and protein reserves. This would maintain the discharge from the glucoreceptor fibers at a subthreshold level for initiating feeding as long as the reserves are above a certain minimum. This subthreshold discharge might be playing an important role in inducing the hormonal changes that gradually switch the metabolism of muscle and adipose tissue from glucose to free fatty acid consumption during intermeal periods.

The lipostatic long-term control of feeding could be easily integrated into this mechanism. There is evidence that adipose tissue produces prostaglandins and that some of these increase food intake while others produce anorexia (Bergstrom, 1967; Shaw & Ramwell, 1968; Martin et al, 1972). We may assume that depleted adipose cells produce a prostaglandin that increases the excitability of the feeding centers and enhance their response to the "hunger discharges" of the hepatic glucoreceptors, or that this substance decreases the sensitivity of the glucoreceptors to glucose or catecholamines which would produce less hyperpolarization and therefore less satiation (fig. 1). On the other hand, overloaded adipose cells would produce a prostaglandin with opposite effects, which would reduce daily food intake. This long-term regulation of the amount of body fat would slowly tend to correct the errors of the short-term control of food intake, and to maintain a certain "set point" of body weight over extended periods of time.

---

this same substance on the hepatocytes membrane potential. Abscissa: average increase in hepatocytes membrane potential, in millivolts. Ordinate: "Corrected anorexia" in grams per kilogram body weight per 10 minutes. Each point represents the value of each bar in the histograms of figs. 2, 4 & 5 and the value of the average membrane potential for the same period of time. (Russek & Grinstein, unpublished). Discontinuous lines: $\pm$ 1 Sy.x.

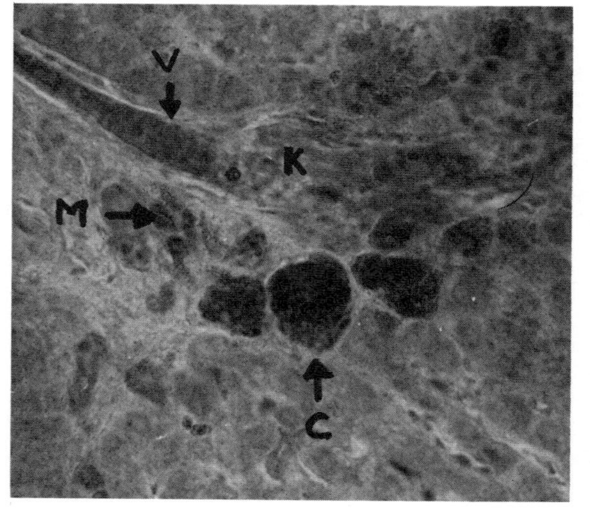

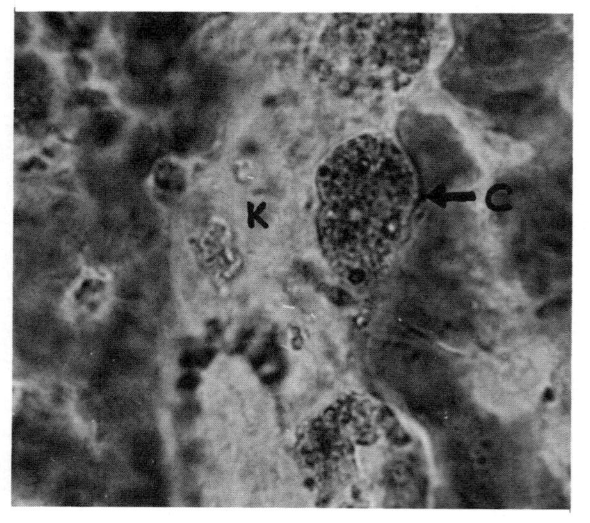

Fig. 7.- Chromaffin cells in the liver parenchyma of rabbit (left; 195x) and rat (right; 250x). The tissues were fixed with Coupland mixture and the adrenaline was identified by Sevki's reaction (dichromate). Adrenal medulla was used as a positive control, and enterochromaffin cells and mast cells as negative ones: True chromaffin cells (C) and adrenal medulla cells take brown to green colorations while enterochromaffin and mast cells (M) take pink to red colorations. The hepatic chromaffin cell clusters are located in the Kiernan spaces (K), near blood vessels (V) which are branches of the portal or hepatic artery (Martinez-Morales, Racotta & Russek, unpublished).

References

Adolph, E. F., 1943, Physiological regulations. Cattell Press, Lancaster, Pa.

Adolph, E. F., 1947, Urges to eat and drink in rats, Am. J. Physiol., 151: 110

Bergstrom S., 1967, Prostaglandins: Members of a new hormonal system, Science, 152: 382.

Booth, D. A., Coons, E. E., and Miller, N. E., 1969, Blood glucose response to electrical stimulation of the hypothalamic feeding area, Physiol. & Behav., 4: 991.

Booth, D. A., Lovett, D., Simson, P. C., 1970, Subcutaneous dialysis in the study of effects of nutrients on feeding, Physiol. Behav., 5: 1201.

Brobeck, J. R., 1947, Food intake as a mechanism of temperature regulation, Yale J. Biol. Med., 20: 545.

Brobeck, J. R., 1960, Food and temperature, in "Recent progress in Hormone Research" G. Pincus, ed. p. 439, Academic Press, New York.

Dean, P. M. and Matthews, E. K., 1970, Glucose- induced electrical activity in pancreatic islet cells, J. Physiol., 210: 255.

Daniel, E. E. Paton, P. M., Taylor, G. S. and Hodgson, B. J., 1970, Adrenergic receptors for catecholamine effects on tissue electrolytes, Fed. Proc., 29: 1410.

Friedmann, N., Solyo, A. V. and Solyo, A. P., 1971, Cyclic adenosine and guanosine monophosphates and glucagon: Effect on liver membrane potentials, Science, 171: 400.

Hamilton, C. L., 1963, Interactions of food intake and temperature regulation in the rat, J. comp. physiol. Psychol., 56: 476.

Haylett, D. G. and Jenkinson, D. H., 1969, Effects of noradrenaline on the membrane potential and ionic permeability of parenchymal cells in liver of the guinea-pig, Nature (London), 224: 80.

Kakolewski, J. W. and Deaux, E., 1970, Initiation of eating as a function of ingestion of hypoosmotic solutions, Am. J., Physiol., 218:590.

Kennedy G. C., 1952, The role of depot fat in the hypothalamic control of food intake in the rat, Proc. Roy. Soc. (London)., 140: 578.

Kreutziger, G. C., 1968, Freeze-etching of intercellular junctions of mouse liver, in "Procedings of the XXVI Annual Meeting of the Electronmicroscope Society of America" (C. J. Arseneaux, ed.), Claitor's Pub. Div., Baton Rouge, Louis.

Martin, F. H., Baile, C. A., Webb, R. L. and Kingsbury, W., 1972, Prostaglandin inhibition and prostaglandin antagonist induction of feeding following hypothalamic injections into sheep, IV Inter. Cong. Nutrition, Mexico City, p-9.

Mayer, J., 1955, Regulation of energy intake and the body weight; the glucostatic theory and the lipostatic hypothesis, Ann. N. Y. Acad. Sci., 63: 15.

Mayer, J., 1956, Regulation de l' appetit, Proc. Inter. Physiol. Congr., 20: 138.

Mc Cleary, R. A., 1953, Taste and postingestion factors in specific hunger behavior, J. comp. physiol. Psychol., 46: 411.

Nicolaidis, S., 1968, The prandial calorigenic effect. III International Conference on the Regulation of Food and Water intake, Haverford, Pa. (September 1 - 3, 1968).

Nicolescu, J., 1958, An atlas concerning morphological aspects of visceral nerve endings. Editura Medicala, Bucarest, Rumania.

Niijima, A., 1969, Afferent impulse discharges from glucoreceptors in the liver of guinea pig, Ann. N. Y. Acad. Sc., 157: 690.

Prior, R. L., Clifford, A. J., Gibson, G. E. and Visek, W. J., 1971, Effect of insulin on glucose metabolism in hyperammonemic rats, Am. J. Physiol., 221: 432.

Racotta, R., Vega, C. and Russek, M., 1972, Liver catecholamines and preabsorptive satiation, Fed. Proc., 31: 309 abs.

Riegele, L., 1928, Über die feinere Vernalten der Nerven in der Leber von Mensch und Saugetiere, Ztschris. f. mikr. anat. Forsch. Bd., 14: 73.

Rodriguez-Zendejas, A. M., Vega, C., Soto-Mora, L. M. and Russek M., 1968, Some effects of intraperitoneal glucose and intraraportal glucose, and adrenaline, Physiol. Behav., 3: 259.

Russek, M., 1963, An hypothesis on the participation of hepatic glucoreceptors in the control of food-intake., Nature, 197: 79.

Russek, M., 1970a, Gluco-ammonia receptors in the liver, Fed. Proc., 29: 658.

Russek, M., 1970b, Demonstration of the influence of an hepatic glucosensitive mechanism on food-intake, Physiol. Behav., 5: 1207.

Russek, M., 1971, Hepatic receptors and the neurophysiological mechanisms of feeding behavior, in "Neuroscien. Research" Vol. 4. (S. Ehrenpreis, ed.) pp. 213-282 Academic Press, New York.

Russek, M., 1973, "Stevenson Memorial Volume", Toronto Press (in press).

Russek, M. and Stevenson, J. A. F., 1972, Correlation between the effects of several substances on food intake and on the hepatic concentration of reducing sugars, Physiol. Behav., 8: 245.

Russek, M. and Teitelbaum, P., 1968, A further analysis of adrenaline induced anorexia in the rat, Proc. Inter. Cong. Physiol. Sc., 24: 378.

Russek M., Rodriguez-Zendejas., A. M. and Piña, S., 1968, Hypothetical liver receptors and the anorexia caused by adrenaline and glucose, Physiol. & Behav., 3: 249.

Russell, J. A.. 1965, The adrenals in: "Physiology & Biophysics" (T. C. Ruch and H. D. Patton, eds.)p. 1138 Saunders Co., Phila.

Sharma, K. N. and Nasset, E., 1963, Electrical activity in mesenteric nerves after perfusion of gut lumen, Am. J. Physiol., 202: 725.

Sharma, K. N. , Anand, B. K., Dua, S. and Singh, B., 1961, Role of stomach in regulation of activities of hypothalamic feeding centers, Am. J. Physiol., 201: 593.

Shaw, J. E. and Ramwell, P. W., 1968, Release of prostaglandin from rat epididymal fat pad on nervous and hormonal stimulation, J Biol. Chem., 243: 1498.

Shimazu, T. and Fukuda A. 1965, Increased activities of glycogenolytic enzymes in liver after splanchnic-nerve stimulation, Science, 150: 1607.

Shimazu, T., Fukuda, A and Ban, T., 1966, Reciprocal influences of the ventromedial and lateral hypothalamic nuclei on blood glucose level and liver glycogen content, Nature, 210: 1178.

Stellar, E., 1954, The physiology of motivation, Psychol. Rev., 61: 5.

Stevenson, J. A. F., 1964, The hypothalamus in the regulation of energy and water balance, The Physiologist, 7: 305.

Tanikawa, K., 1968, "Ultrastructural aspects of the liver and its disorders". p. 50 Igaku Shoin Tokyo.

ABSTRACT:

About ten years ago we postulated the existence of hepatic glucoreceptors, which were later demonstrated electrophysiologically by Niijima. (Ann. N.Y. Acad. Sc., 157: 690, 1969). Substantial evidence about its participation in the control of feeding and glycemia has been accumulated (Russek, M. Neurosc. Res., 4: 213, 1971). Only recently we have started to study the actual coding of the metabolic information that these receptors convey to the CNS. The working hypothesis is that these receptors are innervated hepatocytes, with electrical coupling between the hepatocyte membrane and the nerve fiber membrane. The discharge frequency of the fiber ("hunger discharge") would depend upon the membrane potential of the hepatocyte.

We have shown that increased glucose transport through the membrane, produced either by glucose or by glycogenolysis, hyperpolarizes the hepatocytes. Therefore, when there is enough hepatic glycogen (high hepatic glucose output), or when glucose is being absorbed in the intestine and entering the hepatocytes, the "hunger discharges" are reduced or absent, and vice-versa. Aminoacids are also monitored by these receptors, perhaps through the glycogenolytic effect of ammonia, or through a direct effect of ammonia upon the membrane. Thus, as long as there are protein reserves in liver, and aminoacids liberated from muscle reserve proteins, the "hunger discharges" are kept at a subthreshold level. In this fashion, hepatic membrane potential would integrate, and the glucoammonia receptor discharge frequency would codify, information about hepatic glycogen reserves and hepatic and muscle protein reserves. This information might be used by the CNS in the control of feeding behavior and of the hormones that participate in blood glucose homeostasis.

Catecholamines hyperpolarize the membrane by an alpha effect which increases the K^+ permeability and by a beta effect which produces glycogenolysis. We have some evidence that catecholamines contained in hepatic chromaphin cells might be liberated reflexly immediately after feeding and be responsible for "preabsorptive" satiation.

NEUROCHEMICAL MECHANISMS OF TEMPERATURE REGULATION AND FOOD INGESTION

R. D. Myers

Laboratory of Neuropsychology, Purdue University

West Lafayette, Indiana

I. INTRODUCTION

Although the actual concept of a neurohumoral code for a physiological control process may be a relatively straightforward one (Miller, 1965; Myers and Sharpe, 1968a), the specific details of any of the postulated coding systems are virtually unknown. The biological meaning of the term coding, which was used originally by the geneticists, refers to a synaptic systematization of chemical events that either trigger or suppress an efferent pathway (Myers, 1974). As such, the process of systematization can often explain how a physiological or behavioral response is activated or how it is inhibited. But what does this mean in terms of a neuronal process?

Essentially, when a set of afferent impulses produced by some sort of functional imbalance or other signal is relayed to a circumscribed region in the forebrain, for example, specific neurons may release one substance that activates the respective pathway that is delegated morphologically to that function. As the balance in the physiological system is restored, the release of another substance from other neurons within the same set of cells acts either to inhibit or to modulate the synaptic release of the first substance.

In this presentation, I shall describe the somewhat remarkable evidence that supports the notion of a humoral coding in the hypothalamus of the primate, which is apparently responsible for (a) the control of body temperature and (b) the initiation of feeding behavior. All of the experiments to be highlighted here have been carried out with the monkey or the cat, principally because of compelling anatomical reasons pertaining to the relatively large dimensions of the diencephalic nuclei of these two species.

II. AMINERGIC MECHANISMS INVOLVED IN THERMOREGULATION

During the past 10 years, a substantial number of investigators have provided support for the theory that the balance in the release of the two monoamines, serotonin (5-HT) and norepinephrine constitutes the hypothalamic mechanism whereby thermoregulation is achieved (Feldberg and Myers, 1964). Basically, the amine theory, so termed now because it also incorporates acetylcholine (ACh) in a vital role, has been generated on the basis of two lines of evidence.

First, a micro-injection of 5-HT into the anterior hypothalamus of a cat or monkey evokes hyperthermia (for review see Myers, 1970, 1974) whereas norepinephrine injected at precisely the same locus produces a fall in the same animal's temperature. Because ACh or a cholinomimetic injected at sites distributed diffusely from the anterior hypothalamus through the mesencephalon evokes a sharp rise in temperature of short duration, ACh is believed to transmit signals for heat production along an efferent pathway which extends caudal-ward through these structures (Myers and Yaksh, 1969; Avery, 1971).

Second, the resting output of spontaneously released 5-HT, norepinephrine or ACh can be altered by either peripheral warming or cooling of the cat or monkey. What is so crucial, however, is the fact that the enhanced release of each amine can be triggered _differentially_ by a brief change in the animal's ambient temperature. As we shall see momentarily, these findings are pivotal to the verification of the postulated concepts of neurohumoral coding.

III. INTERACTIONS WITHIN THE CATECHOLAMINERGIC SYSTEM FOR FEEDING

Of particular fascination to our group was the extension of the original observation in the rat (Grossman, 1960) to the primate with respect to the feeding behavior induced by a catecholamine. In the rhesus monkey, a micro-injection of norepinephrine into certain areas of the hypothalamus causes the animal to eat (Myers, 1969), a response which is dose dependent (Sharpe and Myers, 1969). At a small percentage of sites, as little as 10 μg of the catecholamine evoke a simultaneous decline in body temperature at the same time that the animal feeds voraciously (Yaksh and Myers, 1972a). These two responses together with the prandial drinking that accompanies the consumption of dry biscuits are portrayed in

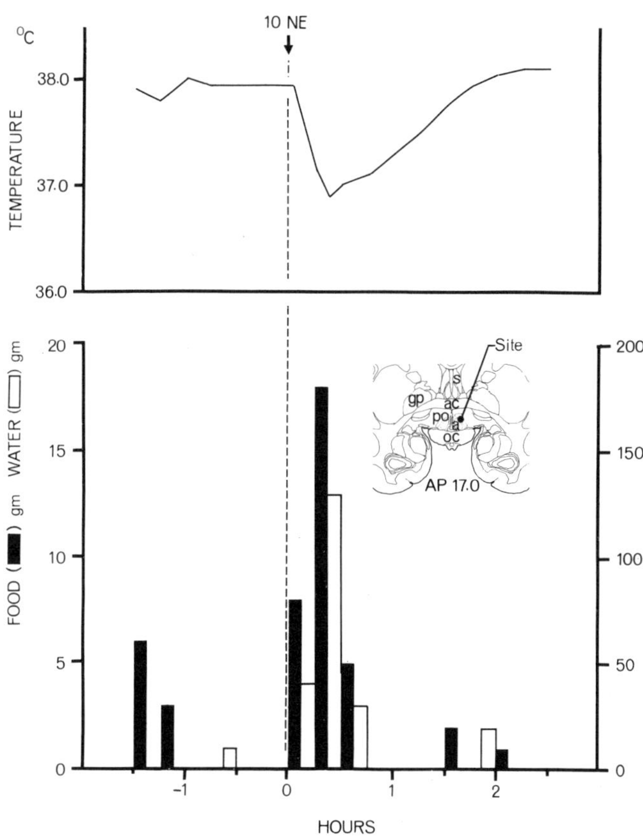

Figure 1. Eating, drinking and temperature response of an unanesthetized monkey following injection at the arrow of NE in a dose of 10 μg base into the anterior hypothalamus at AP 17.0 (*inset*) (from Yaksh and Myers, 1972a).

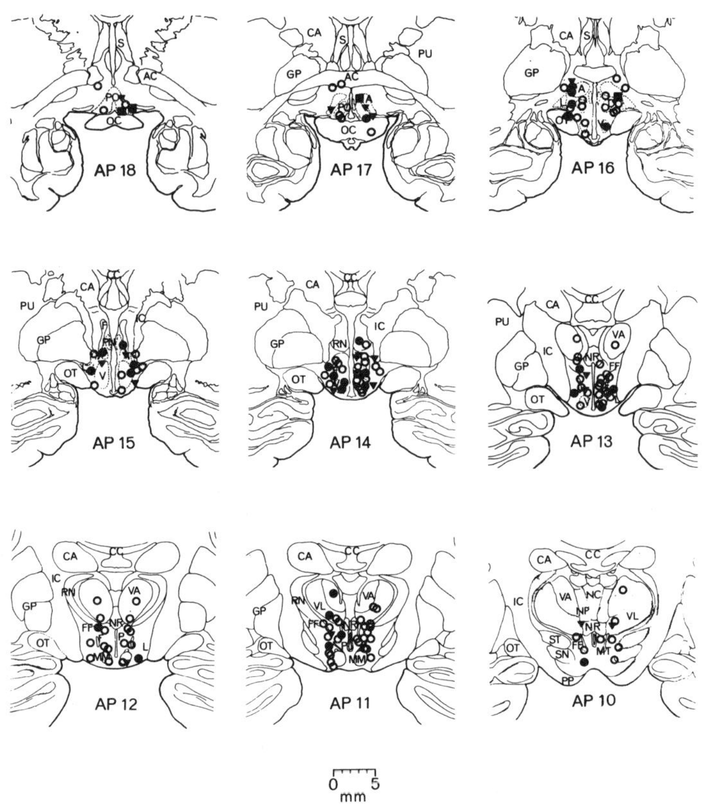

Figure 2. Anatomical mapping at 9 coronal (AP) levels in the hypothalamus at which NE micro-injected in a dose of 5 to 12.5 μg base produces either a hypothermia of more than 0.4°C (▼) or feeding of more than 20 gms with a latency of less than 15 min (●) or both responses simultaneously (■). Sites which are associated with no response are also indicated (o). For anatomical abbreviations see Yaksh and Myers, 1972a.

Figure 1. Interestingly enough, the fall in temperature abates at about the same period as the intake of food. This would reflect a similar duration of action of this catecholamine on the neurons of the anterior hypothalamus, preoptic area designated in the inset in Figure 1, which subserves both functions - feeding and temperature.

The anatomical characteristics of the noradrenergic coding of eating and hypothermia are illustrated by the morphological map shown in Figure 2. Whereas the greatest density of sites mediating hypothermia is in the anterior-preoptic region, the preponderance of noradrenergic sites facilitating feeding are scattered far more diffusely. As shown in the Figure, the region of functional overlap at which both feeding and hypothermia are elicited by a micro-injection of norepinephrine is the anterior-preoptic area. On the basis of the intimate relationship between food intake, energy metabolism and the maintenance of body heat, these elements of functional integration are not unexpected. Even so, the vast majority of sites delegated to ingestive behavior and thermoregulation are morphologically differentiated probably on the basis of catecholaminergic nerve terminals which in other species have their origin in separate diencephalic nuclei, fiber bundles or peripheral pathways (Fuxe _et al._, 1970; Jones _et al._, 1969).

IV. RELEASE OF HUMORAL SUBSTANCES IN PUSH-PULL PERFUSATE

Going hand in hand with the results of these experiments in which the concentration of an endogenous substance is artificially elevated by a micro-injection is the evidence obtained by experiments in which the release of the self-same substance has been enhanced. Before examining this work, however, it is worthwhile to consider a direct way that has been devised for determining whether an active factor can be liberated from neurons by a physiological stimulus.

In this Symposium, Professor M. Monnier and Dr. R. R. Drucker-Colín describe the very promising research devoted to the examination of an unknown humoral factor which arises from the brain of a sleep-deprived animal. When transfused to the brain of a normal animal, the factor causes the recipient to go to sleep. This transfusion paradigm which was originated in Piéron's laboratory in the early part of this century (Legendre and Piéron, 1910) has been utilized in the monkey to demonstrate the presence of a thermogenic factor in the anterior hypothalamus as well as a factor that affects feeding behavior.

A. THE BRAIN TRANSFUSION PARADIGM

To find out whether the release from neurons of a neurohumoral factor is activated by a physiological stimulus, CSF or perfusate is collected, and then its action is tested on homologous neurons in a recipient animal or in the same animal at a later time (Myers, 1967). For example, if perfusate is collected from the anterior hypothalamus of the monkey by means of push-pull cannulae, and then transfused directly to an homologous site in a recipient monkey, a change in body temperature occurs. When the donor is exposed to an ambient temperature of $0^{o}C$ while the transfusion is carried out, the recipient begins to shiver and its temperature rises. However, if the donor's chamber is warmed to $40^{o}C$ or above, and the procedure is repeated, the temperature of the recipient monkey declines (Myers and Sharpe, 1968b). In each instance, the hypothalamus of the second monkey senses and responds to the hypothalamic factor liberated from the donor's hypothalamus in a compensatory manner, as if its own body had been subjected to the two extremes of ambient temperature.

The same sort of physiological compensation is also observed when a monkey is fasted or satiated (Myers, 1969). Perfusate collected from sites located primarily in the perifornical region, generally more caudal than the thermally sensitive region, possesses a potent action on the eating behavior of the primate. When a donor monkey is deprived of food for 18 hours and the perfusate, again collected by means of push-pull cannulae, is transfused to the homologous region of a fully-fed monkey, the recipient begins to eat.

Similarly, if the brain transfusion method is modified in such a way that the donor is used as its own recipient, a surprisingly similar result is obtained. That is, when the monkey is deprived of food for 5 or 18 hours or is instead satiated, and then the effluent that is collected under one of these conditions is re-perfused at a later time, the animal's subsequent ingestion of food and water depends solely on its state of satiety or hunger (Yaksh and Myers, 1972b). Figure 3 illustrates a set of experiments in which the perfusate was collected from the monkey's lateral hypothalamus (site A). After the monkey had been satiated and its own effluent, which was stored on ice during the interim, was re-perfused at the same site about two hours later, the animal consumed food pellets. As shown in the Figure, the duration and magnitude of this perfusate-induced eating depends entirely upon the period of deprivation.

When the experimental situation is reversed and the donor is satiated while the recipient remains hungry, the perfusate collected from the satiated donor monkey can attenuate feeding in the food-deprived recipient (Yaksh and Myers, 1972b).

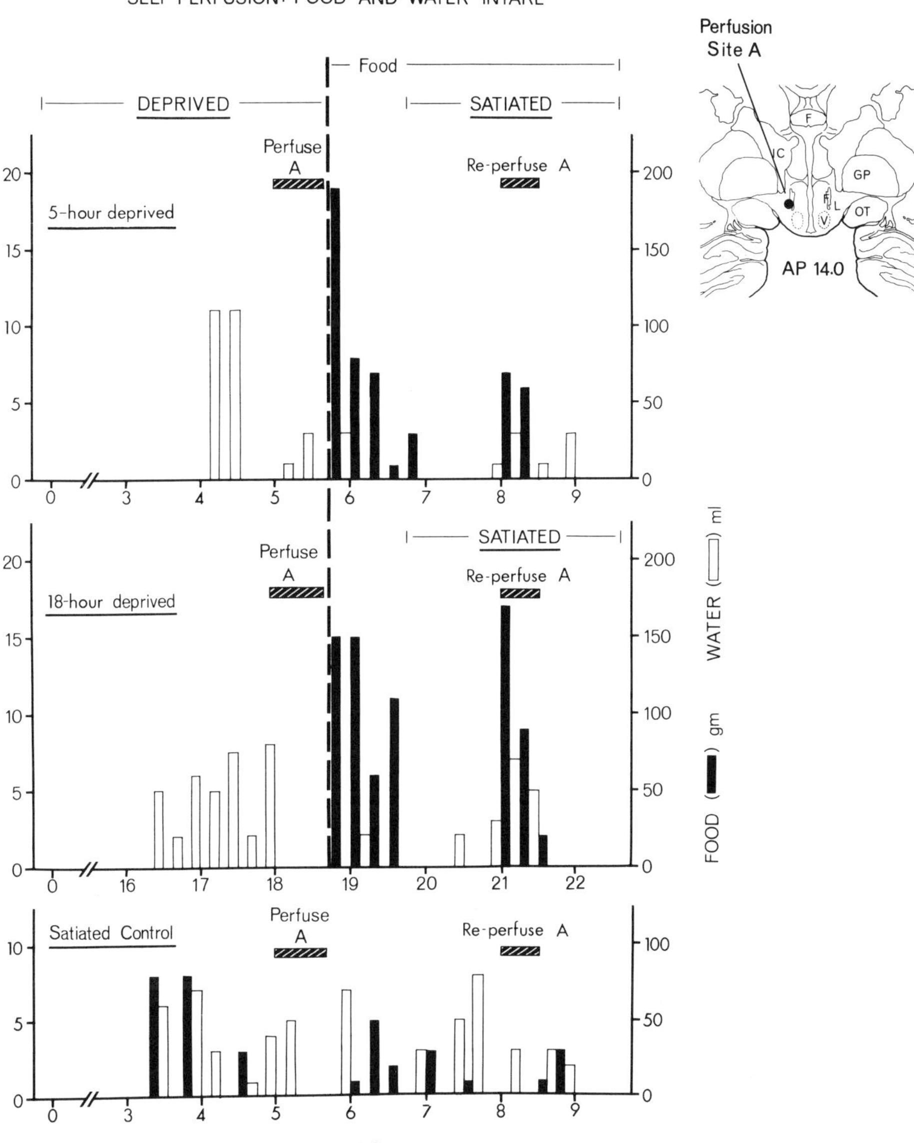

Figure 3. Feeding (▬) and drinking (▭) responses of an animal during three self-perfusion experiments at site A (inset) in which the donor perfusion (Perfusate A) occurred when the animal was 5 hr deprived (top panel), 18 hr deprived (middle panel) or satiated (bottom panel). At the vertical dashed line, the food lever was activated and the animal allowed to satiate after which the animal was again perfused (Re-perfuse A) at site A with the perfusate obtained earlier. For anatomical abbreviations see Yaksh and Myers, 1972b.

Taken as a whole, these experiments show that several distinct humoral factors are released by a physiological stimulus. The factors are present within circumscribed anatomical regions, their chemical characterization has not been accomplished, and they are exceedingly potent.

B. EVOKED RELEASE OF 5-HT OR NOREPINEPHRINE

Does an injection of 5-HT, ACh or norepinephrine actually mimic the synaptic events that take place in the brain-stem during ingestive behavior, the absence of feeding, or a rise or fall in body temperature? Turning this difficult question around experimentally, we have tried to determine whether the regional output of the amines is altered in response to hunger or a thermal challenge.

In a preliminary study carried out in our Laboratory in which the blood pressure of the pithed rat was used as an assay preparation for the content of norepinephrine in hypothalamic effluent, we have found some evidence that the output of a catecholamine-like pressor substance in a push-pull perfusate almost doubles when the monkey is deprived of food. The site of perfusion is again the perifornical area at which such a hunger-induced change occurs (Yaksh and Myers, 1972a). The interesting feature of these active releasing sites is their selective reactivity to norepinephrine which evokes feeding when it is micro-injected in each of the loci.

Curiously enough, it is in the field of thermoregulation that a clear-cut demonstration of the evoked release of three of the biogenic amines has been presented. Coinciding with the shivering and hyperthermia produced when 5-HT is injected into the anterior hypothalamus is the significant change in 5-HT output from the same region, which is increased when the normal monkey is exposed to the cold (Myers and Beleslin, 1971). Using an isolated strip of the rat's stomach fundus, which is highly sensitive to 5-HT, we have observed an enhanced liberation of the indoleamine that is induced as the animal thermoregulates against a cold challenge. That this release is locus-specific is illustrated in Figure 4 which shows the anatomical regions in which this stimulated release occurs.

Alternatively, when the monkey is exposed to a warm environmental temperature, the 5-HT level remains essentially unchanged or is even slightly suppressed. However, external warming does provoke an identical sort of release of ^{3}H- or ^{14}C-norepinephrine in another species (Myers and Chinn, 1973). In the cat, the locus of enhanced catecholamine activity is the anterior hypothalamus and again the sites are identical to those at which a micro-injection

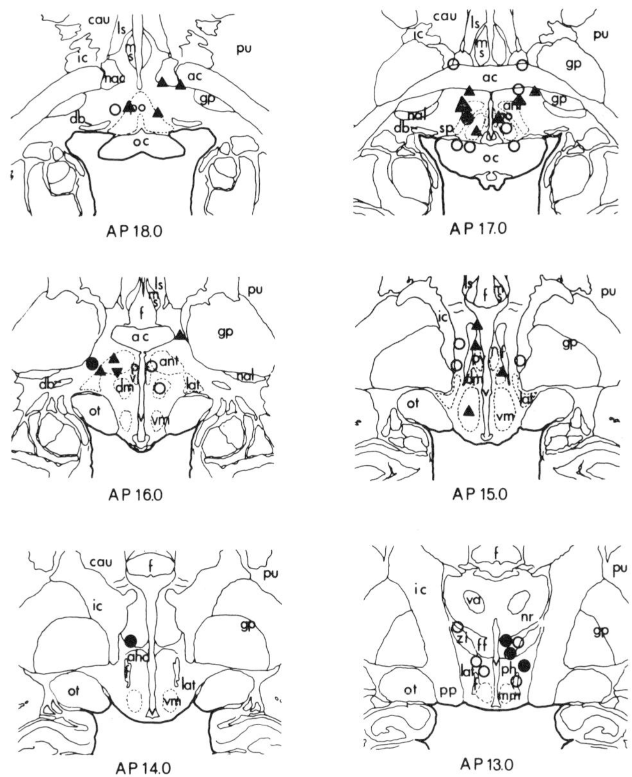

Figure 4. Anatomical mapping at six coronal levels from AP 13.0 to 18.0 of sites in and adjacent to the hypothalamus of the unanesthetized rhesus monkey at which isolated push-pull perfusions were carried out. As a result of heating or cooling the monkey, the release of 5-HT either increased (▲), decreased (▼) or remained unchanged (●). An open circle (o) denotes the absence of 5-HT in the perfusate. For anatomical abbreviations see Myers and Beleslin, 1971.

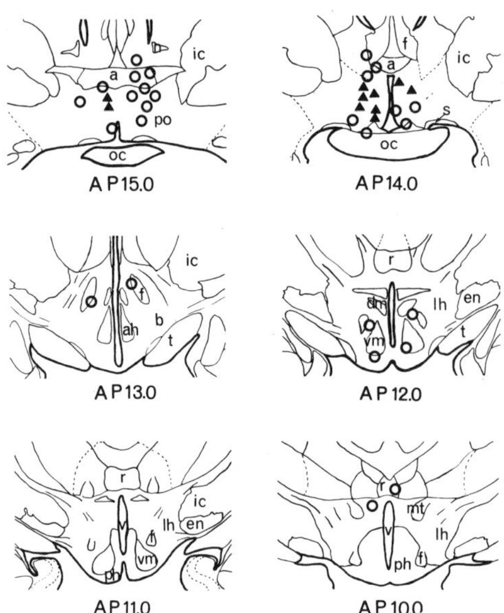

Figure 5. Composite anatomical mapping at six coronal (AP) planes of diencephalic sites, in 11 unrestrained cats at which isolated perfusions were carried out by means of push-pull cannulae. During warming of the animal, the release of ^{14}C- or ^{3}H-NE either increased (▲) or remained unchanged (o). For anatomical abbreviations see Myers and Chinn, 1973.

of norepinephrine produces a fall in temperature (Rudy and Wolf, 1971). Thus, the endogenous activity of norepinephrine, shown in Figure 5, corresponds to the activation of the heat loss mechanism in the rostral hypothalamus.

C. EVOKED RELEASE OF ACETYLCHOLINE

Going hand in hand again with the experiments in which a micro-injection into the hypothalamus of ACh or a cholinomimetic agent evokes the production of heat are the results of a push-pull perfusion study in which we searched for ACh in the effluent. Considering only the sites at which the release of ACh could be detected by assay of the effluent on the eserinized guinea pig ileum preparation, cooling of the monkey was found to enhance the output of ACh at 88% of the sites in the anterior, preoptic area. Conversely, warming of the same monkey attenuates the release of ACh at 80% of the perfusion loci (Myers and Waller, 1973). Caudal to this region, the output of ACh is elevated at about 50% of the active releasing sites when the monkey is exposed to a cold environment. However, warming the monkey suppresses ACh output at nearly all of the more caudal perfusion loci. Within the mammillary region and the mesencephalon, the output of ACh is augmented by cooling at two-thirds of the sites and is suppressed by warming at one-third of the loci.

These results strongly suggest that a cholinergic pathway arising within the anterior hypothalamus traverses the more caudal portion of this structure and extends through the mesencephalon. The principal function of this fiber system apparently is to convey the signals for heat production. In attempting to characterize the cholinoceptive properties of this pathway, Hall and Myers (1972) have provided evidence to suggest that in the primate, thermogenesis is mediated at least partially by means of nicotinic receptors. When either nicotine or ACh is micro-injected directly into the hypothalamus, a carbon-copy hyperthermia is evoked of similar magnitude, latency and duration.

V. HYPOTHALAMIC CODING

Overall then, the balance in the release of 5-HT and norepinephrine seemingly would trigger or inhibit the signals for heat production which in turn are transmitted along a diffuse efferent system by means of ACh. This fine balance in monoamine release, within the rostral hypothalamus of the primate, is established by a cellular mechanism which is not understood at the moment. For instance, thermosensitive cells known to exist in the rostral hypothalamus of this and other species could be those which release 5-HT and norepinephrine. In view of the findings of Jell (1973), this possibility seems very unlikely, because cold and warm sensitive neurons in the cat discharge in response to iontophoretically applied monoamines in a random rather than in a systematic fashion. Therefore, the origin of 5-HT and norepinephrine activity in all likelihood rests in the afferent aminergic nerve terminals which ascend from the periphery, carry the impulses from cold and warm detectors and terminate in the anterior, preoptic area.

The feeding system in the primate is still very much of a mystery in that the hypothalamic basis of a noradrenergic feeding response is simply not understood. A speculative position is that neuronal signals reflecting an imbalance in plasma glucose, lipids or other nutrients are detected by hypothalamic neurons which in turn activate a feeding pathway.

The issue of whether 5-HT, norepinephrine or acetylcholine can be considered as neurotransmitters for the control mechanisms of thermoregulation and possibly ingestive behavior is exceedingly controversial (e.g., Bloom, 1974; Karczmar, 1974; this symposium). For example, it is my opinion that at this stage of our knowledge only ACh can be considered to be a transmitter in the basal and other areas of the forebrain (Myers, 1974). There are several important reasons for this viewpoint. First, ACh obeys all of the traditional criteria (Werman, 1972) set forth for a neurotransmitter including the presence of synthesizing and catabolizing enzymes. Second, ACh is the only substance, at least in the forebrain and midbrain of the mammal, which meets three additional criteria, elaborated upon by Myers (1974). These criteria are: (1) the rapidity of the rise and decay time of the membrane depolarization induced by ACh; (2) the ubiquitous morphological distribution of ACh in the cerebrum in terms of its equal availability to transmit both excitatory or inhibitory impulses in every functional system; and (3) the ultrastructural and neurochemical uniqueness of the cholinergic nerve ending particularly in terms of its local depolarizing action and the almost instantaneous degradation of this choline ester by acetylcholinesterase.

Other substances including the monoamines would have to be classified quite differently depending on whether they are: (a) found in nerve terminals (e.g., 5-HT and norepinephrine); (b) synthesized extraneuronally but nevertheless exert an action on nerve cells (e.g., angiotensin); or (c) synthesized in the soma of the neuron (peptide releasing factors). Interestingly, all of these other substances are morphologically restricted, possess different and perhaps more sluggish modes of inactivation, and ostensibly subserve functions that do not require milli-second responsiveness (Myers, 1974).

Finally, the classification of an endogenous substance that has a neural function and its presumed role in the intricate mechanism by which a process is coded depends upon all aspects of its action and the totality of its *in vivo* characteristics.

ACKNOWLEDGEMENTS

The research reported here has been supported, in large part, over the years, by National Science Foundation Grant GB 24592, Office of Naval Research Contract N00014-67-A-0226-0003, National Institute of Mental Health Training Grant MH10267-08 and from the National Institutes of Health, Health Sciences Advancement Award FR-06013.

REFERENCES

Avery, D. D., 1971, Intrahypothalamic adrenergic and cholinergic injection effects on temperature and ingestive behavior in the rat, *Neuropharmacol.* 10:753-763.

Feldberg, W. and Myers, R. D., 1964, Effects on temperature of amines injected into the cerebral ventricles. A new concept of temperature regulation, *J. Physiol.* 173:226-237.

Fuxe, K., Hökfelt, T. and Ungerstedt, U., 1970, Morphological and functional aspects of central monoamine neurons, *Int. Rev. Neurobiol.* 13:93-126.

Grossman, S. P., 1960, Eating or drinking elicited by direct adrenergic or cholinergic stimulation of hypothalamus, *Science* 132: 301-302.

Hall, D. H. and Myers, R. D., 1972, Temperature changes produced by nicotine injected into the hypothalamus of the conscious monkey, *Brain Res.* 37:241-251.

Jell, R. M., 1973, Responses of hypothalamic neurones to local temperature and to acetylcholine, noradrenaline and 5-hydroxytryptamine, *Brain Res.* 55:123-134.

Jones, B., Bobillier, P. and Jouvet, M., 1969, Effets de la destruction des neurones contenant des catecholamines du mesencephale sur le cycle veille-sommeil du chat, *C. R. Soc. Biol.* (Paris) 163:176-180.

Legendre, R. and Piéron, H., 1910, Des résultats histophysiologiques de l'injection intra-occipito-atlantoïdienne de liquides insomniques. *Cr. Soc. Biol.* (Paris) 1:1108-1109.

Miller, N. E., 1965, Chemical coding of behavior in the brain. Science 148:328-338.

Myers, R. D., 1967, Transfusion of cerebrospinal fluid and tissue bound chemical factors between the brains of conscious monkeys: a new neurobiological assay. Physiol. Behav. 2:373-377.

Myers, R. D., 1969, Chemical mechanisms in the hypothalamus mediating eating and drinking in the monkey. Ann. N. Y. Acad. Sci. 157:918-933.

Myers, R. D., 1970, The role of hypothalamic transmitter factors in the control of body temperature, in "Physiological and Behavioral Temperature Regulation," (J. D. Hardy, ed.), pp. 648-666, Charles Thomas, Springfield.

Myers, R. D., 1974, "Chemical Stimulation of the Brain," Van Nostrand Reinhold, New York.

Myers, R. D. and Beleslin, D. B., 1971, Changes in serotonin release in hypothalamus during cooling or warming of the monkey, Amer. J. Physiol. 220:1746-1754.

Myers, R. D. and Chinn, C., 1973, Evoked release of hypothalamic norepinephrine during thermoregulation in the cat, Amer. J. Physiol. 224:230-236.

Myers, R. D. and Sharpe, L. G., 1968a, Chemical activation of ingestive and other hypothalamic regulatory mechanisms, Physiol. Behav. 3:987-995.

Myers, R. D. and Sharpe, L. G., 1968b, Temperature in the monkey: transmitter factors released from the brain during thermoregulation, Science 161:572-573.

Myers, R. D. and Waller, M. B., 1973, Differential release of acetylcholine from the hypothalamus and mesencephalon of the monkey during thermoregulation, J. Physiol. 230:273-293.

Myers, R. D. and Yaksh, T. L., 1969, Control of body temperature in the unanaesthetized monkey by cholinergic and aminergic systems in the hypothalamus, J. Physiol. 202:483-500.

Rudy, T. and Wolf, H., 1971, The effect of intrahypothalamically injected sympathomimetic amines on temperature regulation in the cat, J. Pharmacol. Exp. Ther. 179:218-235.

Sharpe, L. G. and Myers, R. D., 1969, Feeding and drinking following stimulation of the diencephalon of the monkey with amines and other substances, Exp. Brain Res. 8:295-310.

Werman, R., 1972, CNS cellular level: membranes, Ann. Rev. Physiol. 34:337-374.

Yaksh, T. L. and Myers, R. D., 1972a, Hypothalamic "coding" in the unanesthetized monkey of noradrenergic sites mediating feeding and thermoregulation, Physiol. Behav. 8:251-257.

Yaksh, T. L. and Myers, R. D., 1972b, Neurohumoral substances released from hypothalamus of the monkey during hunger and satiety, Amer. J. Physiol. 222:503-515.

ABSTRACT:

Experimental evidence suggests that the functions of both temperature regulation and feeding behavior, which are intimately related and vital to survival, are subserved by differential neurohumoral codes within restricted regions of the hypothalamus and other parts of the limbic system. By the method of perfusing isolated regions of the brain of the cat and monkey with push-pull cannulae, we have tried to correlate the localized release of a neurohumoral substance with the response to a microinjection of the same substance at the same locus. From neurons within the anterior hypothalamus, serotonin and norepinephrine are apparently released in balanced opposition to one another when heat production or heat loss, respectively, is required. Efferent signals for thermogenesis are apparently transmitted by a cholinergic pathway which has been traced from the preoptic area through the mesencephalon.

The modulation of food intake seems to involve noradrenergic synapses in the near lateral or the perifornical regions of the monkey's hypothalamus. A pharmacologically potent "hunger factor" as well as a "satiety factor" are released from this hypothalamic area; each is capable of inducing or preventing eating, respectively, when re-introduced into the same region. Part of the active principle which stimulates feeding appears to be norepinephrine. This corresponds well with the pharmacological effects on eating produced by the local injection of the catecholamine and also the inhibition of feeding activity by a chemical lesion produced by 6-hydroxydopamine of the catecholaminergic nerve terminals in the hypothalamus. Thus, this monoamine may be used as both an inhibitory substance in the anterior hypothalamus (to block heat production) and as an excitatory factor in the lateral area (to stimulate feeding).

Finally, the view is presented that whereas acetylcholine may be used universally in the brain-stem as a neurotransmitter underlying all vegetative functions, the other biogenic amines including serotonin and norepinephrine are not CNS transmitters in the classical sense.

FUNCTIONAL CHANGES OF SYNAPSES

Victor ALEMAN, Alejandro BAYON and Jorge MOLINA

Sección de Neurobiología, Centro de Investigación y de Estudios Avanzados del IPN
México, D.F. México

There seems to be some data to suggest that synapses are capable of changing their efficacy (Cragg, 1972; Eccles, 1964; Lømo, 1969), either increasing or decreasing it depending on the number and frequency of the electric impulses to which they are subjected.

This synaptic plasticity may be the basis for more important and complex functions of the nervous tissue. Therefore it is important to ascertain how these synaptic functional changes occur at the molecular level.

It is known that the cyclic AMP system (adenyl cyclase, protein kinase, cyclic 3',5'-nucleotide phosphodiesterase and their substrates) shows a higher activity in the nervous tissue than that of other tissues (Sutherland et al., 1962; Butcher and Sutherland, 1962). Furthermore this system seems to be particularly active in the nerve end terminals (De Robertis et al., 1967; Gaballah and Popoff, 1971a; Gaballah and Popoff, 1971b).

Since it has been demonstrated that electrical stimulation (Kakiuchi et al., 1969) and some putative neurotransmitters (Chou et al., 1971; Kakiuchi and Rall, 1968) increase cyclic AMP levels in nervous tissue, a model system is proposed that could explain synaptic efficiency changes, that last for several minutes. In order to test our proposal we have used the superior cervical ganglion of the rat because it is not an excessively complex system, and furthermore it has been shown to exhibit, under certain conditions, synaptic facilitation (Dunant, 1969).

Determination of cyclic AMP Levels, Protein Kinase and Phosphorylation Activities.

Ganglia from 250g female rats were dissected and decapsulated in bicarbonate saline, pH 7.4 at 37°C [NaCl, 136 mM; KCl, 5.6 mM; $NaHCO_3$, 16.2 mM; NaH_2PO_4, 1.2 mM; $MgCl_2$, 1.2 mM; $CaCl_2$, 1.0 mM; Glucose 5.5 mM; gassed with O_2 + CO_2 (95:5)], and mounted in the stimulation chambers. Ganglia were handled to a minimum and gently.

For the study of the changes in cyclic AMP, ganglia were preincubated for 20 min in bicarbonate saline solution kept at 37°C, stimulated *in vitro* (with a frequency of 50/sec, 5 ms pulse duration, intensity of 4 volts and total train duration of 20 sec in all cases), incubated and, at the indicated times, groups of them were frozen and homogenized. Controls were similarly handled but they were not stimulated. In other experiments ganglia were stimulated *in vivo*. Cyclic AMP was determined by the method of Albano et al. (1971).

For the study of the temporal patterns of protein kinase activity, ganglia were stimulated *in vivo*, removed, decapsulated in bicarbonate saline solution at 37°C and homogenized at the indicated times. Enzyme activity was measured by a modification of the method of Miyamoto et al. (1969). Controls were unstimulated.

For the changes in protein phosphorylation following stimulation, ganglia were dissected, decapsulated in bicarbonate saline solution at 37°C, preincubated for 20 min in the same solution without phosphate, stimulated under the same conditions (in order to deplete it from ATP), followed by a 50 min incubation period in the presence of 25 μc of ^{32}P. After this time they were stimulated, ^{32}P excess removed and homogenized at the indicated times. The nuclear fraction was separated by centrifugation at 700 x g for 10 min and the supernatant centrifuged at 200,000 x g for 15 min; the pellet thus obtained is the crude mitochondrial fraction.

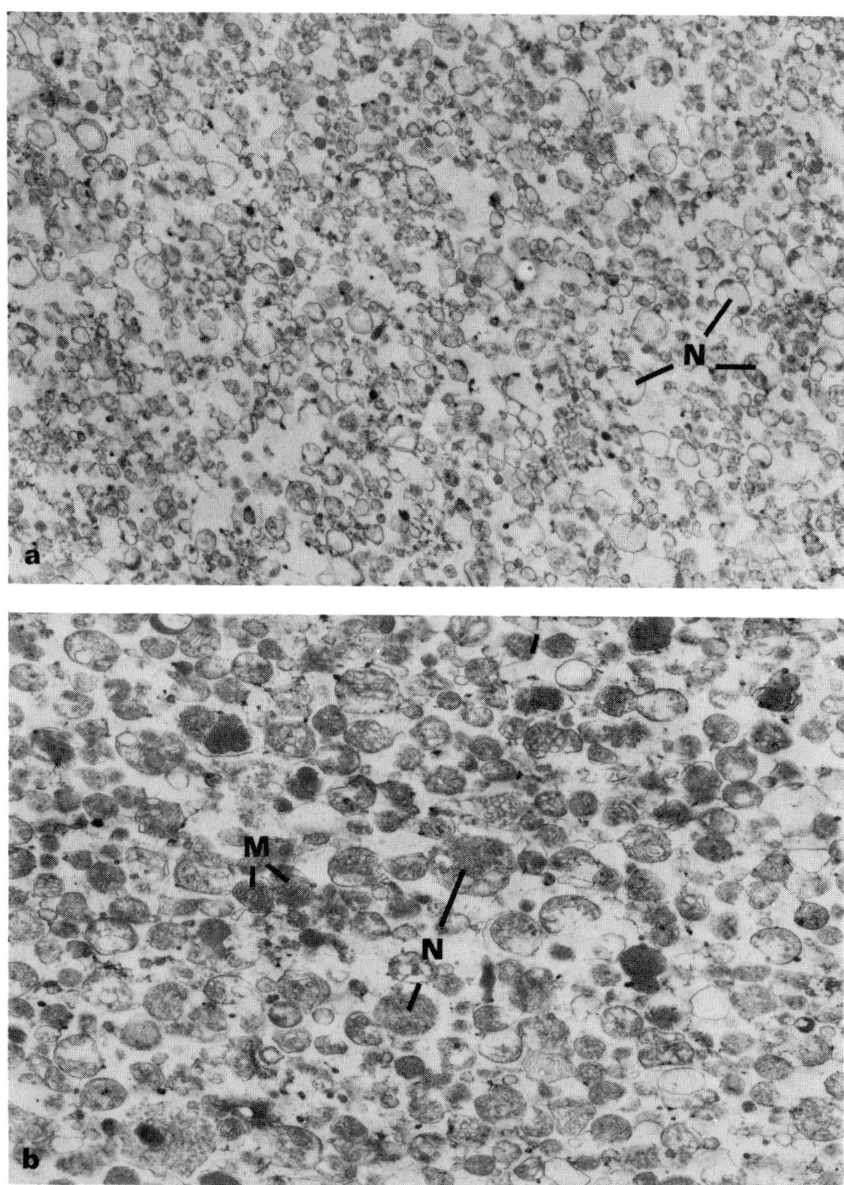

Fig. 1. Electron micrographs of nerve end terminals from the rat superior cervical ganglion. Subcellular fractionation was carried out as indicated in the text. After separation, the nerve end terminals were pelleted, fixed in glutaraldehyde and processed for E.M.; (a) C fraction (De Robertis nomenclature); (b) D fraction. Nerve end terminals (N), mitochondria (M). Magnification x 8,100.

The separation of the nerve end terminal fractions was accomplished by the use of discontinuous sucrose density gradients in a micro scale range and then centrifuged at 65,000 x g for 150 min. The purity of each fraction thus obtained was checked by electron microscopy.

The nerve end terminals thus obtained were aliquoted on filter paper discs and washed 4 times in 5% (w/v) TCA containing 1.0 M monosodium phosphate, one time with ethanol-ether (1:1 v/v) at 37°C repeated once more at room temperature and two more times with ether made at room temperature. Discs were dryed and the radioactivity counted in a Packard Tri-Carb Spectrometer.

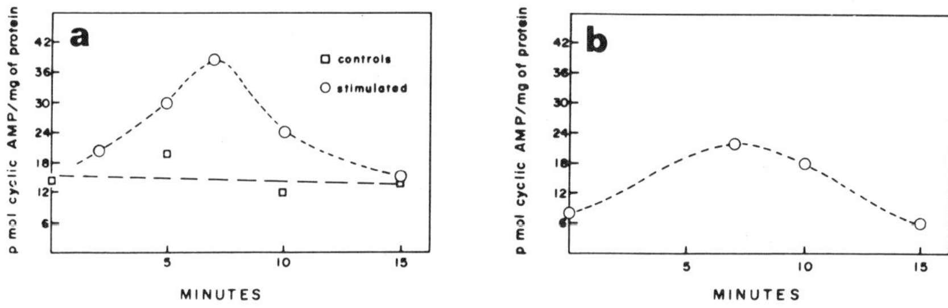

Fig. 2. Levels of cyclic AMP in stimulated and non-stimulated sympathetic ganglia. (a) Ganglia were stimulated in vitro and stimulating conditions were those described under methods. They were incubated in the stimulating chamber in bicarbonate saline saturated with O_2 - CO_2 (95:5 v/v) at 37°C, frozen and homogenized at the indicated times. Ganglia used for control values were incubated under the same conditions, but not stimulated; (b) All conditions were as in Fig. 2a except that ganglia were stimulated in vivo, decapsulated at 37°C, frozen in liquid nitrogen and homogenized at the indicated times. 0---0, stimulated; zero control time gave a similar value than zero stimulated time.

Results

After subcellular fractionation of the ganglia, two nerve end terminal fractions were isolated with a good degree of purity, as can be seen in Fig. 1a and 1b. Besides, it was possible to obtain nuclear, mitochondrial and supernatant fractions.

The maximum amount of cyclic AMP was obtained seven minutes after stimulation, and it showed 300% increase above the non-stimulated control value. Similar results were obtained when the ganglia were stimulated either in vivo or in vitro (Fig. 2a and 2b).

Non-stimulated control ganglia homogenized at the same times to those of stimulated ganglia, gave

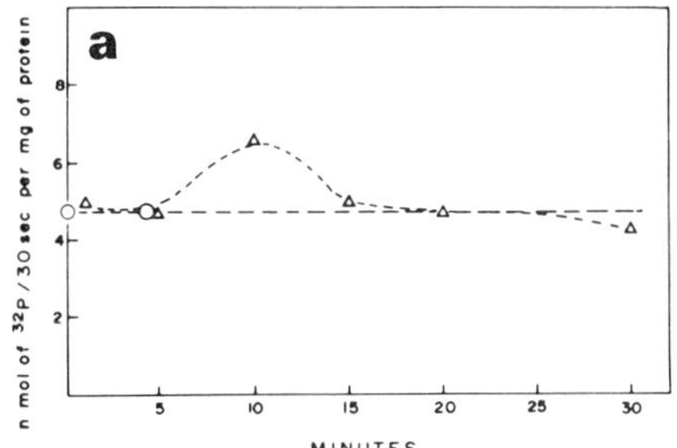

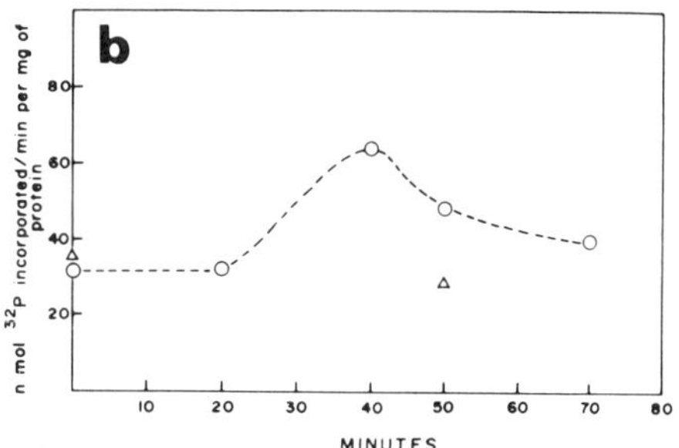

Fig. 3 (a). The effect of synaptic transmission of protein kinase activity. Ganglia were stimulated in vivo, excised, decapsulated and homogenized at the indicated times. Δ---Δ, stimulated ganglia; 0---0, non-stimulated ganglia. Protein kinase activity was assayed by a modification of the method of Miyamoto et al. (1971). (b) The effect of synaptic activity on protein phosphorylation of nerve endings. Ganglia were stimulated and incubated in vitro and homogenized at the indicated times. After gradient separation, nerve endings were collected, diluted 2 times, pelleted at 111,000 x g for 30 min, resuspended in 0.1n NaOH and sampled in filter paper discs. Other conditions are specified under methods. 0---0, stimulated; Δ---Δ, non-stimulated.

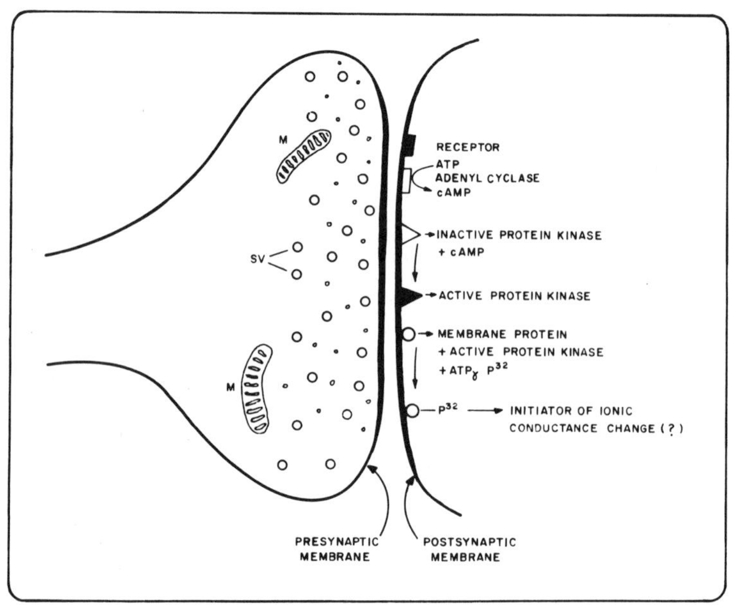

Fig. 4. A model for phosphorylation of the proteins of the postsynaptic membrane.

sufficient degree of purity, as it is shown in Figs. 1a and 1b. Although at this time our results are at a preliminary stage, Fig. 3b shows that, in the excised ganglion, protein phosphorylation occurs in the nerve end terminal fraction, and furthermore we found that it is dependent of the degree of synaptic activity.

In connection with the second question, it is difficult to establish a relationship between the protein phosphorylation and the cyclic AMP system. However, in our analysis of cyclic AMP levels and protein kinase activity at different times, we found a sequential time correlation that is in agreement with the relationship between the cyclic AMP system and protein phosphorylation. Also, cyclic AMP levels and protein kinase activity are dependent on synaptic activity. Thus, this seems to give an answer to our third question.

Based on the above mentioned findings and ideas, we would like to propose that medium term synaptic modulation can be explained by phosphorylation of protein(s) from the postsynaptic membrane through values similar to the zero time control value. This last value is close to that already reported for the rabbit superior cervical ganglion (McAfee et al., 1971).

The maximum activity of protein kinase was reached at about ten minutes after stimulation, and this increment was 40% above the unstimulated zero time control value (Fig. 3a).

Maximum protein phosphorylation of the less dense nerve ending fraction (C fraction) was obtained at about 40 minutes and the increment at this point was 200% above the zero time control value (Fig. 3b); control ganglia nerve ending fractions at the indicated times gave values comparables to that of zero time.

Discussion

In the present investigation there are two questions that are directly interconnected. They are: Is the cyclic AMP system in some way involved with synaptic transmission? Is the medium term synaptic modification (minutes) mediated through the cyclic AMP system? Strictly speaking, if one is able to answer the last question, at the same time one is answering at least in part, the first question.

It is known that protein phosphorylation is the resultant of the cyclic AMP system activation in the cell. Therefore if phosphorylation occurs at the synaptic membranes and furthermore both effects are mediated through synaptic activity, it is important to know if long lasting synaptic structural modifications, as this phosphorylation happens to be, could possibly lead to lasting synaptic functional changes.

Thus we decided first to answer the following questions: Does protein phosphorylation occur in nerve end terminals? Second, if it does, is it or not mediated through the cyclic AMP system? Finally, is the cyclic AMP system dependent on synaptic activity?

To answer the first question, we made a subcellular fractionation of the ganglia and subsequently we obtained a nerve end terminal fraction with a simplified model that is shown in Fig. 4. The present model does not explain how protein phosphorylation of postsynaptic membranes can end up in a temporal synaptic facilitation. That is, we do not give any idea of the molecular events which occur after protein phosphorylation and that give rise to synaptic facilitation.

The concept of protein phosphorylation that we are using here, does not exclude that this protein could be associated with another type of molecules. It even does not exclude the possibility that a non-protein associated molecule such as a phospholipid, be the one directly phosphorylated, provided that this association is essential for the specific function of such protein. As it is indicated in Fig. 4, one possibility is that this protein may be connected with the mechanisms of selected ionic permeability, which could be modified through this phosphorylation.

Although the cyclic AMP system could be directly involved in the mechanism of synaptic transmission by itself, it is proposed that it should be more probably related to post-transmission modulatory effects, regulating the efficiency of subsequent interneuron communication. If that is the case, any agent blocking exclusively the cyclic AMP system will not block the synaptic transmission but will interfere with the modulatory functions of this system.

References

Albano, J.D.M., Brown, B.L., Elnis, R.P., Sghersi, A.M. Tampion, W., 1971, Saturation analysis of cyclic AMP, J. Endocrinol. 49: 12.

Butcher, R.W. and Sutherland, E.W., 1962, Adenosine 3' 5'-phosphate in biological materials. I. Purification and properties of cyclic 3',5'-nucleotide phosphodiesterase and use of this enzyme to characterize adenosine 3',5'-phosphate in human urine, J. Biol. Chem. 237: 1244.

Chou, W.A., Ho, A.K. and Loh, H.H., 1971, Neurohormones on brain adenyl cyclase activity in vivo, Nature New Biol. 233: 280.

Cragg, B.G., 1972, Plasticity of synapses, in "The Structure and Function of Nervous Tissue", (G.H. Bourne, ed.) pp.1-60, Academic Press, N.Y.

De Robertis, E., Arnaiz, G.D.R.L., Alberici, M., Butcher, R.W. and Sutherland, E.W., 1967, Subcellular distribution of adenyl cyclase and cyclic phosphodiesterase in rat brain cortex, J. Biol. Chem. 242: 3487.

Dunant, Y., 1969, Presynaptic spike and excitatory postsynaptic potential in sympathetic ganglion. Their modification by pharmacological agents in "Mechanisms of Synaptic Transmission", (K. Akert and P.G. Waser, eds.) pp.131-139, Elsevier Publishing Co., Amsterdam.

Eccles, J.C., 1964, The physiology of the synapses (Chapters 6,7 and 16) Springer: Berlin-Göttingen-Heidelberg.

Gaballah, S. and Popoff, C., 1971 (a), Localization of adenosine 3',5'-monophosphate dependent protein kinase in brain, J. Neurochem. 18: 1795.

Gaballah, S. and Popoff, C., 1971(b), Cyclic 3',5'-nucleotide phosphodiesterase in nerve endings in developing rat brain, Brain Res. 25: 220.

Kakiuchi, S. and Rall, T.W., 1968, The influence of chemical agents on the accumulation of adenosine 3',5'-phosphate in slices of rabbit cerebellum, Mol. Pharmacol. 4: 367.

Kakiuchi, S., Rall, T.W. and McIlwain, H., 1969, The effect of electrical stimulation upon the accumulation of adenosine 3',5'-phosphate in isolated cerebral tissue, J. Neurochem. 16: 485.

Lømo, T., 1969, Some properties of a cortical excitatory synapse, in "Excitatory Synaptic Mechanisms", (P. Andersen and J. Jansen Jr., eds.), Oslo University Press, Oslo.

McAfee, D.A., Schorderet, M. and Greengard, P., 1971, Adenosine 3',5'-monophosphate in nervous tissue: Increase associated with synaptic transmission, Science, 1971: 1156.

Miyamoto, E., Kuo, J.F. and Greengard, P., 1969, Cyclic nucleotide-dependent protein kinases. III. Purification and properties of adenosine 3',5'-monophosphate-dependent protein kinase from bovine brain, J. Biol. Chem. 244: 6395.

Sutherland, E.W., Rall, T.W. and Menon, T., 1962, Adenyl cyclase. I. Distribution preparation and properties, J. Biol. Chem. 237: 1220.

THE CHEMICAL CODING OF AGGRESSION IN BRAIN

Donald J. Reis

Laboratory of Neurobiology, Department of Neurology
Cornell University Medical College
1300 York Avenue, New York, New York 10021

In the past several years there has been increasing evidence that several compounds of low molecular weight in the brain believed to function in the central as well as in the peripheral nervous system as neurotransmitters may serve to modulate the expression of aggressive behavior. These agents are synthesized, stored, and released by specific and discrete neural systems in the brain. They include the catecholamines norepinephrine (NE) and dopamine (DA), the indole-5-hydroxytryptamine (5-HT, serotonin) and the amine acetylcholine (ACh). In the following paper I would like to present evidence suggesting that each of these agents may have selective and independent actions in mediating the expression of two neurologically distinct classes of aggression: affective and predatory aggression.

I. TWO NEURAL SYSTEMS MEDIATING AGGRESSIVE BEHAVIOR

Aggressive behavior has been defined as behavior which leads to damage or destruction of some goal entity (52). In general, the different classes of aggression have been distinguished on the basis of the stimuli evoking the response. Thus, Moyer (52) has identified several classes of aggression including predatory, intermale, fear-induced, irritable, territorial, maternal, and instrumental. He has suggested that the central neural mechanisms and neuroendocrine modulations of these various aggressive states, in part, may differ.

In contrast, when aggressive behavior is classified by the type of response evoked, a simpler picture of the behavior emerges.

Indeed, there is substantial evidence that there are primarily two neurologically distinctive patterns underlying the expression of aggressive or attack behaviors. These may be termed affective aggression [affective attack (25)] and predatory aggression [biting attack (25)]. These two classes of aggression differ in their mode of expression, the provoking stimuli, the neuroanatomical substrate, and as will be developed later in this essay, their neurochemical modulation. Many of the differences between these two classes of aggression were recognized in the pioneering studies of Flynn and his associates (25). Some of the characteristics of these two types of aggression are outlined in Table I and will be briefly characterized.

Table I - Some Characteristics of Affective and Predatory Aggression

AFFECTIVE AGGRESSION	PREDATORY AGGRESSION
a. Intensive patterned autonomic activation, esp. sympatho-adrenal.	a. Little autonomic activation.
b. Threatening and defensive postures.	b. Stalking postures.
c. Menacing vocalization.	c. Little vocalization.
d. Attack is with claws, frenzied, mutilating.	d. Attacks by biting, lethally directed to back of neck.
e. Initiated by somatic (esp. nociceptive) as well as exteroceptive stimuli.	e. Initiated by exteroceptive stimuli, usually visual.
f. Generally lowered threshold for aggression (irritability), not always goal directed.	f. No irritability.
g. Intra- as well as inter-specific.	g. Inter-specific.
h. May be used only for display.	h. Always aims at success.
i. Not usually related to feeding.	i. Related to feeding.
j. Probably quite hormonally responsive.	j. Probably less hormonally responsive.

[From Reis (61)]

A. Affective Aggression

Affective aggression has been the most intensely studied class. In cat, the prototype of this behavior has been called sham rage (8), the defense reaction (1), or affective attack (31). It can be evoked in cat and other species by brain lesions [e.g. (8, 65)], by electrical stimulation of the brain (25, 31, 66, 91), as a natural response to real or threatened attack (2, 3, 18), or to the infliction of a painful stimulus (shock-induced aggression) (33).

Affective aggression is the mode of display seen in states characterized as "irritable", intermale, territorial or maternal. It is probably a ubiquitous mode of aggressive behavior in vertebrates.

B. Predatory Aggression

Predatory aggression is that class of aggression which leads to destruction of a natural prey, usually for food. In the laboratory, several paradigms have been used including the natural attack by rats against mice [muricidal killing (36)] or frogs (7). Predatory behavior can also be evoked by brain stimulation in cat (91) and rat (38, 56).

C. Neural Systems Mediating Aggressive Behaviors

The two modes of aggression are probably served by different neural pathways. In support of this contention are the facts that electrical stimulation of discrete sites within the hypothalamus or lower brainstem may evoke only one of the two types of behavior in several different species and that lesions placed at sites from which only one of the types of aggression can be evoked by electrical stimulation result in degeneration along different pathways within the brain (14).

The neural pathway mediating affective aggression probably is widely distributed in the brain and spinal cord. The principal pathways as defined by electrical stimulation include a series of neuronal structures running from the amygdala caudally through its two major projections, the ventral-amygdalo-fugal pathway and the stria terminalis, through the lateral hypothalamus and periaqueductal gray matter into the caudal brainstem (24, 35). The representation of affective aggression is redundant. Portions of the behavior can be evoked at many levels of the neuraxis. It is a neural network which is anatomically closely related to the spinal-thalamic tract, possibly explaining its close relationship to pain. The predatory responses are anatomically less well characterized.

From sites in the hypothalamus they appear to project into the ventral midbrain tegmentum (13).

The neural networks from which attack behavior can be evoked by focal electrical stimulation are themselves under modulatory control of other brain regions. The modulatory regions are not themselves critical for the expression of the behaviors but can be demonstrated with excitation to inhibit or facilitate evoked or natural aggression.

II. METHODS FOR STUDYING CENTRAL NEUROTRANSMITTERS IN BEHAVIOR

A. Criteria for Establishing the Participation of a Central Neurotransmitter in Aggression

To determine that a compound functions as a neurotransmitter within the brain critical in the mediation of a specific behavior, a set of criteria needs to be met (60, 61). These are listed in Table II. These criteria take into account the facts that behavior in the brain is organized within extensive neuronal networks with regional localization of function and that different neurotransmitter systems in the brain are regionally distributed. Implicit in this set of criteria is the fact that no single methodology is sufficient to establish the role of neurotransmitter function for a single compound. Biochemical, pharmacological, anatomical, and neurophysiological methods must be utilized.

B. Methods for Studying the Relationship of Neurotransmitters and Aggression

A number of research strategies have been employed in an attempt to determine the role of specific neurotransmitters in different classes of aggression (61, 89). These techniques include:

1. The demonstration by biochemical methods of differences in the concentration or turnover of specific neurotransmitters in the brain between strains of a species differing in aggressivity, i.e. natural [spontaneous (89)] aggression. This includes strains of mice with a naturally reduced threshold for affective aggression (40) or strains of rats with higher probability of killing a specified prey such as a mouse (36).

2. The demonstration of altered concentration or turnover of these agents or their metabolites in the brains of animals made aggressive by a variety of manipulations including isolation, pain, electrical stimulation, or selective lesions of the brain. Such

Table II - Criteria for Establishing the Essential Participation of a Central Neurotransmitter in a Specific Behavior

A. The neurotransmitter should be present in neurons within the central neural networks identified by other methods as necessary for expression of the specified behavior.

B. The specific neural system harboring the transmitter should also contain precursors, intermediate metabolites, and the enzymic machinery required for its synthesis and rapid inactivation, either by degradation or re-uptake of the transmitter.

C. Increased local release of the transmitter should be be demonstrated when the behavior occurs spontaneously and is evoked by focal electrical stimulation and/or brain lesions.

D. Local application of transmitter at appropriate sites within the neuronal network should evoke the behavior.

E. Chronic destruction of the specific neuronal systems harboring the neurotransmitter should abolish the specified behavior and result in disappearance of the neurotransmitter in regions to which the neurons project and which are essential to the behavior. Replacement of the transmitter should reverse the effects of such destruction.

F. Pharmacological agents which interact with the transmitter or its receptors to enhance or reduce its action should affect the behavior in a parallel manner.

[From Reis (61)]

aggressive states may be referred to as <u>evoked aggression</u> [induced aggression (89)].

3. The modification of natural or evoked aggression by drugs which facilitate or inhibit the availability of neurotransmitters at their receptors.

4. Elicitation of the behavior by intraventricular or cerebral injections of the neurotransmitters or their agonists.

5. Alteration of the behavior in the predicted direction by damage of individual chemical pathways by intraventricular or intracerebral administration of drugs with a selective toxicity for specific transmitter systems. The principal example of this is the drug 6-hydroxydopamine (6-OHDA) which will selectively destroy catecholamine pathways in the brain (82) or 6-hydroxydopa which has a more selective action on noradrenergic neurons (86).

III. NORADRENERGIC MECHANISMS

The neuronal systems in the brain which synthesize, store and release the neurotransmitter NE appear to play an important role in the elaboration of aggressive behavior (60). The principal evidence supports the view that the release of NE in brain facilitates affective and inhibits predatory aggression.

A. Noradrenergic Systems in the Brain

The cell bodies of noradrenergic neurons reside in the mesencephalon, pons, and medulla oblongata and send widely ramified axonal processes throughout the brain and spinal cord. On the basis of lesion studies Ungerstedt (87) has defined two principal path-

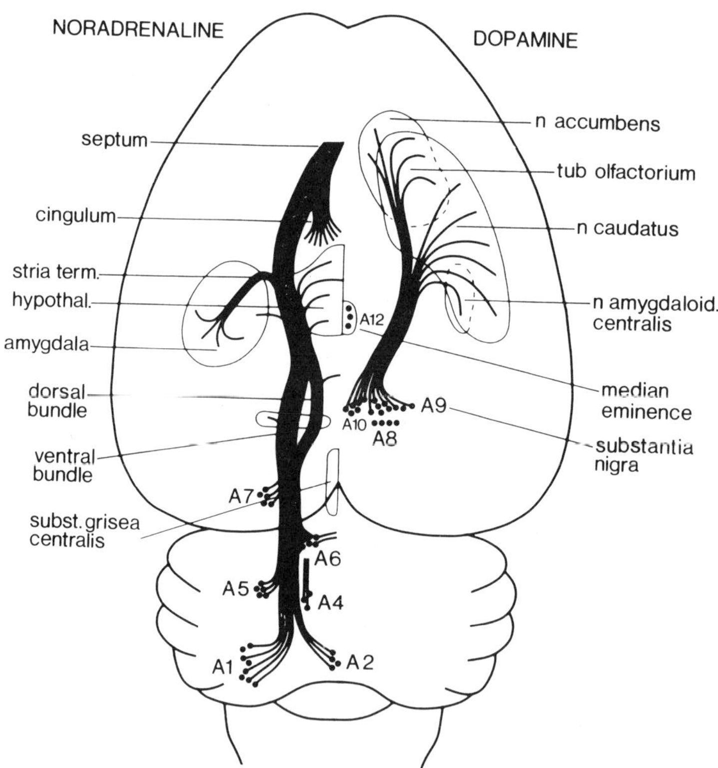

Figure 1. The noradrenergic and dopaminergic innervation of brain. Cell bodies of noradrenergic neurons are localized in lower brainstem (identified as A1-A7) and widely innervate the remainder of brain. The dopamine neurons are localized in mesencephalon (A8-A10) and also in the median eminence (A12). (From 87).

ways: a ventral NE system arising from cell groups in the medulla oblongata and pons, and a dorsal NE system arising from cell bodies primarily localized in the nucleus locus coeruleus. The dorsal system ascends in the medial reticular formation and enters the median forebrain bundle. It primarily innervates cellular groups in the lower brainstem, mesencephalon, and hypothalamus. It heavily innervates the bed nucleus of the stria terminalis, an area integrally involved in the expression of affective aggression (35). The dorsal pathway arises from the relatively few cells of the locus coeruleus. The processes of these cells are widely ramified, innervating cerebellum, neocortex, several portions of the limbic system including hippocampus and amygdala, and also, in part, the hypothalamus.

B. NE in Affective Aggression

The bulk of evidence suggests that neuronal release of NE facilitates or possibly initiates affective aggression. The evidence includes the demonstration of: (i) increased concentrations or turnover of NE in brains of animals in whom affective aggression has been induced; (ii) the initiation of several types of aggression by drugs which facilitate the action of NE within the brain; (iii) evidence that release of brain NE is closely linked to the affective display of sham rage behavior in the cat.

1. Brain NE in induced aggression. Rats in whom affective aggression is induced by isolation (89,93), the delivery of painful shocks (37, 53) or by forced restraint (44) appear to have increased concentrations or turnover of NE. One of the more interesting experiments in this regard has been recently reported by Lamprecht *et al* (44). In this study rats were repeatedly immobilized daily for four weeks. At the termination of the stress period, the rats showed an increase in the amount of shock induced fighting as well as heightened activity of the peripheral sympathetic nervous system. Four weeks following cessation of daily periods of restraint, sympathetic nerve activity as evidenced by the presence of the enzyme dopamine-β-hydroxylase in the peripheral blood had returned to normal. However, these animals still retained an increased propensity for fighting in association with a continued elevation in the hypothalamus of the activity of the enzyme tyrosine hydroxylase. The experiment suggests that the persistence of aggression was in some manner linked to the exalted capacity for synthesis of NE.

2. Drug-induced aggression. Aggressive behavior has been induced in several species by the systemic administration of drugs or pharmacologically-active precursors of NE, all of which serve to increase the efficacy of central noradrenergic transmission. The nature of the aggression induced by noradrenergic-mimetic drugs

seems to be due to increased irritability. Animals are thus more prone to fight in a group housing situation or when handled, and can thus be considered as expressing a form of affective aggression. Several examples may be cited: first, the precursor of NE, L-DOPA, which can increase the amount of available transmitter, increases fighting or rage in groups of caged animals when administered by itself (21, 58, 90), after inhibition of MAO (22, 66, 69), or following blockade of 5HT synthesis by the drug pCPA (47). Whether the effect of L-DOPA is related to the synthesis of DA, or NE, or is secondary to the capacity of L-DOPA to release amines from nerve endings (55) has not yet been determined. Second, amphetamine, which excites central adrenergic receptors, increases spontaneous or isolation-induced fighting in rodents (41, 42, 45, 59, 94) and facilitates aggressivity of rats during withdrawal from morphine (43). Third, tricyclic antidepressants or MAO inhibitors which increase NE availability centrally, either respectively by blocking the uptake inactivation or by inhibiting metabolic inactivation by deamination, also increase affective aggression (22, 71).

3. Noradrenergic mechanisms in sham rage behavior. The evidence for the participation of central noradrenergic pathways in the display of rage in cat has been reviewed in detail elsewhere (60). The evidence can be summarized as follows:

Electrical stimulation of the amygdala or hypothalamus which produces sham rage behavior in the cat is associated with a fall in the concentration of NE but not DA or 5HT in the brain (28, 64). Electrical stimulation at adjacent sites failing to produce the behavior does not result in any change of the brain amines. Thus electrical stimulation of the brain can selectively decrease brain NE concentrations but only when such stimulation can evoke the specific behavior.

A high decerebrating lesion which produces spontaneous recurrent attacks of rage (Figure 2) also results, after several hours, in a fall in the concentration of NE but not 5HT in the lower brainstem (65). A decerebrating lesion placed slightly caudal at the intercollicular level, and which fails to result in spontaneous rage (but leaves intact the display of rage to noxious stimulation) does not result in any change in the amine concentration in the same areas of the lower brainstem. This finding, taken together with the observation of selective decrease of brain NE when sham rage is evoked by electrical brain stimulation is further evidence that the fall of NE concentration is specific for the behavior.

The fall of brain NE is the result of increased activity of noradrenergic neurons and not the consequence of diminished amine synthesis. NE concentrations fall because the neurally mediated release of the transmitter is increased (26, 62, 83) and is in

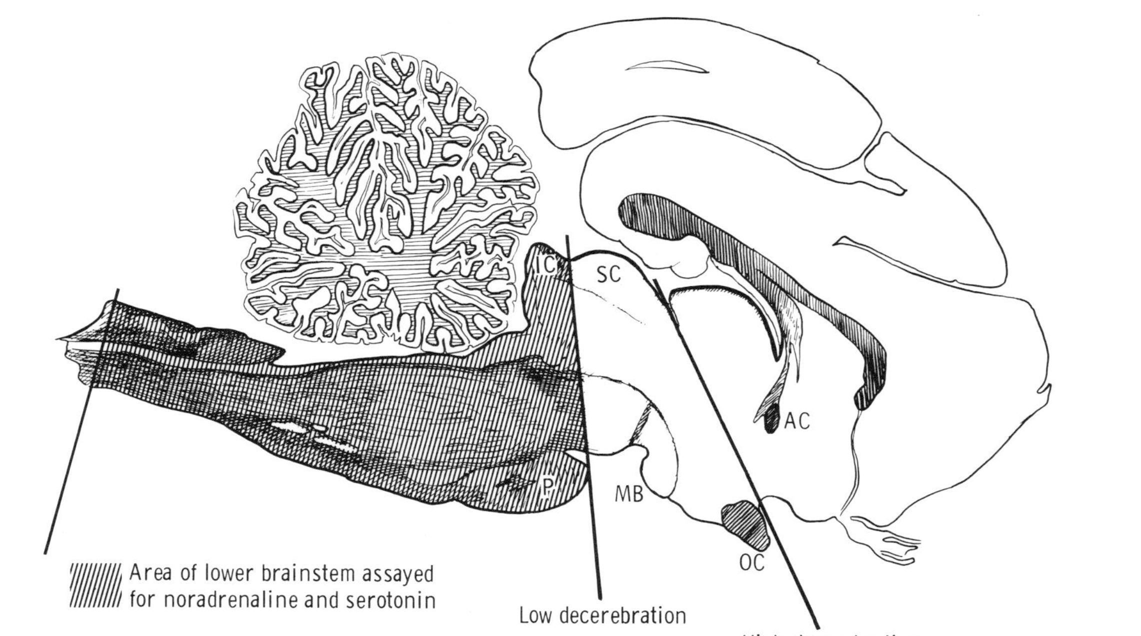

Figure 2. Schematic representation of mid-saggital section of cat brain showing plane of transection producing sham rage (high decerebration) or quiet behavior (low decerebration). The area of brainstem (cross hatched) assayed for norepinephrine and 5HT lies between line to left (at first cervical segment) and plane of low decerebration. Cerebellum and remainder of brain were not assayed. AC, anterior commissure; IC, interior colliculus; MB, mammillary bodies; OC, optic chiasm; P, pons. High decerebration resulted in fall of NE but not 5HT in lower brainstem. Low decerebration had no effect. (From 63).

excess of the rate of neuronal resynthesis. It appears to occur in most, if not all, of the NE terminals throughout the brainstem (26, 62). The augmented neural activity of noradrenergic neurons is not the result of body movement or of changes in cardiovascular function associated with the induced behavior since it occurs even after spinal cord transection abolishes all movement and reflex cardiovascular activity (65). The noradrenergic neuronal activity therefore probably represents a release of brainstem noradrenergic neurons from suprasegmental inhibition.

The release of NE in the brain is proportional to the intensity of the evoked behavior (63). Because drugs which facilitate the action of NE centrally enhance the behavior while drugs which reduce the action of the amine attenuate it, we have proposed that the central neuronal release of NE may trigger parts or all of the behavior (60, 63). Furthermore, drug studies have suggested that it is only a small percentage of the NE in the brain, primarily that which is newly synthesized, which is required to trigger the attacks of rage (60, 66). The precise localization of regions in the brain harboring the critical noradrenergic terminals remains unknown.

When attack behavior evoked by electrical stimulation of hypothalamus is prolonged by three days of intermittent stimulation, the activity of the enzymes tyrosine hydroxylase, the rate-limiting step in the formation of catecholamines, and phenylethanolamine-N-methyl-transferase (PNMT) (the enzyme converting NE to epinephrine) is increased in the adrenal gland (67). The increase in adrenal enzyme activity, probably due to an increase in enzyme molecules, is specific for the behavior and is neurogenically mediated. Stimulation at adjacent sites failing to produce the behavior fails to change the enzyme activity. There is also evidence that tyrosine hydroxylase activity is also increased by such stimulation in the brain. The importance of these observations is that they suggest that when attack behavior is prolonged it can produce more enduring changes in the biosynthetic machinery required for the production of the very neurotransmitter which is critical in the expression of the same behavior.

The mechanism by which NE produces its effect on the behavior is not entirely known. The transmitter may be acting to inhibit or excite neurons in specific areas of the brain. We have proposed (60, 63) that at least some parts of the behavior may result from inhibition by NE of neurons in brainstem regions which are themselves inhibitory to all or parts of the behavior.

C. Brain NE in Predatory Aggression

While there is ample evidence that the release of NE is facilitatory for affective aggression, central adrenergic mechanisms appear to inhibit predatory killing. The conclusion is primarily based on the observations that those drugs which presumably facilitate the action of NE centrally such as amphetamines, L-DOPA, or tricyclic antidepressants, are very effective in blocking muricidal killing in rat (32, 39, 57, 68). The suppression of predation occurs in spite of the fact that these drugs initiate affective aggression in the same species.

Attempts to explain in biochemical terms the opposing actions of the adrenergicomimetic agents in the two classes of aggression have not been altogether satisfactory. The "paradoxical" action of the tricyclic antidepressants in facilitating affective and inhibiting predatory aggression has been interpreted as related to their ability to suppress NE turnover selectively in muricidal rats (68). Others have suggested that drugs may, under some circumstances, act like a neuroleptic agent blocking noradrenergic transmission post-synaptically (95). I believe, however, that the contradictory findings can best be reconciled by assuming that it is the site at which NE neurons terminate rather than the NE neurons viewed as a uniform system which determines the role of the amine in specific behavior.

D. Some Contradictory Data

While a large body of evidence supports the hypothesis that the release of NE from central noradrenergic neurons facilitates or even triggers affective aggression and suppresses predatory aggression, there are several experimental findings which are not easily reconciled with this view.

First are the complex changes in aggressive behavior reported in several recent studies (10, 34, 50, 85) in which the drug 6-OHDA, which selectively destroys catecholaminergic neurons (11, 12, 88), or 6-OH-DOPA, a compound with more selective effects on noradrenergic neurons (86), was introduced into the brain. It would be predicted that if NE facilitates affective and inhibits predatory aggression, selective destruction of the NE neuron should result in opposing results, i.e. inhibition of affective and facilitation of predatory killing. The results, however, do not follow this simple prediction. Eichelman, Thoa and associates have found that after the intraventricular injection of 6-OHDA or 6-OH-DOPA (19, 85, 86) there was a decrease in shock-induced fighting behavior in rats, a a mode of affective aggression. The increase in fighting, however, appeared only gradually after a latency of several days, then continued to increase over several weeks. The delay in fighting did not correlate with the fall of brain NE which occurred much sooner (86), indicating the reduction of the neurotransmitter was probably not itself the cause of the augmented aggression. It is conceivable that the delay in the increase of affective aggression was due to the development of a denervation hypersensitivity of central noradrenergic receptors. Stimulation of these receptors could be by a small pool of remaining NE fibers or possibly by overflow from dopaminergic neurons in the critical area. If such is the mechanism of the augmented aggression in these animals, the findings would then be consistent with the NE hypothesis.

The effects of 6-OHDA on predatory aggression have recently been investigated by Jimerson and Reis (34). In these studies the effects of 6-OHDA bilaterally injected into the lateral hypothalamus of rats on the latencies for attack and killing of a natural prey, frog, were observed. The drug resulted in a reduction in the probability that an individual rat would attack or kill a frog, in the latencies for attack and kill, and in the intensity of the execution of the act of killing itself. The impaired behavior persisted but was characterized by daily fluctuations. On some days treated animals killed most frogs; on others, none. While the latencies in treated animals for attack and kill were often increased, on other days these latencies were close to control. Motor activity and feeding and drinking behavior remained depressed throughout. The results of the study would suggest that destruction of catecholamine pathways passing through or terminating in the lateral hypothalamus would impair predatory aggression by interfering with the ability of the rat to initiate sustained killing (a motivational deficit) and/or in recognizing the frog as prey (a perceptual defect) and not in the ability to execute and perform the motor act of killing.

A second series of contradictory observations relates to the long-standing observations that when NE is introduced intraventricularly it usually fails to produce aggression and indeed may even elicit sedation or sleep (50). Only rarely, when infused in very low (73) or extremely high (16) concentrations has any sort of behavior been elicited. These findings stand in contrast with the fact that intraventricularly administered ACh or carbachol will often produce affective aggression, thereby demonstrating that such behavior can be evoked through the intraventricular route.

The failure of NE, in most circumstances, to produce affective aggression when introduced intraventricularly may reflect the problem of achieving physiological levels of the amine at the appropriate site. For example, NE when microinjected into restricted areas of the upper brainstem has produced outbursts of affective attack in cat (74).

However, another interpretation is that NE has multiple and possibly opposing actions in the brain. At some sites the amine produced behavioral inhibition, while in others it lead to behavioral excitement and aggressive behavior. These contrasting actions possibly could be mediated by different types of adrenergic receptors, alpha and beta. Thus, the diffuse application of NE to central neurons through the ventricles might result in a preponderance of behavioral inhibition masking any excitatory actions of the amine. Conversely, local application by microinjection of the amine might produce contradictory responses depending upon which receptors within the brain it was stimulating. Thus, it is probably the site at which the NE neuron terminates rather than the NE neuron of the uniform system which determines the role of the neurotransmitter in any specific behavior.

IV. DOPAMINE IN AGGRESSION

A. Dopamine Systems in Brain

The cell bodies of the neurons which synthesize, store and release the neurotransmitter DA reside within the substantia nigra and the interpeduncular nucleus (17, 87). The axons of these cells project into the striatum and basal forebrain, thereby exhibiting a far more restricted field of innervation than NE nerve cells. (Fig. 1).

B. Dopamine in Affective Aggression

The possibility that central dopaminergic like noradrenergic neurons facilitate the expression of affective aggression has been suggested by several recent investigations. First, it has been demonstrated that rats treated with apomorphine, a drug which is believed to act as a DA agonist (20), produces interspecific fighting (51, 74, 75). The behavior, which is only seen in adult males (51), is facilitated by isolation (75) or by pain (51) and hence is clearly influenced by the environment. However, it is possible that apomorphine may also activate noradrenergic neurons (49). Second, Lycke *et al* (47) have demonstrated that mice infected with herpes simplex encephalitis have an increased synthesis of brain DA and a reduced synthesis of brain 5HT. The infected animals are exceptionally aggressive and fight intensively between themselves. Since the aggressivity of these mice is reduced by inhibition of DA synthesis with alpha-methylparatyrosine, it seems likely that the behavior is related to the increased availability of DA, possibly facilitated by reduction of 5HT. However, the possibility that changes in NE release may also influence the behavior has not been excluded conclusively.

The third line of evidence for the possible participation of

DA in facilitating affective aggression is derived from the fact that many of the studies designed to demonstrate the role of NE in aggressive behavior do not exclude the possibility that DA is also involved. Thus, the finding that the production of sham rage by electrical stimulation of brain results in a fall of brain NE but not DA does not rule out the possibility of augmented turnover of DA. Likewise, the excited and aggressive behaviors induced by the administration of L-DOPA or amphetamine to rats or cats conceivably could involve dopaminergic as well as noradrenergic mechanisms. Scheel-Kruger and Randrup (70), for example, have shown that the aggressivity produced in rats by the MAO inhibitor pargyline is facilitated by inhibiting the synthesis of NE by blockade of dopamine-β-hydroxylase. Everett and Borcherding (21) have demonstrated that the aggressivity in mice produced by L-DOPA is best correlated with levels of newly formed DA rather than NE. In addition, amphetamines, which evoke affective aggression, also increase the availability of DA as well as NE and their respective receptors (10). The amphetamine-induced aggression can be effectively blocked by neuroleptic agents believed more effective in blocking DA than NE (29).

Dopaminergic mechanisms, on the other hand, cannot account for all of the actions of affective aggression attributed to NE. For example, recurrent attacks of sham rage (which are facilitated by tricyclic antidepressants) can be evoked by decerebrating lesions (65) which destroy most of the DA terminals in basal ganglia and limbic forebrain (87).

C. Dopamine in Predatory Aggression

There have been no studies designed to examine the role of DA in predatory aggression. However, all drugs which would be expected to enhance the availability of DA at their receptors, such as amphetamine, L-DOPA, and monoamine oxidase inhibitors, block predatory aggression (32, 39). Thus, if DA plays a role in modulating this class of aggressive behavior, it would be inhibitory.

V. CHOLINERGIC MECHANISMS

A. Cholinergic Systems

Those neurons synthesizing, storing and releasing the neurotransmitter acetylcholine (ACh) in the brain have a much wider distribution than those harboring the monoamine neurotransmitters. The absence of a specific histochemical method for identifying ACh has not permitted an extensive and specific mapping of the cholinergic pathways (46, 80). These are much more extensively organized within the brain than are the monoaminergic projections and include intersegmental projections in reticular and limbic systems as well as locally restricted interneurons.

B. ACh and Affective Aggression

There is considerable pharmacological evidence that in addition to noradrenergic neurons, central cholinergic mechanisms participate in the neural organization of affective aggression. When introduced into the third or lateral ventricles of cat, ACh or its congener carbachol evoke a fully-developed display of angry behavior (9, 48) and affective attack. In some studies (9, 27, 48) the behavior has been attributed to the development of seizure discharges in the hippocampus or amygdala. It has also been proposed that the behavior is triggered by the diffusion by the agent to critical sites located near the walls of the third ventricle or the periaqueductal gray matter (48). However, the local injection of cholinergic agents into specific intercranial sites has produced affective display in cat and rat (30, 54) without the appearance of seizure activity in simultaneously monitored EEG's (74).

C. ACh in Predatory Aggression

While ACh can clearly elicit affective attack behavior, the most compelling evidence for the participation of central cholinergic networks in aggression is the demonstration of its role in the regulation of predatory killing particularly in the elegant studies of Bandler. Using as a model predatory frog-killing behavior in the rat (7), Bandler demonstrated (5, 6) that a lethal attack on a frog could be facilitated by intracerebral injection of carbachol or ACh (in conjunction with a cholinesterase inhibitor) at specific sites in the lateral hypothalamus, medial or mid-line thalamic nuclei, and ventral midbrain tegmentum. The cholinergic effect appeared specifically related to evocation of predatory aggression by ACh because of the stereotyped nature of the killing response, the lack of increased irritability of the animals, and the correspondence of sensitive sites to those areas in the rat or cat from which electrical stimulation could elicit or facilitate predatory aggression. Significantly, local injection of the cholinesterase inhibitor neostigmine at active hypothalamic sites also facilitated the behavior while systemic injection of atropine depressed it. These findings are important in that they suggest that it is the naturally-occurring synaptic release of ACh which facilitates predatory killing. It also indicates that the effects of the topical application of cholinergic agents to the hypothalamus does not act indirectly by releasing other amines, particularly catecholamines (10).

More provocatively, Smith *et al* (46) have described the production of mouse-killing behavior in rats which were not constitutionally killers by intracerebral injection of carbachol into lateral hypothalamus (81). Neostigmine also produced killing in non-killers while atropine blocked predatory killing in "killer" rats. These experiments differed from Bandler's in that Bandler was unable, within the dose range that he used, to convert non mouse-killing rats to mouse killers by cholinomimetic drugs.

The detailed anatomy of the cholinergic pathways subserving the predatory aggressive response are as yet unknown. However, it is probable that they lie in part within the trajectories of the descending pathways which subserve predatory aggression as described by Chi and Flynn (13, 14), and overlap, in part, the central cholinergic systems described by Shute and Lewis (46, 80).

VI. SEROTONERGIC MECHANISMS

The neural systems synthesizing, storing and releasing the neurotransmitter 5HT are, within the brain, relatively restricted. The cell bodies of these neurons are localized at the raphé nuclei sending their axons rostrally and caudally to widely innervate the brain and spinal cord (87).

The present evidence suggests that 5HT is preponderantly inhibitory to several modes of aggressive behavior. Most of the data is pharmacological and based on experiments in which the synthesis of 5HT is blocked by inhibiting the enzyme tryptophan hydroxylase by pCPA. In this manner, it has been demonstrated that shock-induced spontaneous fighting or predatory killing in the rat (15, 76, 77, 78, 84), or aggressiveness and rat killing in cat (23) are increased after treatment with pCPA. Sheard (76) has shown that an increase in aggressivity in the rat is associated with a fall of brain 5HT and of its principal metabolite 5-hydroxyindole acetic acid (5-HIAA), and may be returned to normal levels by treatment with a precursor 5-hydroxytryptophan. These findings would indicate that the behavioral action of pCPA is due to its blockade of the synthesis of 5HT. It has also been postulated that the anti-aggressive action of lithium in rat and man (79, 92) is due to an enhancement by the drug of the availability of 5HT.

Drug-induced aggressivity is also facilitated by inhibition of 5HT synthesis. For example, in mice, a dose of L-DOPA, which by itself fails to affect behavior will, after pCPA pre-treatment (47), evoke aggressivity. Likewise, aggressive pecking behavior in the chick can be evoked by amphetamines only after pCPA pre-treatment (72).

There is little evidence that direct excitation of serotonergic neurons will inhibit predatory aggression. Kulkarni's observation that 5HT will block muricidal behavior in rat is consistent with the view that 5HT release is inhibitory for predatory as well as affective aggression (39).

VII. SUMMARY AND CONCLUSIONS

The expression of aggressive behavior in higher vertebrates is primarily mediated by two neurologically distinct systems. The first, ubiquitously distributed in vertebrates, subserves the expression of affective aggression. The second, possibly only developed in predators, subserves predatory aggression. The two classes of aggression differ in their autonomic, somatic and behavioral manifestations, the evoking stimuli, and their sensitivity to hormones. Large portions of the neuronal network subserving each type of aggression also differ.

All available evidence suggests that several compounds including NE, DA, ACh and 5HT, believed to act as neurotransmitters in the CNS, may have characteristic actions on each class of aggression, (Fig. 3) implying, thereby, that specific neuronal systems which synthesize, store and release these agents are differently engaged. At present, the largest body of data suggests that the release of the catecholamine NE facilitates and in some cases may trigger the expression of affective aggression. Whether the amine acts to modulate all or merely portions of the behavior, as for example the intense alerting, is unknown. On the other hand, the same transmitter appears to inhibit predatory aggression. The mechanism whereby a single transmitter facilitates one and inhibits another closely allied behavior is unknown. Conceivably, it could relate to either the selective action of different neuronal groups releasing an identical transmitter or to differences in adrenergic receptor mechanisms in different regions of the brain.

Cholinergic mechanisms, on the other hand, appear primarily involved in the facilitation or triggering of the central neural networks mediating predatory aggression. ACh also appears to facilitate affective aggression. The roles of DA and 5HT are uncertain. On the basis of the limited available data, DA appears to act like NE, facilitating affective and inhibiting predatory aggression. The action of 5HT appears to be inhibitory for both classes of aggression.

The foregoing considerations lead to several conclusions. First, no single transmitter serves a unique codon for either class of aggression. Several and possibly many other neurotransmitters

Neurotransmitter	Class of aggression	
	Affective	Predatory
Norepinephrine -	↑↑	↓
Dopamine -------	↑?	↓?
Acetylcholine ---	↑	↑↑
Serotonin -------	↓	↓

Figure 3. Different actions of some central neurotransmitters on two classes of aggression. (From 61).

appear to interact in their expression. Second, the behavioral action of the various neural transmitters is not uniquely limited to aggression. Each of these agents has been considered to participate in other behaviors as well. For example, NE and ACh both have been implicated as neurotransmitters subserving the hypothalamic regulation of eating and drinking. Thus, each transmitter serves multiple behaviors while several behaviors are subserved by multiple transmitters.

Any consideration of the neurochemistry of aggression ultimately leads to reflections on the possible chemical control of violence and aggressive behavior in man. At present so little is known of neural substrates and the molecular basis of human behavior that extrapolations from animal investigations are at best dangerous and premature. There are, however, two considerations arising from animal studies which bear on the problem of the control of human aggression by drugs which act upon the central neurotransmitter systems. First, the plurality of neural transmitters mediating aggression suggests that blockade of the action of only one transmitter would be unlikely to control all types of aggressive behavior. Secondly, the multiplicity of behaviors subserved by any one transmitter raises the question of whether any drug would be able to have a selective action on aggression without affecting other types of behaviors.

Further elaboration of the problem must await more detailed studies of the neural pathways that subserve different types of aggressive behavior, the identification of the transmitters within each network and the critical applications of neurochemical, neurophysiological and pharmacological methods to identify the dynamic chemical changes which occur during the aggressive acts themselves.

REFERENCES

1. Abrahams, V.C., Hilton, S.M. and Zbrozyna, A., 1960, Active muscle vasodilitation produced by stimulation of the brainstem: Its significance in the defense reaction, J.Physiol. 154:491.

2. Adams, D.B., 1968, The activity of single cells in the midbrain and hypothalamus of the cat during affective defense behavior, Arch.Ital.Biol. 106:243.

3. Adams, D.B., Baccelli, G., Mancia, G. and Zanchetti, A., 1969, Cardiovascular changes during naturally elicited fighting behavior in the cat, Am.J.Physiol. 216:1226.

4. Alexander, R.S., 1946, Tonic and reflex functions of medullary sympathetic cardiovascular centers, J.Neurophysiol. 9:205.

5. Bandler, R.J., 1970, Cholinergic synapses in the lateral hypothalamus for the control of predatory aggression in the rat, Brain Res. 20:409.

6. Bandler, R.J., 1971, Direct chemical stimulation of the thalamus: effects on aggressive behavior in the rat, Brain Res. 26:81.

7. Bandler, R.J. and Moyer, K.E., 1970, Animals spontaneously attacked by rats, Commun.Behav.Biol. 5:177.

8. Bard, P., 1928, A diencephalic mechanism for the expression of rage with special reference to the sympathetic nervous system, Am.J.Physiol. 84:490.

9. Baxter, B.L., 1967, Comparison of behavioral effects of electrical or chemical stimulation applied at the same brain loci, Exp.Neurol. 19:412.

10. Besson, M.J., Cheramy, A., Feltz, P. and Glowinski, J., 1971, Dopamine: spontaneous and drug-induced release from the caudate nucleus in the cat, Brain Res. 32:407.

11. Bloom, F.E., Algeri, S., Groppetti, A., Revuelta, A. and Costa, E., 1969, Lesions of central norepinephrine terminals with 6-OH-dopamine: biochemistry and fine structure, Science 166:1284.

12. Breese, G.R. and Traylor, T.D., 1970, Effects of 6-hydroxy-dopamine on brain norepinephrine and dopamine; evidence for selective degeneration of catecholamine neurons, J.Pharmacol. Exp.Ther. 174:413.

13. Chi, C. and Flynn, J.P., 1971, Neural pathways associated with hypothalamically elicited attack behavior in cats, Science 171: 703.

14. Chi, C. and Flynn, J.P., 1971, Neuroanatomic projections related to biting attack elicited from hypothalamus in cats, Brain Res. 35: 49.

15. Conner, R.L., Stolk, J.M., Barchas, J., Dement, W.C. and Levine, S., 1970, The effect of parachlorophenylalanine (pCPA) on shock-induced fighting behavior in rats, Physiology and Behavior 5: 1221.

16. Cordeau, J.P., de Champlain, J. and Jacks, N., 1971, Excitation and prolonged waking produced by catecholamines injected into the ventricular system of cats, Can. J. Physiol. Pharmacol. 49: 627.

17. Dahlström, A. and Fuxe, K., 1965, Evidence for the existence of monoamine neurons in the central nervous system. II. Experimentally induced changes in the intraneuronal amine levels of bulbospinal neuron systems and IV. Distribution of monoamine nerve terminals in the central nervous system, Acta Physiol. Scand., Suppl. 247, 64: 1.

18. Delgado, J.M.R., 1964, Free behavior and brain stimulation, Int. Rev. Neurobiol. 6: 349.

19. Eichelman, B.S., Thoa, N.B. and Ng, L.K.Y., 1972, Facilitated aggression in the rat following 6-hydroxydopamine administration, Physiol. Behav. 8: 1.

20. Ernst, A.M., 1967, Mode of action of apomorphine and dexamphetamine on gnawing compulsion in rats, Psychopharmacologia 10: 316.

21. Everett, G.M. and Borcherding, J.W., 1970, Effect on concentrations of dopamine, norepinephrine and serotonin in brains of mice, Science 168: 849.

22. Everett, G.M. and Wiegand, R.G., 1961, Non-hydrazide monoamine oxidase inhibitors and their effects on central amines and motor behavior, Biochem. Pharmacol. 8: 163.

23. Ferguson, J., Henriksen, S., Cohen, J., Mitchell, G., Barchas, J. and Dement, W., 1970, "Hypersexuality" and behavioral changes in cats caused by administration of p-chlorophenylalanine, Science 168: 499.

24. Fernandez de Molina, A. and Hunsperger, R.W., 1962, Organization of the subcortical system governing defense and flight reactions in the cat, J. Physiol. (Lond.) 160: 200.

25. Flynn, J.P., Vanegas, H., Foote, W. and Edwards, S., 1970, Neural mechanisms involved in a cat's attack on a rat, in "The Neural Control of Behavior" (R.E. Whalen, R.F. Thompson, M. Verzeano and N.F. Weinberger, eds.), pp. 135-173, Academic Press, New York.

26. Fuxe, K. and Gunne, L-M., 1966, Depletion of the amine stores in brain catecholamine terminals in amygdaloid stimulation, Acta Physiol. Scand. 62: 493.

27. Grossman, S.P., 1963, Chemically induced epileptiform seizures in the cat, Science 142: 409.

28. Gunne, L-M. and Lewander, T., 1966, Monoamines in brain and adrenal glands of cat after electrically induced defense reaction, Acta Physiol. Scand. 67: 405.

29. Hasselager, E., Rolinsky, Z. and Randrup, A., 1972, Specific antagonism by dopamine inhibitors of items of amphetamine-induced aggressive behavior, Psychopharmacologia 24: 485.

30. Hernandez-Peon, R., Chavez-Ibarra, G., Morgane, P.J. and Timo-Iaria, C., 1963, Limbic cholinergic pathways involved in sleep and emotional behavior, Exp. Neurol. 8: 93.

31. Hess, W.R., 1957, "The Functional Organization of the Diencephalon," Grune and Stratton, New York.

32. Horovitz, Z.P., Ragozzino, P.W. and Leaf, R.C., 1965, Selective block of rat mouse-killing by antidepressants, Life Sci. 4: 1909.

33. Hutchinson, R.R., Ulrich, R. and Azrin, N.H., 1965, Effects of age and related factors on the pain-aggression reaction, J. Comp. Physiol. Psychol. 59: 365.

34. Jimerson, D. and Reis, D.J., 1973, Effects of intrahypothalamic 6-hydroxydopamine on predatory aggression in rat, Brain Res. (in press).

35. Kaada, B., 1967, Brain mechanisms related to aggressive behavior, in "Aggression and Defense: Neural Mechanisms and Social Patterns" (C.D. Clemente and D.B. Lindsley, eds.), pp. 95-133, University of California Press, Los Angeles.

36. Karli, P., Vergnes, M. and Didiergerges, F., 1969, Rat-mouse interspecific aggressive behavior and its manipulation by brain ablation and brain stimulation, in "Aggressive Behavior" (S. Garattini and E.B. Sigg, eds.), pp. 47-55, Excerpta Medicia Foundation, Amsterdam.

37. Kety, S.S., Javoy, F., Thierry, A-M., Julou, L. and Glowinski, J., 1967, A sustained effect of electroconvulsive shock on the turnover of norepinephrine in the central nervous system of rat, Proc. Natl. Acad. Sci. 58: 1249.

38. King, M.B. and Hoebel, B.G., 1968, Killing elicited by brain stimulation in rat, Commun. Behav. Biol. 2: 173.

39. Kulkarni, A.S., 1968, Muricidal block produced by 5-hydroxytryptophan and various drugs, Life Sci. 7: 125.

40. Lagerspitz, K.Y.H., Tirri, R. and Lagerspitz, K.M.J., 1967, Neurochemical and endocrinological studies of mice selectively bred for aggressiveness, Rep. Inst. Psychol. (Turku) 29: 1.

41. Lal, H., De Feo, J.J. and Thut, P., 1968, Effect of amphetamine on pain-induced aggression, Commun. Behav. Biol. 1: 333.

42. Lal, H., Nesson, B. and Smith, N., 1970, Amphetamine-induced aggression in mice pretreated with dihydroxyphenylalanine (DOPA) and/or reserpine, Biol. Psychiat. 2: 299.

43. Lal, H., O'Brien, J. and Puri, S.K., 1971, Morphine withdrawal aggression: sensitization by amphetamines, Psychopharmacologia 22: 217.

44. Lamprecht, F., Eichelman, B.S., Thoa, N.B., Williams, R.B. and Kopin, I.J., 1972, Rat fighting behavior: serum dopamine-β-hydroxylase and hypothalamic tyrosine hydroxylase, Science 177: 1214.

45. Lewander, T., 1968, Urinary excretion and tissue levels of catecholamines during chronic amphetamine intoxication, Psychopharmacologia 13: 394.

46. Lewis, P.R. and Shute, C.C.D., 1967, The cholinergic limbic system: Projections to hippocampal formation, medial cortex, nuclei of the ascending cholinergic reticular system and the subfornical organ and supraoptic crest, Brain 90: 521.

47. Lycke, E., Modigh, K. and Roos, B.E., 1969, Aggression in mice associated with changes in the monoamine-metabolism of the brain, Experientia 25: 951.

48. MacLean, P.D., 1957, Chemical and electrical stimulation of hippocampus in unrestrained animals. II. Behavioral findings, AMA Arch. Neurol. 78: 128.

49. Maj, J., Grabowska, M. and Gajda, L., 1972, Effect of apomorphine on motility in rats, European J. Pharmacol. 17: 208.

50. Marley, E., 1966, Behavioral and electrophysical effects of catecholamines, Pharmacol. Rev. 18: 753.

51. McKenzie, G.M., 1971, Apomorphine-induced aggression in the rat, Brain Res. 34: 323.

52. Moyer, K.E., 1968, Kinds of aggression and their physiological basis, Commun. Behav. Biol. Part A. 2: 65.

53. Musacchio, J.M., Julou, L., Kety, S.S. and Glowinski, J., 1969, Increase in rat brain tyrosine hydroxylase activity produced by electroconvulsive shock, Proc. Natl. Acad. Sci. 63: 1117.

54. Myers, R.D., 1964, Emotional and autonomic responses following hypothalamic chemical stimulation, Can. J. Psychol. 18: 6.

55. Ng, K.Y., Chase, T.N., Colburn, R.W. and Kopin, I.J., 1970, L-DOPA-induced release of cerebral monoamines, Science 170: 76.

56. Panksepp, J., 1971, Aggression elicited by electrical stimulation of the hypothalamus in albino rats, Physiology and Behavior 6: 321.

57. Panksepp, J., 1971, Effect of hypothalamic lesions on mouse-killing and shock-induced fighting in rats, Physiology and Behavior 6: 311.

58. Randrup, A. and Munkvad, I., 1966, DOPA and other naturally occurring substances as causes of stereotypy and rage in rats, Acta Psychiat. Scand. 191: (Suppl. 42) 193.

59. Randrup, A. and Munkvad, I., 1967, Stereotyped activities produced by amphetamine in several animal species and man, Psychopharmacologia 11: 300.

60. Reis, D.J., 1972, The relationship between brain norepinephrine and aggressive behavior, Res. Publ. Ass. of Res. Nerv. Ment. Dis. 50: 266.

61. Reis, D.J., 1973, Central neurotransmitters in aggression, Res. Publ. Ass. of Res. Nerv. Ment. Dis. (in press)

62. Reis, D.J. and Fuxe, K., 1968, Depletion of noradrenaline in brainstem neurons during sham rage behavior produced by acute brainstem transection in cat, Brain Res. 7: 448.

63. Reis, D.J. and Fuxe, K., 1969, Brain norepinephrine: evidence that neuronal release is essential for sham rage behavior following brainstem transection in cat, Proc. Natl. Acad. Sci. U.S.A. 64: 108.

64. Reis, D.J. and Gunne, L-M., 1965, Brain catecholamines: relation to the defense reaction evoked by amygdaloid stimulation in the cat, Science 149: 450.

65. Reis, D.J., Miura, M., Weinbren, M. and Gunne, L-M., 1967, Brain catecholamines: relation to defense reaction evoked by acute brainstem transection in cat, Science 156: 1768.

66. Reis, D.J., Moorhead, D.T. II and Merlino, N., 1970, DOPA-induced excitement in cat and its relationship to brain norepinephrine concentrations, Arch. Neurol. 22: 31.

67. Reis, D.J., Moorhead, D.T. II, Rifkin, M., Joh, T. and Goldstein, M., 1971, Changes in enzymes synthesizing catecholamines in attack behavior evoked by hypothalamic stimulation in cat, Nature 229: 562.

68. Salama, A. and Goldberg, M.E., 1970, Neurochemical effects of imipramine and amphetamine in aggressive mouse-killing (muricidal rats), Biochem. Pharmacol. 19: 2023.

69. Scheel-Kruger, J. and Randrup, A., 1967, Stereotyped hyperactive behaviour produced by dopamine in the absence of noradrenaline, Life Sci. 6: 1389.

70. Scheel-Kruger, J. and Randrup, A., 1968, Aggressive behavior produced by pargyline in rats pretreated with diethyldithiocarbamate, J. Pharm. Pharmacol. 20: 948.

71. Schrold, J., 1970, Aggressive behavior in chicks induced by tricyclic antidepressants, Psychopharmacologia 17: 225.

72. Schrold, J. and Squires, R.F., 1971, Behavioral effects of a-amphetamine in young chicks treated with p-Cl-phenylalanine, Psychopharmacologia 20: 85.

73. Segal, D.S. and Mandell, A.J., 1970, Behavioral activation of rats during intraventricular infusion of norepinephrine, Proc. Natl. Acad. Sci. U.S.A. 66: 289.

74. Senault, B., 1970, Comportement d'aggressivité intraspécifique induit par l'apomorphine chez le rat, Psychopharmacologia 18: 271.

75. Senault, B., 1971, Influence d'isolement sur le comportement d'aggressivité intraspécifique induit par l'apomorphine chez le rat, Psychopharmacologia 20: 389.

76. Sheard, M.H., 1969, The effect of p-chlorophenylalanine on behaviour in rats: relation to 5-hydroxytryptamine and 5-hydroxyindoleacetic acid, Brain Res. 15: 524.

77. Sheard, M.H., 1970, Behavioural effects of p-chlorophenylalanine in rats: Inhibition by lithium, Commun. Behav. Biol. 5: 71.

78. Sheard, M.H., 1970, Effect of lithium on foot shock aggression in rats, Nature 228: 284.

79. Sheard, M.H., 1971, Effect of lithium on human aggression, Nature 230: 113.

80. Shute, C.C.D. and Lewis, P.R., 1967, The ascending cholinergic reticular system: neocortical, olfactory and subcortical projections, Brain 90: 497.

81. Smith, D.E., King, M.B. and Hoebel, B.G., 1970, Lateral hypothalamic control of killing: evidence for a cholinoceptive mechanism, Science 167: 900.

82. Smith, G.P., Strohmayer, A.J. and Reis, D.J., 1972, Effect of lateral hypothalamic injections of 6-hydroxydopamine on food and water intake in rats, Nature New Biol. 235: 27.

83. Sweet, R.D. and Reis, D.J., 1971, Collection of [^{3}H] norepinephrine in ventriculo-cisternal perfusate during hypothalamic stimulation in cat, Brain Res. 33: 584.

84. Tagliomente, A., Tagliomente, P., Gessa, G.L. and Brodie, B.B., 1969, Compulsive sexual activity induced by p-chlorophenylalanine in normal and pinealectomized male rats, Science 166: 1433.

85. Thoa, N.B., Eichelman, B. and Ng, L.K.Y., 1972, Shock-induced aggression: effects of 6-hydroxydopamine and other pharmacological agents, Brain Res. 43: 467.

86. Thoa, N.B., Eichelman, B., Richardson, J.S. and Jacobowitz, D., 1972, 6-Hydroxydopa depletion of brain norepinephrine and the facilitation of aggressive behavior, Science 178: 75.

87. Ungerstedt, U., 1971, I. Stereotaxic mapping of the monoamine pathways in the rat brain, Acta Physiol. Scand. Suppl. 367: 1.

88. Uretsky, N.J. and Iversen, L.L., 1970, Effects of 6-hydroxydopamine on catecholamine containing neurons in the rat brain, J. Neurochem. 17: 269.

89. Valzelli, L., 1967, Drugs and aggressiveness, Advances in Pharmacol. 5: 79.

90. Vander Wende, C. and Spoerlein, M.T., 1962, Psychotic symptoms induced in mice by the intravenous administration of solutions 3,4, dihydroxyphenylalanine (DOPA), Arch. Int. Pharmacodyn. 137: 145.

91. Wasman, M. and Flynn, J.P., 1962, Directed attack elicited from hypothalamus, Arch. Neurol. 6: 220.

92. Weischer, M.L., 1969, Uber die antiaggressive Wirkung von Lithium, Psychopharmacologia 15: 245.

93. Welch, B.L. and Welch, A.S., 1965, Effects of grouping on the level of brain norepinephrine in white Swiss mice, Life Sci. 4: 1011.

94. Welch, B.L. and Welch, A.S., 1969, Aggression and the biogenic amine neurohumors, in "Aggressive Behavior" (S. Garattini and E.B. Sigg, eds), pp. 179-189, Excerpta Medicia Foundation, Amsterdam.

95. Yen, H.C.Y., Katz, M.H. and Krop, S., 1970, Effects of various drugs on 3,4-dihydroxyphenylalanine (DL-DOPA)-induced excitation (aggressive behaviour) in mice, Toxicol. Appl. Pharmacol. 17: 597.

REGULATION OF HIPPOCAMPAL INFORMATION PROCESSING BY K+ RELEASE AND (K+)o ACCUMULATION: POSSIBLE ROLE IN LEARNING

Iván Izquierdo

Departamento de Fisiologia, Farmacologia e Biofísica, Instituto de Biociencias, Universidade Federal de Rio Grande do Sul (UFRGS), Porto Alegre, RS, Brazil

The hippocampus has a peculiarly restricted extracellular space, not so much because its extracellular clefts are specially narrow, but rather because they are very little ramified (Green and Maxwell, 1961). Beatriz Vasquez and I, in 1967 (unpublished) obtained inulin space values of 7.9 $\pm$ 1.2 % (n=12) for rat hippocampus, against 18.3 $\pm$ 2.1 % (n=4; p below 0.0005) for the remainder of the cerebral hemisphere. Measurements of the relative surface occupied by extracellular clefts in electron micrographs of rat stratum pyramidale and radiale (Green and Maxwell, 1961) gave in our hands even smaller values, of 2.0% or less. In these layers of the hippocampus, neurones are opposed to each other for long stretches, separated only by the intercellular clefts, with no interposed glial processes (Green, 1964; Green and Maxwell, 1961), and the situation is clearly like an extreme case of those theoretically studied by Lebovitz (1970) in which neuronal activity causes long-lasting (K+)o accumulations.

There is much indirect evidence for such (K+)o increases in the hippocampus during electrical activity. It has been reviewed elsewhere (Izquierdo, Nasello and Marichich, 1971; Izquierdo, 1972) and will only be commented briefly and incidentally here.

ELECTROPHYSIOLOGICAL CONSEQUENCES OF HIPPOCAMPAL K+ RELEASE AND (K+)o ACCUMULATION.

Following single afferent volleys, hippocampal (K+)o builds up for a period of up to several hundred milliseconds (Vasquez et al.,

1969; Izquierdo and Nasello, 1970; see Lebovitz, 1970) and succeeding evoked responses to further stimuli become enhanced, which is known as facilitation. When both pulses are to the same afferent pathway, it is called homosynaptic; if they are to different pathways, facilitation is heterosynaptic (Vasquez et al., 1969; Izquierdo and Nasello, 1970). High (K+)o increases transmitter release at synapses, and this has been proposed to be the cause of facilitation (see Izquierdo and Nasello, 1970; Izquierdo, 1972). Hippocampal perfusion with a high K+ fluid causes a widespread depletion of synaptic vesicles (unpublished) and increases the amplitude of hippocampal evoked responses (Izquierdo et al., 1970). Hippocampal facilitation is enhanced by a topical application of veratrine and depressed by one of tetraethylammonium (Izquierdo and Nasello, 1970), drugs known to, respectively, increase and depress the K+ efflux of spikes (Nasello et al., 1972; Izquierdo, 1972).

The incidence of hippocampal facilitation depends on which are the afferent pathways stimulated: homosynaptic facilitation is practically a constant phenomenon, whereas heterosynaptic facilitation is more frequent to fornix-commisural or commisural-fornix stimulus pairs than to stimulus combinations involving the subiculum (Vasquez et al., 1969). This may be related to the distance between the synaptic endings of each group of fibers: afferent fornix fibers (Petsche et al., 1966) and at least part of the commisural ones (Blackstad, 1956; Green, 1964) terminate on the basal dendrites of pyramidal cells, whereas subicular fibers terminate on the apical dendrites (Blackstad, 1956; Green, 1964). This further argues in favor of the K+ mechanism, since obviously the lower the distance between two groups of fibers firing in close succession, the higher the probability that the K+ released by one will affect the other (Izquierdo and Nasello, 1970; Izquierdo et al., 1971; Izquierdo, 1972).

When afferent stimulation is made repetitive, hippocampal-evoked responses build up gradually because of successive facilitation; after a certain number of pulses, however, the evoked responses become deformed by sharp waves called "spike complications" (Green, 1964; Izquierdo, 1972), and then they give way to haphazard, undriven waves constituting seizure discharges, which go on for a variable amount of time and are followed by spreading depression (Euler et al., 1958; Green and Adey, 1956; Green, 1964). This sequence of events occurring on faradic stimulation is mimicked by high K+ perfusion (Zuckermann and Glaser, 1963; Izquierdo et al., 1970), and the (K+)o concentration at which seizures either occur spontaneously or are caused even by single shocks is, in the rat, 34 mEq/l (Izquierdo et al., 1970: Izquierdo and Orsingher, 1972). The process triggered at that (K+)o concentration is not known, although there is evidence that it involves synaptic mechanisms (Izquierdo et al., 1970). The number of stimuli needed to cause "spike complications" (Izquierdo, 1967) or seizures (Nasello et al., 1969) is reduced by veratrine and increased by tetraethylammonium. Further evidence on the K+ mechanism of seizures may be found elsewhere (Fertziger and Ranck, 1970: Green, 1964: Izquierdo, 1972).

POSSIBLE INFORMATIONAL CONSEQUENCES OF HIPPOCAMPAL K+ RELEASE AND (K+)o ACCUMULATION

Hippocampal pyramidal cells as a whole are sensitive to all sensory modalities, but the responses of individual cells to each type of stimulus vary considerably. Thus, certain sensory modalities are unable to excite pyramidal cells up to their firing level, even if they cause EPSPs, whereas other modalities cause firing very readily (Yokota et al., 1967); some units respond to certain external stimuli but not to others (Green and Machne, 1955), etc. It is possible that this wide variety of individual or mass cell responses to sensory stimuli is due to the fact that each of these reach the hippocampus through different relay stations, albeit by the same two or three final afferent pathways (Yokota et al., 1967; Izquierdo, 1972, 1973).

Taking this into account, it is possible to consider how the informational contents of the diverse sensory responses may be changed by facilitation and by (K+)o levels, in relation to the messages emitted by the hippocampus to projecting structures upon each response. Parenthetically, even if studies on facilitation and on the effect of high K+ have been conducted mostly by observations on evoked responses, it seems licit to extrapolate conclusions to unit activity. Normal hippocampal-evoked responses occur synchronously with (Izquierdo and Vasquez, 1968) and their amplitude bears a linear relation to (Vasquez et al., 1969) compound action potentials travelling hippocampofugally in the fornix, and which doubtlessly represent trains of spikes originating in pyramidal cells. Normally, hippocampal-evoked responses are synchronous with one such compound spike (Izquierdo and Vasquez, 1968), but when "spike complications" appear, fornix compound spikes become double or multiple (Vasquez et al., 1969).

Let us consider a few examples:

a) A stimulus normally causing only pyramidal cell EPSPs below firing level (Yokota et al., 1967) is preceded by another, facilitating pulse. The transient (K+)o increase caused by this pulse increases synaptic efficiency so that the formerly ineffective stimulus will now cause larger EPSPs and induce firing. In this case, facilitation converts an afferent stimulus unable to inform hippocampal projections into one that may do so.

b) A stimulus causing a large evoked response is preceded by a facilitating pulse. The already large evoked response caused by the formerly isolated stimulus may build up still further to the point where "spike complications" occur. In consequence, at the output of the hippocampus, now double or multiple bursts of spikes will appear instead of one (Vasquez et al., 1969), and to hippocampal projections this will be as if they received paired or multiple volleys instead of single ones. This may be of extreme importance for such projections: as much as 47% of preoptic, and 61% of hypothalamic units responding to hippocampal stimulation will not do so unless this is double or multiple (Poletti et al., 1973).

c) A stimulus causing an already large response with "spike complications" is preceded by a facilitating pulse at a "background" hippocampal (K+)o level of, say, 30 mEq/l. The facilitated stimulus may now cause a seizure discharge instead of an evoked potential (Izquierdo et al., 1970), and instead of double or multiple bursts of spikes at the hippocampal output, a rather sustained and probably haphazard firing of the diverse axons will be observed, which will be followed, when spreading depression appears, by an electrical silence (Kandel and Spencer, 1961). The informational contents of the message received by hippocampal projections will be thus dramatically changed by facilitation: instead of a "staccato" group of two or a few bursts, they will receive a prolonged and temporally- and spatially-dispersed firing, followed by another long period of silence. By the way, it is interesting to note here that certain units in the basal forebrain, preoptic area and hypothalamus seem to be specifically responsive to hippocampal afterdischarges (Poletti et al., 1973).

The number of such examples could be easily multiplied, but perhaps these three will suffice to illustrate the vast informational possibilities that facilitation introduces with regard to hippocampal efferent connections. A detailed electrophysiological study of these connections has appeared recently (Poletti et al., 1973). We would like to point out here one implication of the third example given above, namely, that the informational consequences of any given stimulus will depend on the hippocampal (K+)o level "against" which it is given. When (K+)o is low, it will cause just a normal evoked response and a single synchronous efferent volley at the fornix. When (K+)o is moderately high, the same afferent stimulus may cause instead a response deformed by "spike complications" and two or more efferent fornix volleys. When (K+)o is near 34 mEq/l, the same stimulus may elicit a seizure discharge eventually followed by spreading depression.

HIPPOCAMPAL RNA, ELECTRICAL ACTIVITY AND LEARNING

If stimulation parameters are adjusted continuously so as to avoid evoked response build-up, it is possible to produce prolonged repetitive stimulations of afferent pathways with hippocampal-evoked responses of a constant amplitude (Marichich and Izquierdo, 1970, 1972; Izquierdo et al., 1970; Izquierdo and Orsingher, 1972; Izquierdo, Orsingher and Levin, 1972; Izquierdo, Orsingher and Berardi, 1973; Izquierdo, Tannhauser and Tannhauser, 1973). After 25 min of 2 to 7.7/sec stimulation of the fornix with 0.6 mV hippocampal surface-negative evoked responses throughout, a frequency-dependent increase of hippocampal RNA concentration is obtained. The optimum frequency is 4/sec, and 15.9/sec has an opposite effect (Marichich and Izquierdo, 1970). These RNA changes are accompanied by a loss of intracellular K+ (and presumably by a parallel raise of (K+)o which increases monotonically with frequency (Marichich and Izquierdo, 1972).

As was the case with evoked responses, again here high K+ perfusion mimicks the effect of electrical stimulation. After 35 min of high K+ perfusion an increase of hippocampal RNA concentration is obtained at moderate (K+)o levels, 11 mEq/l being the optimum, and a fall of hippocampal RNA occurs with too high (K+)o (Izquierdo et al., 1969). The effect of high K+ is antagonized by Mg++, which suggests that it depends on increased synaptic efficiency.

There is a close parallelism between the effect of high K+ on evoked responses (Izquierdo et al., 1970) and that on RNA: 11 mEq/l is the optimum level for both effects. It was proposed that at "theta-like" fornix stimulation rates of 4 or 5/sec, (K+)o rises to about 11 mEq/l, and that with epileptogenic stimulation it rises to 34 mEq/l (Izquierdo et al., 1970; Marichich and Izquierdo, 1970, 1972). Since this proposition was based on experiments on anesthetized rats, whose body temperature is below normal, it is possible that an extrapolation to awake rats should contemplate somewhat higher "theta" frequencies, such as are typical of this species (Bennett, 1971; Whishaw and Vanderwolf, 1973). Rat "theta" frequency varies with body temperature (Whishaw and Vanderwolf, 1973).

Further evidence for this parallelism between electrical rhythms and (K+)o levels is the following: a) the K+ loss from the hippocampus upon 25 min of epileptogenic stimulation is about three times as high as that which occurs after 25 min of 5/sec stimulation with 0.6 mV evoked responses throughout (Marichich and Izquierdo, 1972), which agrees well with the proposed (K+)o levels of 34 and 11 mEq/l, respectively; b) the number of 10/sec pulses to an afferent pathway needed to trigger a seizure discharge is also roughly three times that needed to cause a maximum enhancement of evoked responses without "spike complications" (see illustrations of Green and Adey, 1956, and Izquierdo, 1972).

Learning is known to be accompanied by a quite persistent hippocampal "theta" rhythm (see Green, 1964 and Izquierdo, 1972 for references), which is known to be caused by a 4 to 8/sec bombardment of pyramidal cells by trains of afferent pulses coming down the fornix (Green, 1964). Parenthetically, it seems likely that the intensity of this bombardment is continuously regulated by some sort of feedback, since "theta" waves do not build up or degenerate into seizures as happens with stimulation of constant strength. Several drugs which enhance learning, such as amphetamine, nicotine and eserine (Stumpf, 1965: Izquierdo, 1972) also cause "theta" rhythm; others, such as strychnine, which also enhances learning, do not (Izquierdo, 1972).

A 25 min session of avoidance conditioning (Nasello and Izquierdo, 1969: Gattoni, 1973) or the injection of amphetamine, nicotine (Nasello and Izquierdo, 1969) or eserine (Daroqui and Orsingher, 1972), but not a 25 min session of pseudoconditioning (Nasello and Izquierdo, 1969) or an injection of strychnine (Nasello and Izquierdo, 1969; Izquierdo, 1972, 1973), cause an increase of hippocampal RNA concentration of magnitude comparable to the one caused by 25 min of 4/sec stimulation or 35 min of perfusion with 11 mEq/l K+ (Izquierdo et al., 1971; Izquierdo, 1972, 1973).

The hippocampal RNA increase of learning seems due to increased synthesis, since an increased uptake of labelled uridine by nuclear and cytoplasmic RNA is observed (Gattoni, 1973; Bowman and Strobel, 1969). A much higher uridine uptake occurs upon pseudoconditioning, but it is accompanied by a highly increased acid RNAse activity (Gattoni, 1973), which probably explains why no RNA concentration increase can be detected (Nasello and Izquierdo, 1969; Gattoni, 1973). The effect of pseudoconditioning is suggestive of overstimulation; seizures, for example, lower whole brain RNA transitorily also because of increased turnover (Chitre et al., 1964).

Both the RNA concentration and synthesis increases which occur in the hippocampus during learning are short-lived, since they cannot be detected beyond 15 min from termination of the training session (Gattoni, 1973). This last observation confirms an early report of Zemp et al. (1966) on uridine uptake by whole-brain RNA.

The increase of hippocampal RNA seems to be a correlate and not just a mere concomitant of learning. Treatments which interfere with the RNA increase, such as a local injection of actinomycin D or puromycin (Nakajima, 1969), or systemic injections of alpha-methyl-tyrosine (Daroqui and Orsingher, 1972), also interfere with learning (Nakajima, 1969; Orsingher and Fulginiti, 1971). The effect of puromycin seems to depend more on the induction of abnormal hippocampal electrical activity than on an actual interference with RNA (see Nakajima, 1969; Izquierdo, 1972). The effect of alpha-methyl-tyrosine is attributable to a blockade of reticulo-septal adrenergic connections presumed to be essential for the production of "theta" rhythm in the intact animal (Daroqui and Orsingher, 1972).

Increased protein synthesis has also been reported to occur in the hippocampus during training (Hydén and Lange, 1968, 1970; Beach et al., 1969), a phenomenon which might be related to the RNA increase.

POSSIBLE ROLE OF HIPPOCAMPAL FACILITATION AND OF THE RNA BUILD-UP IN STIMULUS RECOGNITION PROCESSES

It is widely held that the hippocampus plays a key role in learning, although opinions differ as to the nature of such a role (see Green, 1964; Bennett, 1971; Izquierdo, 1972, 1973). In order to have any, however, it must be agreed that it should be able to recognize external stimuli (and, particularly, associations of external stimuli), and to keep a record of this recognition (Izquierdo, 1972, 1973). In a preceding section we have discussed a possible and very sensitive method for recognizing paired from isolated sensory stimuli, and to differentiate pairs from one another, namely, facilitation. Then, we have also discussed an apparent chemical correlate of learning: the RNA build-up (which is accompanied by enhanced protein synthesis). Hippocampal RNA synthesis seems to depend on synaptic currents (Izquierdo et al., 1969; Izquierdo, 1972), such as happens in other neurons (Kernell and Peterson, 1970), and thus should be sensitive to facilitation.

We have proposed that hippocampal facilitation, coupled with an RNA synthesis, plays a major role in the stimulus recognition processes inherent to learning (Izquierdo and Nasello, 1970; Izquierdo et al., 1971: Izquierdo and Orsingher, 1972: Izquierdo, 1972). Since both hippocampal facilitation and the RNA build-up appear to be K+-dependent events, we decided to find out whether some relation between learning and hippocampal K+ release could be demonstrated.

RELATION BETWEEN HIPPOCAMPAL K+ RELEASE AND LEARNING

The first experiments designed to test this relation were done in rats genetically selected by an in-breeding procedure for their low learning capacity (Orsingher and Levin, 1970). They started with the incidental observation that more 10/sec afferent stimuli were needed in these rats to elicit hippocampal seizures than in normal animals (Izquierdo and Orsingher, 1972). This finding could be explained by two possible mechanisms: a) that poor-learner rats had a larger hippocampal extracellular space, so that the released K+ could not accumulate: b) that actually K+ release was lower.

The first alternative was ruled out both by unpublished electron micrographs and by the direct measure of K+ release afforded by K+ assays in stimulated and unstimulated hippocampi from normal and poor-learner rats, which showed that the release was much lower in the latter (Izquierdo and Orsingher, 1972). This finding could not be explained by a higher rate of reuptake of the released K+ in the poor-learners, since on one hand ouabain perfusion had no effect, and on the other, membrane ATPase activity of poor-learner and normal rat hippocampus was similar. The lower hippocampal K+ release of poor-learners could not be attributed to a lower K+ gradient across cell membranes, since K+ concentration measurements in whole hippocampal tissue and in cerebrospinal fluid were no different in these animals when compared to normal rats. Thus, we were left with two main alternatives to explain the low K+ release of poor-learners: a) a biochemical defect of hippocampal neural membranes whereby K+ permeability changes during excitation are lower: b) some difference in intra- or extracellular levels of other ions (such as Ca++) that might be involved in K+ release. We found no difference in neocortical, hippocampal or cerebrospinal fluid Na+ concentrations (Izquierdo and Orsingher, 1972: Izquierdo, 1972).

Having thus established that poor-learner rats released less hippocampal K+ upon stimulation than normal rats, we proceeded to investigate hippocampal facilitation and the RNA build-up response in the former, which, according to our hypothesis would be expected to be depressed. This was indeed the case: the incidence of hippocampal facilitation was much lower in poor-learners than in normal animals, and also the characteristic hippocampal RNA build-up in response to 4/sec fornix stimulation was absent in poor-learner rats (Izquierdo, Orsingher and Ogura, 1972).

These data certainly favored the view that hippocampal K+ release, facilitation and the RNA build-up are related to learning. However, at this point we could not tell whether there was a correlation between these phenomena, or they all were just independent consequences of the genetic disturbance inherent in poor-learner animals.

Thus, we proceeded in two directions. One was to explore hippocampal K+ release and performance of a learning task in a large, genetically unselected rat population. Results on this line were very encouraging, since a positive and highly significant linear correlation between both variables was found in 42 pairs of rats from the general stock of our laboratory (Izquierdo, Orsingher and Levin, 1972).

The other approach was to investigate drugs which would impair hippocampal K+ release, and then to test their effect on learning. We decided to start our exploration with anticonvulsants, in virtue of the previous demonstration of the dependence of hippocampal seizure discharges on K+ release (see above). Four drugs were investigated: the time-honoured phenobarbital, diphenylhydantoin, trimethadione, and the newer drug cannabidiol (Izquierdo, Orsingher and Berardi, 1973; Carlini et al., 1973: Izquierdo and Tannhauser, 1973). All four raise hippocampal seizure threshold (Izquierdo, 1974). Diphenylhydantoin (Nasello et al., 1972) and cannabidiol (Izquierdo, Orsingher and Berardi, 1973) were found to block hippocampal K+ release and, accordingly, they were subsequently shown to depress learning, hippocampal facilitation and posttetanic potentiation, and to block the RNA build-up to 4/sec fornix stimulation (Izquierdo and Nasello, 1973). Phenobarbital and trimethadione had no effect on hippocampal K+ release and did not affect acquisition of a conditioned avoidance response even at ataxic doses (Izquierdo, Tannhauser and Tannhauser, 1973). Parenthetically, it seems possible that phenobarbital and trimethadione exert their antiepileptic action by an effect on some further step along the chain of unknown events triggered at high (K+)o concentrations (Izquierdo et al., 1970, 1971: Izquierdo, 1972).

CLOSING COMMENT

The data discussed in the preceding section clearly point out to a definite correlation between hippocampal K+ release, facilitation and the RNA build-up on one hand, and learning capacity on the other. However, it must be admitted that these data still fall short of indicating whether these correlations are expressions of causal relationships, or mere coincidences.

However, the following brief summary of the above data and other evidence suggests that the evidence for a causal relation between hippocampal K+ release, electrical activity, RNA and learning, exceeds by far that collected for any other of the diverse hitherto proposed (see Kandel and Spencer, 1968) learning correlates:

a) interference with on-going hippocampal electrical activity and, therefore, with any type of electrical interaction possibly related to information processing interferes with learning (Olds and Olds, 1961; Bures, 1964);

b) interference with the hippocampal RNA build-up interferes with learning (see above); on the contrary, drugs which increase hippocampal RNA favor learning (Izquierdo, 1972);

c) these drugs which increase hippocampal RNA and favor learning also induce hippocampal "theta" activity and heterosynaptic potentiation of hippocampal evoked responses (Izquierdo et al., 1968);

d) perfusion of the lateral ventricles above the hippocampus with fluids containing moderately high K+ concentrations improves the rate of acquisition of conditioned responses (Sachs, 1963), increases hippocampal RNA (Izquierdo et al., 1969) and enhances hippocampal evoked responses (Izquierdo et al., 1970); perfusion with too high K+ concentrations impairs learning (Sachs, 1963), lowers hippocampal RNA (Izquierdo et al., 1969) and causes seizures (Izquierdo et al., 1970);

e) drugs or genetic variables which block hippocampal K+ release also block hippocampal facilitation and the RNA build-up response to stimulation, and impair learning; other drugs with similar general pharmacologic properties (i.e., anticonvulsants) which do not affect K+ release do not affect learning (see above).

Note: All the personal experiments described here were done in collaboration with Drs. O.A.Orsingher, A.G.Nasello, B.J.Vasquez, E.S. Marichich, R.C. Gattoni and others during the seven happy years we could work together at the Departamento de Farmacología, Facultad de Ciencias Químicas, Universidad Nacional de Córdoba, Argentina.

REFERENCES

Beach, G., Emmans, M., Kimble, D.F., and Lickey, M., 1969, Autoradiographic demonstration of biochemical changes in the limbic system during avoidance training. Proc.Nat.Acad.Sci.U.S. 62:692-697.

Bennett, T.L., 1971, Hippocampal theta activity and behavior - a review. Commun. Behav. Biol. 6:37-48.

Blackstad, T.W., 1956, Commisural connections to the hippocampal region in the rat, with special reference to their mode of termination. J.Comp.Neurol. 105:417-538.

Bowman, R.E., and Strobel, D.A., 1969, Brain RNA metabolism in the rat during learning. J.Comp.Physiol.Psychol. 67: 448-456.

Bures, J., 1964, Spreading depression, in "Animal behaviour and drug action" (H.Steinberg, A.V.S.De Reuck and J. Knight, eds.), pp. 373-377, Churchill: London.

Carlini, E.A., Leite, J.R., Tannhauser, M., and Berardi, A.C.,1973, Cannabidiol and Cannabis Sativa extract protect laboratory animals against convulsive agents. J.Pharm.Pharmacol.(in press).

Chitre, V.S., Chopra, S.P. and Talwar, G.P., 1964, Changes in the ribonucleic acid content of the brain during experimentally induced convulsions. J.Neurochem. 11:439-448.

Daroqui, M.R., and Orsingher, O.A., 1972, Effect of alpha-methyltyrosine pretreatment on the drug-induced increase of hippocampal RNA. Pharmacology 7:366-370.

Euler, C.von, Green, J.D., and Ricci, G., 1958, The role of hippocampal dendrites in evoked responses and afterdischarges. Acta Physiol.Scand. 42:87-111.

Fertziger, A.P., and Ranck, J.B.,Jr., 1970, Potassium accumulation in interstitial space during epileptiform seizures. Exptl.Neurol. 26:571=585.

Gattoni, R.C., 1973, Condicionamiento y pseudocondicionamiento: efectos sobre el metabolismo de RNA en neocorteza e hipocampo. Doctor's Thesis, Medical School, Univ.Córdoba, 97 pp.

Green, J.D., 1964, The hippocampus. Physiol.Rev. 44:561-608.

Green, J.D., and Adey, W.R., 1956, Electrophysiological studies of hippocampal connections and excitability. Electroencephalogr. Clin.Neurophysiol. 8:245-262.

Green, J.D., and Machne, X., 1955, Unit activity of rabbit hippocampus. Amer.J.Physiol. 181:219-224.

Green, J.D., and Maxwell, D.S., 1961, Hippocampal electrical activity. I. Morphological aspects. Electroencephalogr.Clin.Neurophysiol. 13: 337-346.

Hydén, H., and Lange, P.W., 1968, Protein synthesis in the hippocampal pyramidal cells of rats during a behavioral test. Science 159:1370-1373.

Hydén, H., and Lange, P.W., 1970, Protein changes in nerve cells related to learning and conditioning, in "The Neurosciences - second study program" (F.O.Schmitt, ed.), pp. 273-289, Rockefeller Univ.Press: New York.

Izquierdo, I., 1967, Effect of drugs on the spike complication of hippocampal field responses. Exptl.Neurol. 19:1-10.

Izquierdo, I., 1972, Hippocampal physiology: experiments on regulation of its electrical activity, on the mechanism of seizures, and on a hypothesis of learning. Behav.Biol. 7:669-698.

Izquierdo, I., 1973, Hippocampal function in rats with a poor learning capacity, in "The Biochemistry of Learning" (W.B.Essman and S.Nakajima, eds.) (in press).

Izquierdo, I., 1974, Effect of antiepileptic drugs on the number of stimuli needed to cause a hippocampal seizure discharge. Pharmacology (in press).

Izquierdo, I., Marichich, E.S., and Nasello, A.G., 1969, Effect of potassium on hippocampal ribonucleic acid concentration. Exptl. Neurol. 25:626-631.

Izquierdo, I., and Nasello, A.G., 1970, Pharmacological evidence that hippocampal facilitation, posttetanic potentiation and seizures may be due to a common mechanism. Exptl.Neurol. 27:399-409.

Izquierdo, I., and Nasello, A.G., 1972, Pharmacology of the brain: the hippocampus, learning and seizures. Progr.Drug Res. 16:211-228.

Izquierdo, I., and Nasello, A.G., 1973, Effects of cannabidiol and dipheny hydantoin on the hippocampus and on learning. Psychopharmacologia (Berl.) (in press).

Izquierdo, I., Nasello, A.G., and Marichich, E.S., 1970, Effects of potassium on rat hippocampus: the dependence of hippocampal evoked and seizure activity on extracellular potassium levels. Arch. intern.Pharmacodyn. 187:213-228.

Izquierdo, I., Nasello, A.G., and Marichich, E.S., 1971, The dependence of hippocampal function on extracellular potassium levels. Curr. Mod. Biol. 4:35-48.

Izquierdo, I., and Orsingher, O.A., 1972, A physiological difference in the hippocampus of rats with a low inborn learning ability. Psychopharmacologia (Berl.) 23:336-396.

Izquierdo, I., Orsingher, O.A., and Berardi, A.C., 1973, Effect of cannabidiol and of other Cannabis Sativa compounds on hippocampal seizure discharges. Psychopharmacologia (Berl.) 28:95-102.

Izquierdo, I., Orsingher, O.A., and Levin, L.E., 1972, Hippocampal potassium release upon stimulation and performance in a shuttle-box. Behav.Biol. 7:367-371.

Izquierdo, I., Orsingher, O.A. and Ogura, A., 1972, Hippocampal facilitation and RNA build-up in response to stimulation in rats with a low inborn learning ability. Behav.Biol. 7:699-707.

Izquierdo, I., and Tannhauser, M., 1973, The effect of cannabidiol on maximal electroshock seizures in rats. J.Pharm.Pharmacol. (in press).

Izquierdo, I., Tannhauser, M., and Tannhauser, S., 1973, Phenobarbital and trimethadione: two anticonvulsant agents with no effect on hippocampal potassium release and on learning. Ciencia e Cult. (Sao Paulo) (in press).

Izquierdo, I., and Vasquez, B.J., 1968, Field potentials in rat hippocampus: monosynaptic nature and heterosynaptic posttetanic potentiation. Exptl.Neurol. 21:133-146.

Izquierdo, I., Vasquez, B.J., and Nasello, A.G., 1968, Indirect effect of drugs on hippocampal field responses to commissural and subicular stimulation. Pharmacology 1:178-182.

Kandel, E.R., and Spencer,W.A., 1961, Electrophysiological properties of an archicortical neuron. Ann.N.Y.Acad.Sci. 94:570-603.

Kandel, E.R., and Spencer, W.A., 1968, Cellular neurophysiological approaches to the study of learning. Physiol.Rev. 48:65-134.

Kernell, D., and Peterson, R.P., 1970, The effect of spike activity versus synaptic activity on the metabolism of ribonucleic acid in molluscan giant neurone. J.Neurochem. 17:1037-1094.

Lebovitz, R.M., 1970, A theoretical examination of ionic interactions between neural and nonneural membranes. Biophys.J. 10: 423-444.

Marichich, E.S., and Izquierdo, I., 1970, The dependence of hippocampal RNA levels on the frequency of afferent stimulation.

Naturwiss. 57:254.

Marichich, E.S., and Izquierdo, I., 1972, Potassium loss from rat hippocampus during electrical activity. Arch.intern.Pharmacodyn. 196:353-356.

Nasello, A.G., and Izquierdo, I , 1969, Effect of learning and of drugs on the ribonucleic acid concentration of brain structures of the rat. Exptl.Neurol. 23:521-528.

Nasello, A.G., Marichich, E.S., and Izquierdo, I., 1969, Effect of veratrine and tetraethylammonium on hippocampal homosynaptic and heterosynaptic post-tetanic potentiation. Exptl.Neurol. 23:516-520.

Nasello, A.G., Montini, E.E., and Astrada, C.A., 1972, Effects of veratrine, tetraethylammonium and diphenylhydantoin on potassium release by rat hippocampus. Pharmacology 7:89-95.

Nakajima, S., 1969, Interference with relearning in the rat after hippocampal injection of actinomycin D. J.Comp.Physiol.Psychol. 67:457-461.

Olds, J., and Olds, M.E., 1961, Interference in learning in paleocortical systems, in "Brain mechanisms and learning" (A. Fessard, R.W.Gerard, J.Konorski, and J.F.Delafresnaye, eds.), pp. 153-187, Blackwell:Oxford.

Orsingher, O.A., and Fulginiti, S., 1971, Effects of alpha-methyltyrosine and adrenergic blocking agents on the facilitating action of amphetamine and nicotine on learning in rats. Psychopharmacologia (Berl.), 19:231-240.

Orsingher, O.A., and Levin, L.E., 1970, Selección de ratas de bajo nivel de aprendizaje en base a la performance en una "shuttlebox". Proc.3rd.Ann.Meet., Soc.Argent.Farmacol.Exptl., abstracts, p.18.

Petsche, H., Gogolák, G., and Stumpf, C., 1966, Die projektion des Zellen des Schrittmachers für den Thetarhythmus aum den Kaninchen Hippocampus. J.Hirnforsch. 8: 129-136.

Poletti, C.E., Kinnard, M.A., and MacLean, P.D., 1973, Hippocampal influence on unit activity of hypothalamus, preoptic region and basal forebrain in awake, sitting squirrel monkeys. J.Neurophysiol. 36:308-324.

Sachs, E., 1963, cited by E.Roy John, 1965, in "The anatomy of memory" (D.P. Kimble, ed.), pp. 273-275. Science and Behavior Books:Palo Alto.

Stumpf, C., 1965, Drug action on the electrical activity of the hippocampus. Intern.Rev.Neurobiol. 8:77 132.

Vasquez, B.J., Nasello, A.G., and Izquierdo, I., 1969, Hippocampal field potentials: their interaction and their relation to hippocampal output. Exptl.Neurol. 23:435-444.

Whishaw, I.Q., and Vanderwolf, C.H., 1973, Hippocampal EEG anu behavior: changes in amplitude and frequency of RSA (Theta rhythm) associated with spontaneous and learned movement patterns in rats and cats. Behav.Biol. 8:461-484.

Yokota, T., Reeves, A.G., and MacLean, P.D., 1967, Intracellular olfactory response of hippocampal neurons in awake, sitting squirrel monkrys. Science 157:1072-1074.

Zemp, J.W., Wilson, J.E., Schlessinger, K., Boggan, W.O., and Glassman, E., 1966. Brain function and macromolecules. I. Incorporation of uridine into RNA of mouse brain during short-term training experience. Proc.Nat.Acad.Sci.U.S. 55:1423-1431.

Zuckermann, E.C., and Glaser, G.H., 1968, Changes in hippocampal evoked responses induced by localized perfusion with high-potassium cerebrospinal fluid. Exptl.Neurol. 22:96-117.

ABSTRACT:

Potassium released during hippocampal electrical activity accumulates temporarily in the peculiarly restricted local extracellular space. After one or a few afferent volleys the moderate $(K+)_o$ build-up increases synaptic efficiency and causes, respectively, facilitation and post-tetanic potentiation of evoked responses, both homo- and heterosynaptic. After several minutes of sustained afferent stimulation, adjusted continuously so that hippocampal-evoked responses remain constant (a procedure mimicking theta rhythm), there is a $(K+)_o$-dependent RNA increase. If repetitive afferent stimulation is uncontrolled, evoked responses build up gradually at first and then change their shape, until finally seizures occur. This happens when $(K+)_o$ reaches 34 mEq/1.

We have proposed that the hippocampus participates in learning by a mechanism involving stimulus recognition through K+-mediated heterosynaptic interactions coupled with a chemical transduction process initiated by the RNA build-up. The following evidence favors this hypothesis: a) avoidance conditioning (but not pseudoconditioning), or the injection of amphetamine, nicotine or eserine (drugs which enhance learning and induce theta rhythm and heterosynaptic potentiation of hippocampal responses) increase rat hippocampal RNA such as happens after a comparable period of afferent stimulation or high-K+ perfusion; b) pre-treatment with alpha-methyltrosine blocks both the effect of these drugs on learning and that on RNA; c) moderate CSF [K+] increases favor learning; excessive CSF [K+] depresses it, causes hippocampal seizures and an RNA fall; d) rats genetically selected for poor learning ability have a low hippocampal K+ release upon stimulation, practically lack hippocampal heterosynaptic facilitation and have a low incidence of homosynaptic facilitation and heterosynaptic post-tetanic potentiation, show no RNA build-up on stimulation, and require more repetitive stimuli than normal rats in order to trigger hippocampal seizures; e) rats injected with either cannabidiol or diphenylhydantoin, two chemically diverse antiepileptic drugs, also have an impaired hippocampal K+ release, depressed facilitation and post-tetanic potentiation, no RNA response to stimulation, depressed acquisition of learned responses and, of course, an increased hippocampal seizure threshold.

PART III
NEUROHUMORAL MECHANISMS IN BEHAVIORAL REGULATION

INTRODUCTION TO SESSION III

This session is concerned with Neurohumoral Mechanisms in Behavioural Regulation and we welcome our distinguished contributors.

The last decade has witnessed a considerable advance in our understanding of the processes by which behaviour is initiated and controlled. The collaboration of the various disciplines concerned with neuroscience, has been of particular importance in allowing an integrated approach to a common problem. Probably the most important advances have resulted from the development or refinement of techniques for studying the application of drugs, and the release and identification of substances from the mammalian central nervous system. Similarly, our studies of the neurobiological processes regulating behaviour have been aided tremendously by the availability of elegant biochemical techniques, which allow the characterization of the neurohumoral factors with which we are all directly or indirectly concerned. Thus, we no longer consider behaviour in isolation. We observe the behavioural response to a given stimulus, but more important we are now looking at the variety of neurochemical changes accompanying a particular behavioural event. It now seems certain that many of the answers relating to our understanding of the control of behaviour will be found at the cellular and biochemical level.

In the following session, we have a number of contributors, whose presentations are concerned with the neurohumoral mechanisms involved in the regulation of the two opposing states of sleep and vigilance. I think that it is therefore appropriate at this time for us to remember the incentive given to this area of research by the studies of the late Dr. Hernández-Péon. Some of you had the privilege of knowing and working with him, and I am sure that we are all familiar with his studies concerning the involvement of limbic cholinergic systems in the control of sleep and emotional behaviour. Almost 10 years ago, he presented a paper in Switzerland on "Central Neurohumoral Transmission in Sleep and Wakefulness". I think you would agree that this would have been a most appropriate title for a paper in our session this morning. In this paper, Dr. Péon concluded, and I quote, "that there is in the brain an anatomically circumscribed sleep system, the activation of which produces sleep through direct inhibition of the mesodiencephalic vigilance system". We now look forward with interest to hearing about the current "State of the art" with regard to sleep and vigilance mechanisms.

G. H. Hall
Harrogate, England

MODULATORY EFFECTS OF ACETYLCHOLINE AND CATECHOLAMINES IN THE CAUDATE NUCLEUS DURING MOTOR CONDITIONING

H. Brust-Carmona, R. Prado-Alcalá, J. Grinberg-Zylberbaum, J. Alvarez-Leefmans and I. Zarco Coronado

Physiology Department, Faculty of Medicine, National University of Mexico, Mexico 20, D. F.

INTRODUCTION

The caudate nucleus (CN) seems to be an important part of the neuronal circuit which controls the motor conditioned responses (MCR). In fact, lesions of the CN impair the performance of MCR (Thompson, 1959; Chorover and Gross, 1963; Prado *et al.*, 1973). The CN activity is also necessary for the conditioned inhibition of learned motor responses (Brust-Carmona and Zarco-Coronado, 1971a). Therefore, these facts suggest that in the CN exist both the necessary circuit to activate the elements for the performance of MCR and also neurones which are able to inhibit that activity. A logical postulation would be that the "activating system" is controlled by a certain chemical transmitter process and the "inhibitory system" by a different one.

Many authors (Hebb and Silver, 1956; Fahn and Cote, 1968) have pointed out the extraordinarily high concentrations of choline-acetylase and cholinesterase found in the corpus striatum.

Portig and Vogt (1969) have found that sensory stimulation elicits the release of acetylcholine in the CN. Since the MCR is associated with the sensory input, the possibility that the activating system of motor conditioned response being cholinergic was considered. Indeed, the microinjection of atropine in the caudate nuclei of cats impaired two types of MCR (Prado *et al.*, 1972). In this paper the results of ACh microinjections in the CN will be described.

The actions of the peripheral (cholinergic) endings of the autonomic parasympathetic system are balanced by those of the sympathetic system which are adrenergic. Similar balancing mechanisms might be expected in the visceral brain (Bharcava et al., 1972) and in its primitive motor counterpart the extrapyramidal system. The high catecholamine (CA) content of the corpus striatum has been described (Vogt, 1954) as well as the participation of CA in motor regulation (Carlsson, 1959). Thus, the inhibitory function of CN in learning processes could be related to CA. This hypothesis was tested by investigating the effects of varying the concentration of catecholamines in the CN, during motor inhibitory conditioned situations.

EXPERIMENTAL SERIES 1

Method

Twelve adult cats (Ss) of either sex and between 2.5 and 3.5 Kg of body weight were used. The experiments were performed in an electrically insulated, soundproof room, measuring 2x3x2.5 m. The four walls and the ceiling were white. In the room was a table about 1.5 m long, provided at one end with a rewarding device and on the other end an open cage, facing the rewarding device. This rewarding set-up consisted of a revolving disk bearing eight bowls, which the observer could make to appear successively under a hole on the table, handling the device from outside the chamber. At the beginning of the experiment, to habituate the cats to the environment, they were put daily on the table while the bowls, which later on served to present the reward, were being made to appear successively under the hole. Once the animal did not pay any attention to the rewarding device and had learned to remain quietly inside the cage, the conditioning sessions began. The conditioning stimulus (CS) consisted of four flashes (F) 1/sec delivered by a Grass photostimulator (model PS2) set at an arbitrary intensity of 4. The unconditioned stimulus (US) was a piece of fresh meat which the Ss could get leaving the cage and walking 100 cm to reach the bowl. The next trial was performed only when the Ss returned to their place in the cage. When this motor conditioned response (MCR) was greater than 80% correct for three consecutive days, an acoustic stimulus was randomly associated to the flashes. This consisted of 4 clicks (C), 1/sec, produced by delivering square pulses from a Grass S4 stimulator to a Grass AM5 audiomonitor and speaker, and so timed that the first click was simultaneous with the second flash. This type of stimulus (F-C) thus ended with a click and since whenever it was applied the Ss did not receive any food, they learned not to walk towards the rewarding device, staying inside the cage. This was termed the "inhibitory conditioned response" (ICR). Three series of 8 trials of MCR and 15-20 trials of ICR each were performed daily except on Sundays.

After the Ss had 20% to 80% correct responses in the ICR, double walled cannulae (0.5 mm external diameter) were stereotaxically implanted bilaterally, under general anaesthesia (40 mg/Kg nembutal I.P.), in the head of the CN, using the coordinates A 16, L 4.5, H 4.5, after Jasper and Ajmone-Marsan Atlas (1954). The conditioning sessions were resumed 24 hrs after surgery, and following a training period which varied for the different Ss, bilateral injections of either epinephrine hydrochloride E (Servet[D]) or norepinephrine bitartrate NE (Levofed[D]), or isotonic saline solutions were initiated. Seven Ss received, at random, E or saline with a different sequence between the injections. Three had NE or saline and two animals received only saline. The injections were performed in another room, between the first and second series, or between the second and third, and occasionally before the first series. Injections were always bilateral, of 10 or 20 μl in volume, containing 2, 4 or 8 μg and were delivered in 20-30 sec. The conditioning sessions were resumed 4 min after the injections. Because of the irregular pattern of the drug administration, it was considered not valid to apply any statistical analyses to this experimental series.

At the end of the experiments the Ss were deeply anaesthetized and their brains perfused first with saline solution and then with 10% formalin. Histological sections of the brain were subsequently prepared and the location of the cannulae was determined in accordance with the technique of Guzmán-Flores _et al_.

Results

Essentially, the cats acquired and maintained the MCR and the ICR as described in a previous report (Brust-Carmona _et al_., 1971a). However, a reduction in the number of both types of conditioned responses was noted after the implantation of the cannulae. Nevertheless, the Ss reacquired the criterion level of conditioning (80%) for the MCR in 2 to 6 days, but the number of correct ICRs remained lower than before the implantation. This effect was more accentuated when the implantation was performed in animals showing only 20% correct ICRs. The first E injections of 8 μg in 20 μl or 4 μg in 10 μl into the CN markedly decreased the spontaneous movements of the animals which in some instances seemed to fall asleep. These effects were more accentuated following the injection of NE. For this reason the maximum dose of NE used was only 5 μg. A less intense effect was obtained with 2 μg in 10 μl. These doses never produced the sleeping effect, but the animals moved slowly within the reflex chamber, not responding to the conditioned stimulus. However, they ate all the meat offered to them. In subsequent injections the animals showed less inhibition of the spontaneous motor activity and they were able to react to the conditioned stimulus, giving then a number of correct responses very similar to the ones obtained before the injection. In contrast, the number of correct ICRs increased (Fig. 1).

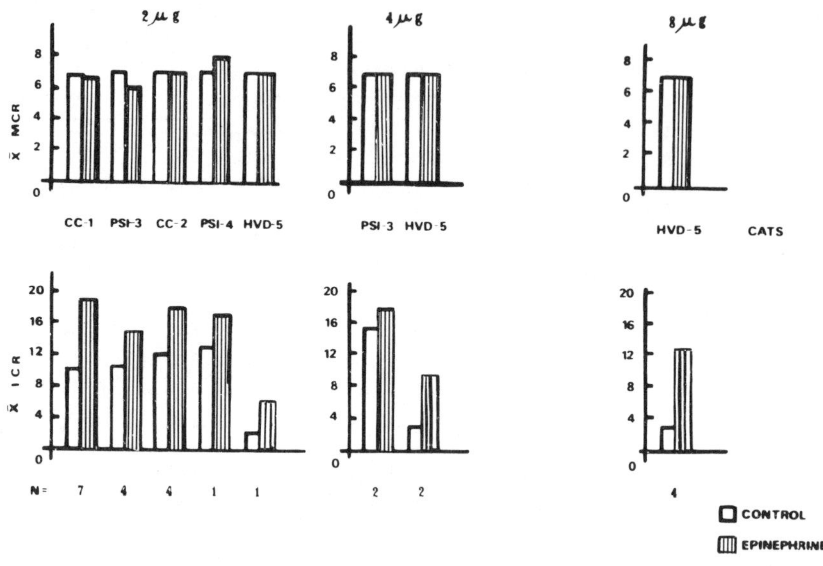

Figure 1

Figure 1 illustrates the total average of motor conditioned responses (MCR) and inhibitory conditioned responses (ICR) before (white bars) and after microinjections into the CN of 2, 4 or 8 μg of epinephrine (dark bars) producing an increase of ICRs. N represents the number of injections in each cat. Observe the enhancement of ICR.

The lower part of Fig. 1 illustrates the average of correct responses obtained in 5 cats, each subjected to a different number of injections (N in the figure). The white bars represent the average of the scores before, and the dark ones after the epinephrine injections, which always were followed by an increase of the ICRs.

The injections of saline solution did not change the learned conditioned responses (Fig. 2).

Figure 2 illustrates the total average of MCR and ICR before (white bars) and after saline solution microinjection in the CN (dark bars). The learned responses remain constant. N represents the number of injections in each cat.

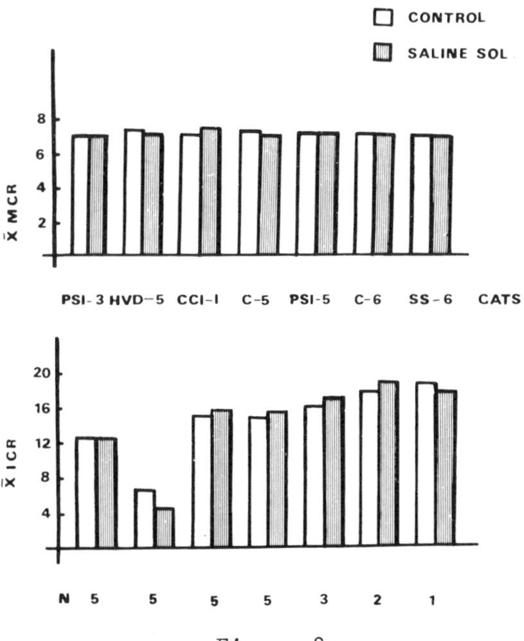

Figure 2

After the first injection of (5 μg) NE, all 3 Ss remained sitting in a resting position, completely immobile for 30-60 min. This effect decreased after the two following injections and from the fourth application on, the animals reacted to the CS, showing an increase in the ICR (50-70%) while the MCR decreased (20-30%).

It is important to note that if the animals were abruptly disturbed while under the effects of either the E or NE injection they were able to jump, showed defensive reactions, and also could eat the meat offered to them.

The histological sections showed that the cannulae were placed in the medial dorsal part of the head of the CN.

EXPERIMENTAL SERIES 2

Method

Eight adult cats (Ss) of either sex and between 2.5 and 3.5 Kg of body weight were used. The two learned responses were very similar to those previously described, except that they were performed in a small soundproof box and that the US was milk instead of meat, as described elsewhere (Prado et al., 1972). The conditioned stimuli (CS) were 4 flashes (1/sec) delivered by a photostimulator (Grass PS2). The CS signalled the Ss to walk (75 cm) from a platform to the cup in order to obtain 2 ml of milk and return to the starting point (MCR). Only after returning to the platform another CS was delivered. The movements of the cats were detected by two photocells placed underneath two holes on the floor of the alley at 16 cm and at 41.5 cm from the door, respectively, and recorded in two channels of a Physiograph Six (Narco-Bio-Systems). The artifacts of the CS were also recorded in another channel of the Physiograph.

After the Ss had reached the criterion level (80%) of responses, cannulae were implanted in the CN as previously described. Conditioning sessions were resumed two or three days after surgery. As in the experimental series 1, when the Ss had reacquired the criterion level they were conditioned to inhibit the motor response whenever four clicks were associated to the flashes signaling that no milk would be delivered (ICR). A daily session consisted of 15 trials of flashes-milk and, randomly distributed among them, 15 trials of flashes-clicks-no milk. Five bilateral injections of 5 μl either of saline (4 cats) or (5 μg) norepinephrine hydrochloride solutions (4 cats), were given. The injections were performed 10 min before the session in another room, every other day, in 20-30 sec. At the end of the experiment, the location of the cannulae was assessed in histological sections of the brains, as already mentioned.

Results

No difference in MCR nor in ICR in both groups after the first injection of saline or NE was observed, but after the second in-injection, the NE group showed a greater number of ICR while the MCR was similar in both groups. The same difference was maintained in the successive injections. The application of the non-parametric Mann-Whitney U Test (Siegel, 1956) indicated that the difference was statistically significant at the *p* level of 0.05 (n1 4, n2 4, U 3). In contrast, the small difference noted in MCR was not statistically significant.

In Figure 3 the upper graph shows the MCR and the lower one the ICR of each daily conditioning session. The average MCR did not change, although in contrast the ICRs average of 4 cats in which NE was injected in the CN (continuous line) is higher than in the other 4 cats in which saline solution was injected (interrupted line). The injections were performed every other day and are represented by the arrows. The vertical lines represent the standard deviation.

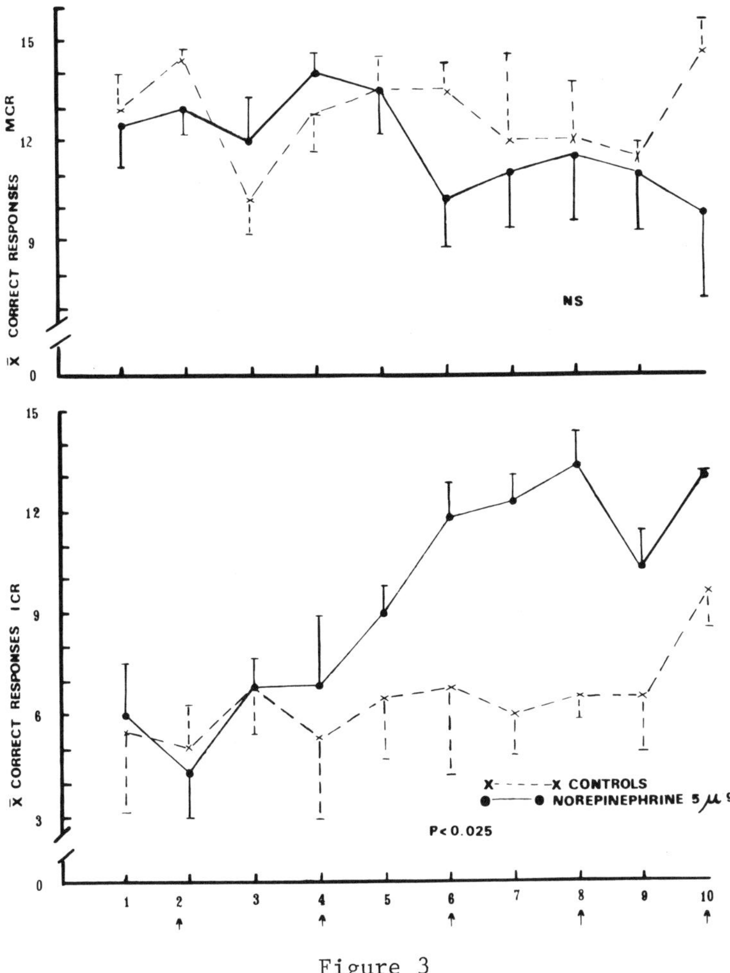

Figure 3

Figure 3 illustrates the number of MCR and ICR obtained in 10 successive sessions. The continuous line represents the MCR and ICR of the 4 animals which received NE, the interrupted line the results of the other 4 cats. The cannulae were placed in the medial part of the head of the CN.

EXPERIMENTAL SERIES 3

Method

Fourteen cats of either sex between 3.0 and 4.0 kg body weight were used. The experiments were performed in a Skinner type box with omnidirectional lever and liquid feeder (Lehigh-Valley Electronics). The Ss were trained according to the following schedule: if a small light above the lever was on, each time the cat pressed the lever 0.5 ml of milk were delivered. When the light was off the rewarding system was also turned off; thus, the animal had to 'inhibit' or 'suppress' the motor response. The lever responses during the light "on" (60 sec) and "off" (20 sec) periods were automatically recorded. This pattern was repeated 6 times (first series). The animal was taken out and handled, and 10 min later returned to the box, where the conditioning pattern was repeated again 6 times (second series). The mean frequency of pressings per min of each series was calculated.

When the Ss showed a sustained rate of pressings during three consecutive days, implantation of cannulae into the CN was performed as described in experimental series 1. Twenty-four to forty-eight hours after surgery the conditioning sessions were resumed until the prior pressing rate was re-established. In 5 cats, on successive days, the following pattern was performed: a) control session; b) microinjection of NE (5 μg in 5 μl) or isotonic saline solution (5 μl) and c) sham injection. Two animals had 3 injections of NE, one had 4 and the other two had 5. In the other 3 cats, after three successive days of an almost constant rate of lever pressing, microinjections of 100 μg of phenpiperazole (Zolertine Miles Laboratories) were performed every other day. Two animals had 3 injections and the other one had 2. The bilateral injections (20-30 sec each) were performed between the two series in another room. In six cats the cannula implantation was performed prior to the conditioned sessions. After complete recuperation the daily training started as previously described; but 10 min before the session 3 cats were microinjected with 5 μl of isotonic saline solution and the other 3 received 10 μg of ACh every other day.

An extra group of cats (9 Ss) which had been used in other experiments (microinjections of NaCl, atropine or an adrenergic blocker) showing a sustained lever pressing rate, were used to test the effect of intraperitoneal injections of reserpine (Ciba[D]) given three hours before the session. Three cats received similar doses (0.1, 0.5 or 1.0 mg/Kg). In order to compare the results the average lever pressing of the two previous sessions was taken as 100%.

The position of the cannulae was determined in the histological sections of the brains, as already mentioned.

Results

The cats increased the rate of lever pressing after the ACh injections compared with the animals which received saline solution (Fig. 4).

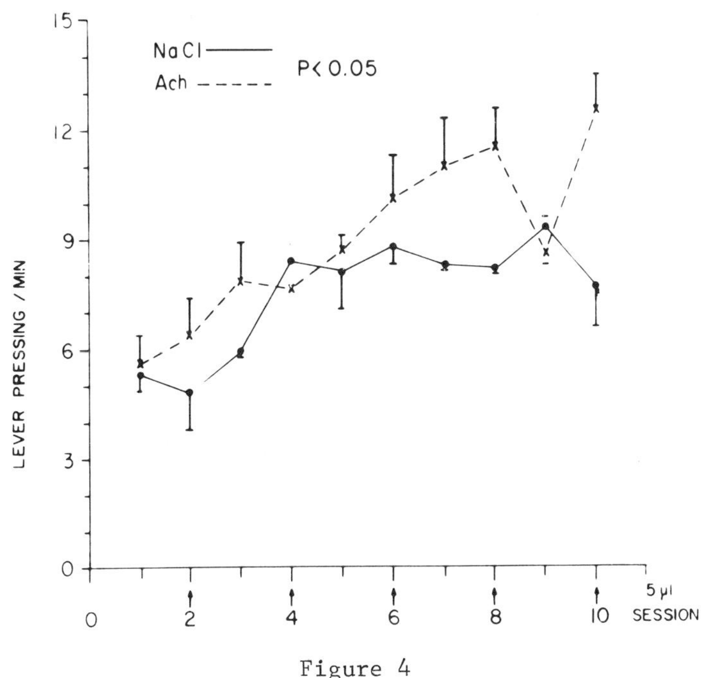

Figure 4

Figure 4 illustrates the daily lever pressing average of cats which were injected 10 µg of ACh (3 Ss) or the same volume (5 µl) of NaCl (3 Ss) in the CN before every other session (represented by the arrows). ACh improved acquisition of the lever pressing response.

The difference observed was statistically significant at the level of $P < 0.05$ (Mann-Whitney U Test). The possible increment of hunger after the ACh injection was tested in 3 Ss by measuring the total amount of ingested milk, which remain unchanged after the sham injection or the ACh injection.

After the implantation of the cannulae the number of lever pressings decreased but in a few days it was completely recovered.

It is interesting to mention that this effect was inversely related to the training time. In overtrained animals the implantation caused almost no change in the lever pressing rate.

Contrasting with the two experiments formerly described, in the present series, only a slight diminution of the spontaneous motor activity after the NE injection was observed. As soon as Ss were placed in the box, they started pressing the lever although at a slower speed than in the control situation. However, the total amount of lever pressing before, compared with those after the NE injections, was not statistically significant (Mann-Whitney U Test). In contrast, the number of lever responses decreased when the light was off, i.e. during the "suppression situation". The difference between the control series and experimental series was statistically significant at the level of $\underline{p} < 0.05$ (n1 5, n2 5, U 4). The average lever pressing during the control series and the experimental ones with the light "on" and "off" is represented in figure 5. Neither the injections of NaCl, nor the sham injections or the control treatments changed the lever pressing rate. Furthermore, no significant difference between them was found. The cannulae placements were in the medial part of the head of the CN.

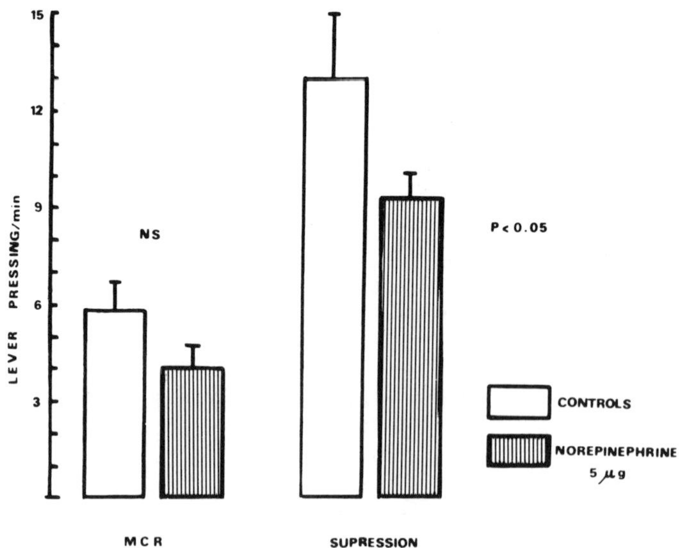

Figure 5

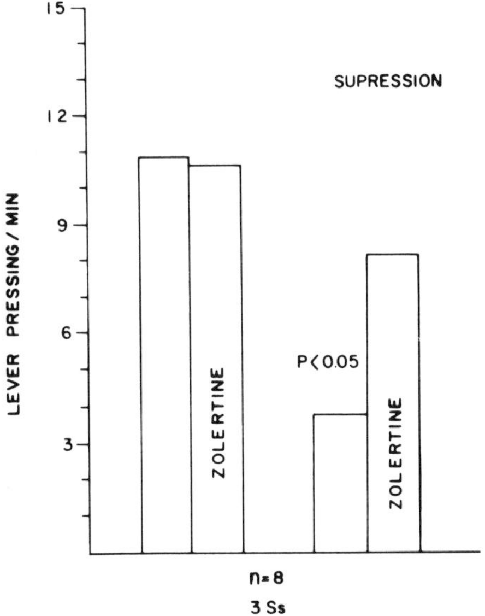

Figure 6

However, the lever pressing rate during the suppression situation increased after the microinjection of zolertine, while no change of the lever pressing rate was observed during the rewarding situation (Fig. 6).

In Figure 5 the bars illustrate the average lever pressing (20 injections in 5 cats) before (white) and after (dark) NE injections in the CN, during the rewarding situation (MCR), and when the lever pressing was not reinforced (suppression). The NE produced a statistically significant improvement of suppression.

In Figure 6 the bars illustrate the average lever pressing (8 injections in 3 cats) prior to and after phenpiperazole injections in the CN during the rewarding situation (left part of the figure) and when the lever pressing was not reinforced (right part) Zolertine produced a statistically significant increment of lever pressing during the suppression situation.

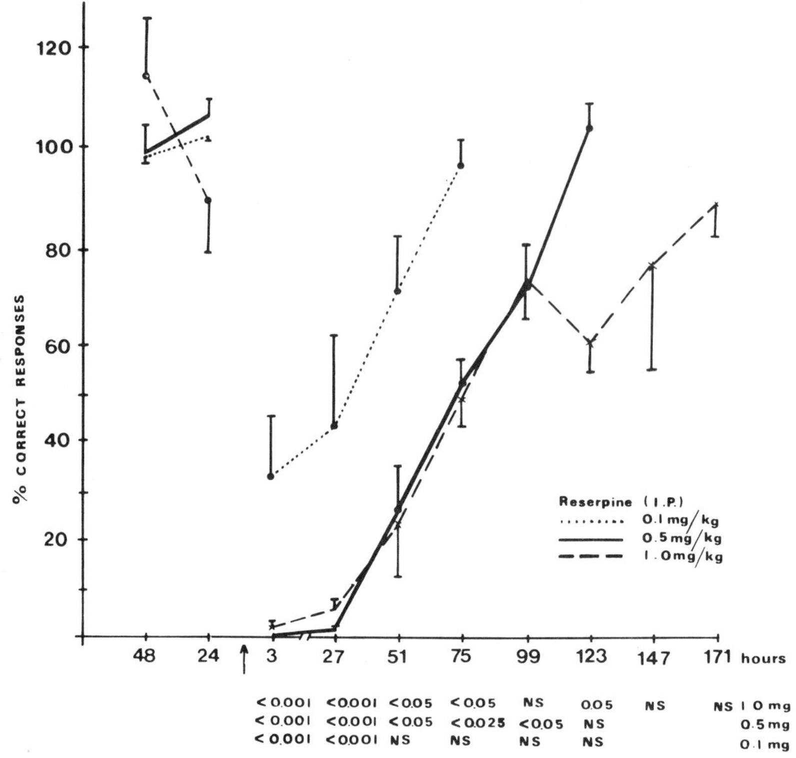

Figure 7

After the application of reserpine, the behavior of the animals was unchanged in their living quarters where they had access to an open space 5 x 3 m. However, when placed in the Skinner box the animals showed less spontaneous activity and their lever pressing decreased in such a way that the highest doses blocked it completely (Fig. 7). In a few days the animals returned to their previous lever pressing rates showing a recuperation time directly related to the doses. The observed changes were statistically significant at different levels of probability which are represented in Fig. 7 (Variance analysis, randomized blocks, Edwards, 1968).

In Figure 7 the graphs show the average diminution of lever pressing (3 cats for each dose) 3 hours after one dose of reserpine IP and its progressive recuperation in the following days. The vertical lines represent the standard deviation.

DISCUSSION

The fact that the conditioned motor responses diminished after the cannula implantation into the head of the CN is in line with the postulation that this structure is an important structure in the neural circuits responsible for conditioning. Furthermore, the recent description of the CN's participation in short term memory processes (Wyers and Deadwyler, 1971), as well as the previous observations showing that well established MCRs are abolished by electrolytical lesions (Thompson, 1959; Chorover and Gross, 1963) and also by pharmacological effects (Brust-Carmona et al., 1971b; Prado et al., 1973), very well support our postulation that the CN could be the locus of the "integration and storage of the engram".

The enhancement of the inhibition of motor conditioned responses (approaching and lever pressing) after the microinjection of E or NE into the CN is in accordance with the assumption that the inhibitory motor functions of the CN are related to the catecholamines. Moreover, the injection of E or NE also decreased the spontaneous motor activity. This effect was more clearly observed when the animals were placed in the conditioning chambers;probably, because of the conditioned situation, the CA were constantly released. The decrement of the learned inhibitory responses after the injection of an adrenergic blocker gives further support to our postulation. Thus, the CA variation in the CN will modify the motor activity. Indeed, the diminution of CA (Ehringer and Hornykiewicz, 1960) and the improvement of Parkinson patients treated with DOPA (Cotzias et al., 1967; Barbeu, 1962) gives further support to the above viewpoint.

On the other hand, the abolition of lever pressing in our cats after the application of reserpine could be explained if an increase in the "free CA" is assumed to occur, during the first hours due to the blocking of the CA reuptake (Glowinski et al., 1966), thus enhancing the inhibitory influences of the CN.

Then the responses which result from motor activity would diminish after reserpine application as in the case of the active avoidance conditioned responses (Ahlenius and Engel, 1971; Seiden and Peterson, 1968) and lever pressing in mice (Butcher et al., 1972). Later on, the depletion of CA in the CN caused by the blocking of the CA storage in the synaptic vesicles by higher doses of reserpine, could be correlated with an increment of lever pressing. However, the stage of CA depletion was probably not reached in our experiments since the doses of reserpine used were 5-10 times lower than the ones used by other authors (Andén et al., 1969; Ahlenius and Engel, 1971). Furthermore, when we tried to repeat the doses of 1 mg/kg 2-3 times, the cats had diarrhea. Thus, we decided that the condition of the animals to test the alimentary motor conditioned responses was not good at all. Besides the peripheral effects of reserpine, the depletion of CA occurring in different brain areas could have different effects; for instance, CA seems to have a facilitatory influence in the mesencephalic reticular formation (Stein, 1968); b) inhibitory influences in hypothalamic structures (Bharcava *et al.*, 1972); c) regulatory actions upon sleep-arousal reactions (Jouvet, 1969). The interpretation of the results obtained after systemic applications of drugs is a very difficult and contradictory task.

The diminution of the MCR by the microinjections of atropine in the CN (Prado *et al.*, 1972) as well as its increment by ACh application on the one hand and, on the other, facilitation of the inhibitory conditioned responses by the E or NE injections in the CN permits one to postulate a possible counterbalance of cholinergic and adrenergic mechanisms in the CN. In our view, the caudate neurones involved in MCR, tonically inhibited by CA, are activated by ACh. In line with this view is the existence of extensive ramifications of CA neurones in the CN (Andén *et al.*, 1966) and the inhibitory effects of CA upon the electrical activity of CN neurones (McLennan and York, 1967).

REFERENCES

1. Andén, N. E., Fuxe, K., Hamberger, B., Hokfelt, T., 1966, Quantitative study of the nigro-neostriatal dopamine neuron system in the rat. *Acta Physiol. Scand.* 67, 306-312.

2. Andén, N. E., Carlsson, A., Haggendal, J., 1969, Adrenergic mechanisms. *Amer. Rev. Pharmacol.* 9, 119-134.

3. Ahlenius, S., Engel, J., 1971, Behavioral and biochemical effects of L-DOPA after inhibition of dopamine-beta-hydroxylase in reserpine pretreated rats. *Naunyn-Schmiedebergs Arch. Pharmak.* 270, 349-360.

4. Barbeu, A., 1962, The pathogenesis of Parkinson's disease: A new hypothesis. *Canad. Med. Assoc. J.* 87, 802-807.

5. Bharcava, K. P., Kulshrestha, V. K., Srivastava, G. B., 1972, Central cholinergic and adrenergic mechanisms in the release of antidiuretic hormone. *Brit. J. Pharmacol.* 44, 617-627.

6. Brust-Carmona, H., Zarco-Coronado, I., 1971a, Instrumental and inhibitory conditioning in cats. II. Effects of paleocortex and caudate nucleus lesions. *Bol. Est. Med. Biol. Méx.* 27, 63-71.

7. Brust-Carmona, H., Prado-Alcalá, R., Grinberg-Zylberbaum, J., 1971b, Bloqueo reversible de respuestas condicionadas motoras por la aplicación de anestésicos locales en el núcleo caudado. Bol. Est. Med. Biol. Méx. 27, 109-114.

8. Butcher, L. L., Rhodes, D. L., Yuwiler, A., 1972, Behavioral and biochemical effects or preferentially protecting monoamines in the brain against the action of reserpine. Eur. J. Pharmacol. 18, 204-212.

9. Carlsson, A., 1959, The occurrence, distribution and physiological role of catecholamines in the nervous systems. Pharmacol. Rev. 2, 490-494.

10. Chorover, S. L., Gross, Ch. G., 1963, Caudate nucleus lesions: Behavioral effects in the rat. Science 141, 826-827.

11. Cotzias, G. C., Van Woert, M. H., Schiffer, L., 1967, Aromatic amino acids and modification of Parkinsonism. New Engl. J. Med. 276, 374-379.

12. Edwards, A. L., 1968, Experimental design in psychological research. 3rd. Edition, pp. 155-167. Holt, Rinehart and Winston Inc.

13. Ehringer, H., Hornykiewicz, O., 1960, Verteilung von Noradrenalin and Dopamin (3-hydroxytyramin) im Gehirn des Menschen und ihr Verhalten bei Erkrankungen des Extrapyra midalen Systems. Klin. Wsch. 38, 1236-1239.

14. Fahn, S., Cote, L. J., 1968, Regional distribution of cholineacetylase in the brain of the rhesus monkey. Brain Res. 7, 323-325.

15. Glowinski, J., Iversen, L. L., Axelrod, J., 1966, Regional studies of catecholamines in the rat brain. IV. Effects of drugs on the disposition and metabolism of H^3 norepinephrine and H^3 dopamine. J. Pharmacol. Exp. Ther. 153, 30-41.

16. Hebb, C. O., Silver, A., 1956, Cholineacetylase in the central nervous system of man and some other mammals. J. Physiol. (Lond) 134, 718-728.

17. Jasper, H. H., Ajmone-Marsan, C., 1954, A stereotaxic atlas of the cat. Nat. Res. Council of Canada, Ottawa.

18. Jouvet, M., 1969, Biogenic amines and the states of sleep. Science 163, 32-41.

19. McLennan, H., York, D. H., 1967, The action of dopamine on neurons of the caudate nucleus. J. Physiol. 189, 393-402.

20. Portig, P. J., Vogt, M., 1969, Release into the cerebral ventricles of substances with possible transmitter function in the caudate nucleus. J. Physiol. (Lond) 204, 687-715.

21. Prado-Alcalá, R., Grinberg-Zylberbaum, J., Alvarez-Leefmans, J., Gómez, A., Singer, S., Brust-Carmona, H., 1972, A possible caudate-cholinergic mechanism in two instrumental conditioned responses. Psychopharmacologia (Berl.) 25, 339-346.

22. Prado-Alcalá, R., Grinberg-Zylberbaum, J., Alvarez-Leefmans, J., Brust-Carmona, H., 1973, Suppression of motor conditioning by the injection of 3 M KCl in the caudate nuclei of cats. Physiol. Behav. 9, 1-6.

23. Seiden, L. S., Peterson, D. D., 1968, Reversal of the reserpine-induced suppression of the conditioned avoidance response by L-DOPA: Correlation of behavioral and biochemical differences in two strains of mice. J. Pharmacol. Exp. Ther. 159, 422-428.

24. Siegel, S., 1956, Nonparametric Statistics for the Behavioral Sciences. New York: McGraw Hill.

25. Stein, L., 1968, Chemistry of reward and punishment. In: Psychopharmacology: A review of progress 1957-1967. D. H. Efron, ed. Public Health Service Publication No. 1836.

26. Thompson, R. L., 1959, Effects of lesions in the caudate nuclei and dorsofrontal cortex on conditioned avoidance behavior in cats. J. comp. physiol. Psychol. 52, 650-659.

27. Vogt, M., 1954, The concentration of sympathin in different parts of the central nervous system under normal conditions and after the administration of drugs. J. Physiol. (Lond) 123, 451-481.

28. Wyers, E. J., Deadwyler, S. A., 1971, Duration and nature of retrograde amnesia produced by stimulation of caudate nucleus. Physiol. Behav. 6, 97-103.

ABSTRACT:

The caudate nucleus (CN) seems to be the integrating and storing place in the circuit responsible for the motor conditioned responses (MCR), since the electrolytical or pharmacological suppression of CN activity blocks the performance of previous MCR as well as the acquisition of a passive avoidance response. The activating process of the CN neurons is probably cholinergic, because the topical application of atropine sulphate blocks MCR (either approaching conditioned response or lever pressing) and acetylcholine (ACh) improves its acquisition. Thus it was postulated that the inhibition could be related to catecholamines (CA). Considering that the neurotransmitter substance would need lower concentrations than those of its precursor, we started by testing the effect of the microinjection of epinephrine (E) or norepinephrine (NE) in the CN, leaving for a subsequent opportunity other substances (DOPA, dopamine). In three series of experiments, cats were trained to inhibit an approaching conditioned response if a series of clicks accompanied the CS (4 flashes) or a lever pressing response if a small light above the lever was off. The bilateral application of E or NE improved the learned inhibitory responses, without changing the MCR. The statistical difference (Mann Whitney U test) of the inhibition of the MCR before and after the CA microinjection was significant at the level of $p < 0.05$. In contrast, neither the MCR nor the inhibition changed when saline solution was injected. A neuronal circuit in which ACh could be the activating transmitter substance and CA responsible for the "inhibitory modulation" is proposed.

THE EFFECTS OF DRUGS AND ELECTRICAL STIMULATION OF THE BRAIN ON MEMORY STORAGE PROCESSES*

James L. McGaugh and Paul E. Gold

Department of Psychobiology, University of California

Irvine, California

The ultimate goal of studies of the neurobiology of memory is to understand how the nervous system processes, stores, and utilizes information. The problem is not one simply of determining the mechanism underlying the neural trace of an experience. We need to know how such traces are produced, where they occur and how they are used when required to control learned behavior. The neuronal basis of memory is undoubtedly highly complex. When new information is acquired many brain systems are probably involved. The particular neuroanatomical and neurochemical systems activated by any new experience will depend upon the type of information to be stored. Visual information, for example, will probably involve at least some neuronal systems and specific cells that are different from those involved in auditory information. Consequently, it does not seem likely that there are unique neuroanatomical systems or unique neurochemical systems that store many different types of memories. But, it may be that some brain systems are involved in certain processes which promote the storage of most if not all information. Thus, although it may not be possible to locate specific "engrams" it should be possible to locate the neural systems which are involved in the processing and storing of information and, eventually, to understand the anatomical interactions and neurochemical bases of such systems.

There is extensive evidence that the systems involved in the storage of memory are active for some period of time following a training experience (Glickman, 1961; McGaugh, 1966). Treatments such as electrical stimulation of the brain and centrally acting drugs influence the retention of experiences if the treatments are administered shortly after learning. In general, the degree of modification of retention produced by the treatments decreases as the interval between the learning experience and the treatment is increased (McGaugh and Herz, 1972). Thus, there is strong support for the view that the treatments affect retention because they modulate time dependent processes involved in memory storage (McGaugh and Dawson, 1971).

As yet, little is known about the bases of the modulating effects of drugs and electrical stimulation of the brain on memory storage processes. An understanding of the specific alterations in neuronal functioning which are associated with impairment or enhancement of memory is essential if such studies are to contribute to progress in discovering the neurobiological bases of memory storage. Our recent research on this problem has attempted to determine: 1) the alterations in CNS function which are related to modulation of memory storage processes, 2) the brain regions in which electrical stimulation is particularly effective in modulating memory storage processes, and 3) whether memory storage processes are modulated by drugs which affect catecholamine biosynthesis. This paper summarizes some of the findings of our studies concerned with these issues. Although there are many treatments which affect memory storage, our research is focused on the effects of direct electrical and chemical brain stimulation and the effects of systemically administered drugs.

EFFECTS OF BRAIN STIMULATION ON MEMORY STORAGE

Until recently most studies of the effects of electrical stimulation of the brain on memory storage used stimulation delivered to rats or mice through pinneal or corneal electrodes. With such procedures retrograde amnesia is produced by current intensities which also produce brain seizure discharges and convulsions (McGaugh, 1973; McGaugh, 1973, in press; McGaugh, Zornetzer, Gold and Landfield, 1972). Since the electrical stimulation used in these studies produced brain seizures, it has generally been assumed that the seizures are responsible for the disruptive effect of the current on memory storage. Further, since brain seizure discharges spread throughout the brain in rats and mice, these earlier findings seemed to suggest brain stimulation would not be a useful technique for localization of specific brain systems involved in the modulation of memory. There is now a great deal of evidence which indicates that this conclusion was unwarranted.

While it is the case that under many experimental conditions the threshold for producing amnesia by electrical stimulation of the brain is very close to the brain seizure threshold (Zornetzer and McGaugh, 1971), our recent findings which are summarized below indicate that the elicitation of brain seizures is neither a necessary nor sufficient condition for producing retrograde amnesia by electrical stimulation of the brain. Our results indicate that the effects of electrical stimulation depend upon the intensity of the current as well as the region of the brain stimulated. Memory is not affected by stimulation of some brain regions even if the stimulation elicits widespread seizure activity. Stimulation of other brain regions produces retrograde amnesia with current intensities well below the brain seizure threshold. The finding that the effects of the stimulation on memory vary with the region stimulated provides the possibility of identifying neural structures which influence memory storage processes (Gold, Zornetzer and McGaugh, in press).

CORTICAL STIMULATION

In our studies of the effect on memory of direct stimulation of the cortex in rats, the rats are first implanted with bilateral cortical electrodes. Then, a week or so later, they are given a single training trial on an inhibitory avoidance task followed by a retention test either a day or several days later. Electrical stimulation is administered after training and EEG activity is recorded from the implanted electrodes before and after the stimulation. On the retention test, the index of retention of the punishment is the latency to make the previously punished response. Long response latencies are considered as evidence that the animals remember the punishment. The use of a single training trial allows us to control rather precisely the interval of time between the training and the administration of brain stimulation.

In the first studies of the effects of cortical stimulation on memory which were conducted in our laboratory (Zornetzer and McGaugh, 1970, 1972) we found that retrograde amnesia could be produced by direct stimulation of the frontal cortex delivered immediately after training and that, over a range from 1.5 to 6.0 mA the degree of amnesia varied directly with current intensity. At current intensities greater than 2.0 mA brain seizures were produced in all animals. In subsequent studies we have investigated the effect of varying the interval between training and the cortical stimulation in order to determine the nature of the gradient of retrograde amnesia produced by cortical stimulation. There has been a great deal of controversy concerning the length of the gradient of retrograde amnesia. The interest is based on the possibility that a gradient might be an index of the time required for the formation or consolidation of the lasting trace of an experience. However, the studies of retrograde amnesia have not produced any uniform "time-constants" for memory storage. Under some conditions (Chorover and Schiller, 1965) amnesia is produced only if the treatments are administered within 20 seconds following training. Under other conditions, amnesia is produced by treatments administered many hours after the training (Kopp, Bohdanecky and Jarvik, 1966; Cherkin, 1969). Our findings indicate that the retrograde amnesia gradient produced by posttrial cortical stimulation varies with a number of conditions, including the current intensity, as well as with the brain region stimulated. In one study, for example (Gold, Macri and McGaugh, 1973), rats were given bilateral stimulation (0.5 sec, 60 Hz) through either frontal or posterior cortical electrodes after they were trained on the one trial inhibitory task. The results of this study are shown in Figure 1. The effects of frontal cortex stimulation are shown on the left. Clearly, the degree of amnesia varied with current intensity: No amnesia was produced with 1.0 mA. With 2.0 mA amnesia was produced only if the current was applied immediately after training; with 4.0 mA a 30 second posttrial treatment was effective; with 8.0 mA amnesia was produced with a treatment administered 15 minutes after training. With posterior cortex stimulation the results were similar. However, for a given stimulation intensity, amnesic gradients were significantly longer with posterior stimulation. The 4.0 mA stimulation produced significant amnesia even if the treatment was administered one hour after training. We interpret these findings as indicating

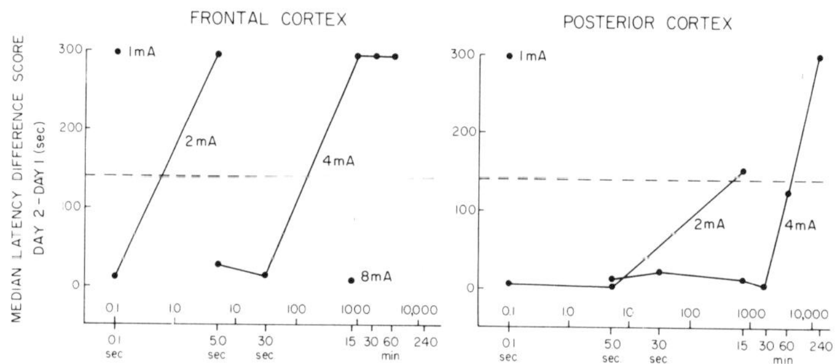

Figure 1. Retention response latencies of rats trained in a one trial inhibitory avoidance task and administered electrical stimulation of the frontal or posterior cortex after the training at the times indicated. The retention score is based on the differences between the response latency on the training trial and the test trial 24 hours later. Groups with median response latencies of less than 140 seconds (indicated by the dotted line) are significantly different ($p < .05$) from groups receiving the training and no subsequent brain stimulation. (From Gold, Macri and McGaugh, 1973.)

that retrograde amnesia gradients reflect time-dependent changes in the susceptibility of memory storage processes to the disruptive effects of electrical stimulation. However, the specific gradient obtained by posttrial cortical stimulation depends upon the current intensity used. The fact that the effect depends upon the brain region stimulated is also of interest.

While these findings might be interpreted as suggesting that posterior brain regions are in some way more involved in memory storage processes, other recent findings suggest that a more complex interpretation is required. For example, in an inhibitory avoidance task which uses slightly different experimental conditioning and training procedures (Gold, Bueno and McGaugh, 1973), the amnesia produced by electrical stimulation of the frontal cortex is greater than that produced by the posterior cortex (Gold, McDonald and McGaugh, 1973). These findings are shown in Figure 2. Note that with 6.0 mA stimulation amnesia is produced by stimulation of frontal cortex administered 15 minutes after training. However, stimulation of posterior cortex 15 minutes

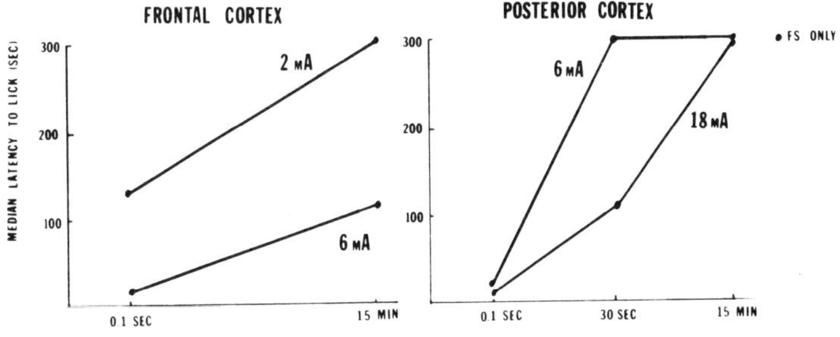

Figure 2. Retention response latencies of rats trained in a modified one-trial inhibitory avoidance task and administered electrical stimulation of the frontal cortex at the times indicated. The rats were first deprived of water and trained to approach a drinking tube and lick for a water reward. On the training trial they were given a footshock during the 10th second of drinking. The scores are the median latency to drink on the 24 hr retention test. With this task and procedures posttrial anterior cortex stimulation is more effective than posterior stimulation in disrupting memory. (From Gold, McDonald and McGaugh, in press).

after training does not affect retention even with a current of 18 mA. Thus, the effectiveness of electrical stimulation of the cortex in producing modulation of memory depends upon many conditions, including the specific training procedures used. It is important to note that in this experiment brain seizures were elicited in all animals. Thus, it is clear that even if the brain stimulation is administered within 30 seconds after training, the elicitation of brain seizure activity is, in this particular case, not a sufficient condition for modulating memory storage processes.

SUBCORTICAL STIMULATION

In view of the evidence provided by the cortical stimulation studies that brain seizures are not a sufficient condition for producing amnesia, we have conducted a series of studies to determine whether brain seizures are necessary for producing amnesia by electrical stimulation of the brain. If the stimulation is delivered through cortical electrodes, amnesia is not produced unless the current is at or above the brain seizure threshold (Gold and McGaugh, 1973). However, evidence from other laboratories suggested that amnesia is produced by stimulation of subcortical regions with current intensities below the threshold for eliciting brain seizures (Wyers et al., 1968; McDonough and Kesner, 1971; Bresnahan and Routtenberg, 1972). These findings suggest that low intensity electrical stimulation can be used to determine whether there are specific brain regions in which posttrial electrical stimulation is particularly effective in modulating memory. To examine this question further we studied the effects on memory, of posttraining electrical stimulation of a number of subcortical regions with current intensities below the seizure threshold (Gold, Macri and McGaugh, in preparation). The animals in these studies were implanted bilaterally with electrodes and then given a single training trial on an inhibitory avoidance task and a subsequent retention test. Immediately following the training the rats received a 10 sec train of 100 Hz, 0.1 ms electrical stimulation through each subcortical electrode. EEG activity was recorded from the cortical and subcortical electrodes immediately after the stimulation. We found several regions in which, with our procedures, subseizure stimulation does not produce amnesia. These regions include the septum, cerebellum, medial and lateral preoptic area of the hypothalamus, dorsomedial thalamus and anteroventral nucleus of the thalamus. However, stimulation of the amygdala was highly effective in producing amnesia. Since the amygdala stimulation was effective when the stimulation was administered immediately after training, we subsequently investigated the effects of varying the training-treatment interval (Gold, Macri and McGaugh, 1973). Our findings are shown in Figure 3. Amnesia was produced by stimulation given 5 sec and 1 hr after training but not with stimulation given 6 hours after training.

These findings provide strikingly clear evidence that elicitation of brain seizure activity is not necessary for producing retrograde amnesia with subcortical stimulation. Moreover, these results indicate that stimulation of the amygdala is a very effective procedure for producing amnesia. The current used in this study was only 25 μA through each electrode. We have also found that in rats amnesia can be produced by unilateral as well as bilateral stimulation (Gold, Edwards and McGaugh, in preparation). The findings of our study of the effects of immediate posttrial unilateral subseizure stimulation are shown in Figure 4. With unilateral stimulation 50 μA current was used. Our findings indicate that the effect of the stimulation varies with the location of the electrode within the amygdaloid complex. Stimulation through electrodes with tips located in the regions marked with Xs was relatively ineffective in influencing memory. In contrast, stimulation through electrodes with tips

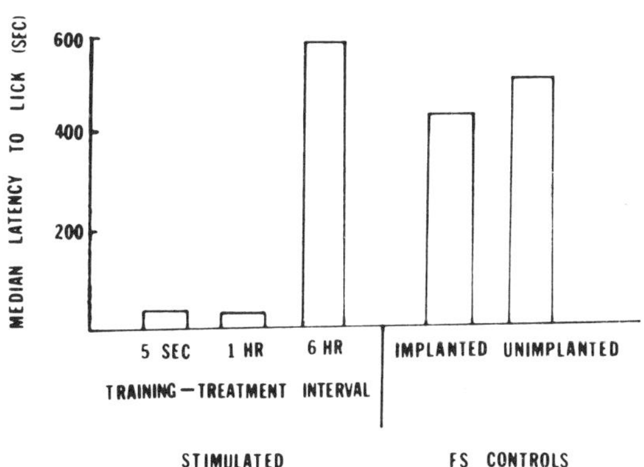

Figure 3. Retention response latencies of rats trained on the one-trial inhibitory avoidance task (same task as that used for experiment summarized in Figure 2), given bilateral stimulation of the amygdala (10 sec train, 25 μA, 100 Hz, 0.1 ms monophasic square wave) at the times indicated, and tested 24 hours later. The groups given posttrial stimulation 5 sec or 1 hr after training were significantly different from the 6 hour groups as well as two control groups. Implanted controls had electrodes in the amygdala but were not stimulated. (From Gold, Macri and McGaugh, in press)

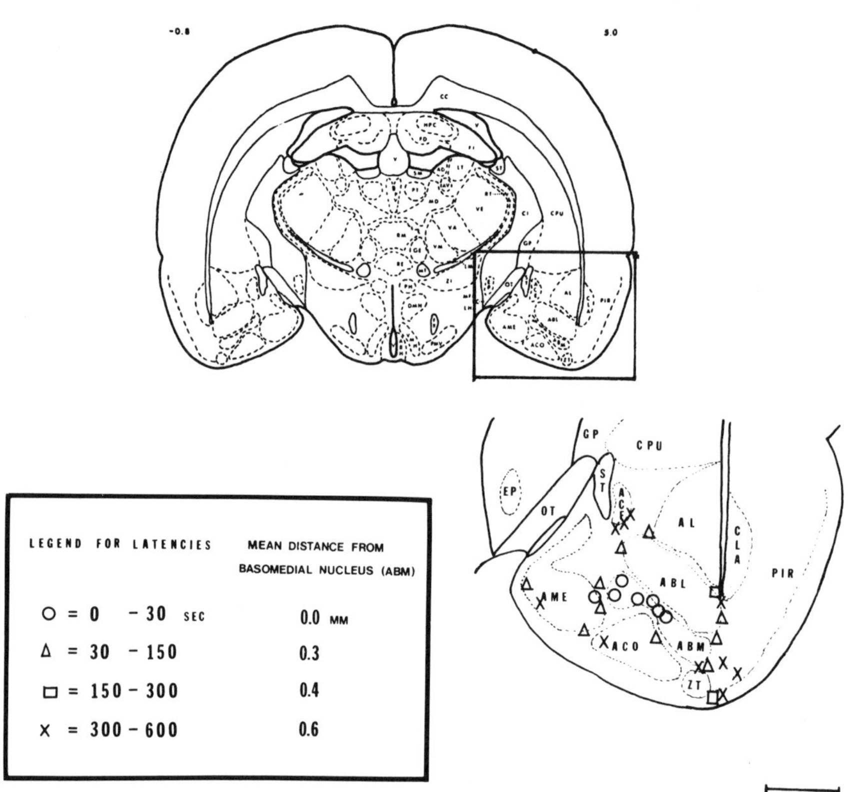

Figure 4. The effects of posttrial unilateral stimulation of the amygdala (10-30 sec train, 50 μA, 100 Hz) on retention of an inhibitory avoidance response (same task as that used for experiments summarized in Figures 2 and 3. Sites of electrode tips within the amygdala are shown on right. The circles, triangles, squares and Xs indicate the response latencies, as shown on the left, of individual animals. Lowest response latencies (i.e., greatest amnesia) were obtained with stimulation of the basomedial nucleus. From Gold, Edwards and McGaugh, in preparation.

in the region marked with small circles was highly effective in producing retrograde amnesia. Most of the highly effective placements are located in the basomedial nucleus of the amygdala. The effectiveness of the stimulation decreased directly with the distance of the electrode from the basomedial nucleus.

These findings indicate quite clearly that the modulating influences of brain stimulation on memory depend upon the specific brain regions stimulated. But, how are such findings to be interpreted? There are several possibilities. One possibility is that the stimulation disrupts memory by disturbing memory systems which are at least in part located in the region of the electrode tip. This interpretation suggests that the basomedial nucleus is critically involved in memory storage. We have found, however, that animals with bilateral amygdala lesions learn the inhibitory avoidance task with little difficulty (Zornetzer and McGaugh, 1970; Gold and McGaugh, in preparation). Thus, at the very least, the amygdala is not required for learning of this task. It should also be noted that, in the rat, brain seizures elicited by cortical stimulation most likely disrupt organized neural activity in the amygdala and, as we reported above, brain seizures are not a sufficient condition for producing memory deficits. Thus, it seems to us to be more likely that electrical stimulation affects memory because the direct stimulation of this region produces changes which alter neuronal functioning in other regions of the brain. Some structures, such as the amygdala, may be merely more effective sites for producing such effects. It is possible, for example, that there are brain states which are optimal for efficient memory storage. Stimulation which produces shift away from the optimal state might interfere with memory storage while treatments which produce shifts toward optimal states might facilitate storage. Thus, it might be that amygdala stimulation is highly effective as an amnesic treatment simply because the stimulation modulates brain states and the changes alter memory storage in neural systems located elsewhere in the brain. Stimulation of the amygdala does, in fact, produce widespread changes in neural activity (Kaada, 1972). We need to know which specific changes if any, are critical for producing the modulating influences on memory storage.

Thus, we believe that our findings have not located "engrams" in the amygdala. Rather, we believe that these findings may contribute to an understanding of neural systems involved in the modulation of memory storage.

Our recent research has focused on impairment of memory storage. However, there is also evidence that learning can be facilitated by low intensity (subseizure) stimulation of subcortical brain regions including the mesencephalic reticular formation (Bloch, 1970; Denti et al., 1970) and the hippocampus (Landfield, Tusa and McGaugh, 1973; Destrade, Soumireu-Mourat and

Cardo, 1973). It should also be noted that other experiments have shown that amnesia can be produced by subseizure stimulation of structures other than the amygdala, including in particular, the caudate nucleus and hippocampus (Zornetzer, Chronister and Ross, 1973; Peeke and Herz, 1971; Haycock et al., in press). These findings suggest that several neural structures may be involved in modulating memory storage processes. The amygdala may be an important brain region in this regard but it is probably not unique.

DRUG MODIFICATION OF MEMORY STORAGE PROCESSES

Drugs have also been used extensively in experiments investigating the neurobiological bases of memory storage. Memory can be impaired or enhanced by posttrial administration of many drugs. While drugs can be used to modulate memory storage processes, the experimental findings have not, as yet, provided unambiguous conclusions concerning the neurochemical systems underlying the behavioral effects. The problem is that any particular drug has more than one effect and it is difficult to determine which of the neurobiological effects is responsible for the drug's influence on memory.

In a series of experiments we have shown that retention is enhanced by posttraining administration of central nervous system stimulants, including strychnine, pentylenetetrazol, picrotoxin and amphetamine. Our findings are summarized in several papers (McGaugh, 1968, 1973; McGaugh and Herz, 1972). Comparable effects have been obtained in numerous other laboratories (cf. McGaugh, 1973). The type of findings we have obtained are illustrated by the results shown in Figures 5 and 6. In these experiments mice were given training trials each day in a two-alley visual discrimination maze. Correct responses were rewarded with food. Training was continued until the mice reached a criterion of 9 out of 10 correct choices. In the experiment summarized in Figure 5 the mice received either posttrial injections of saline or injections of strychnine sulphate (1.0 mg/kg I.P.) at one of several intervals before or after the daily training session (McGaugh and Krivanek, 1970). As can be seen, highly significant facilitation was produced by injections given either before training or up to 1 hour following training. The degree of enhancement decreased as the interval between training and drug administration was increased. Figure 6 shows the results of a similar study of the effect of d-amphetamine on learning (Krivanek and McGaugh, 1969). The results on the left show the effects of varying the dose of the drug when the injections were administered immediately after the daily training. Up to a dose of 2.0 mg/kg the degree of facilitation increased with the dose. The results on the right indicate the effect of varying the time of administration of a low (0.5 mg/kg) and

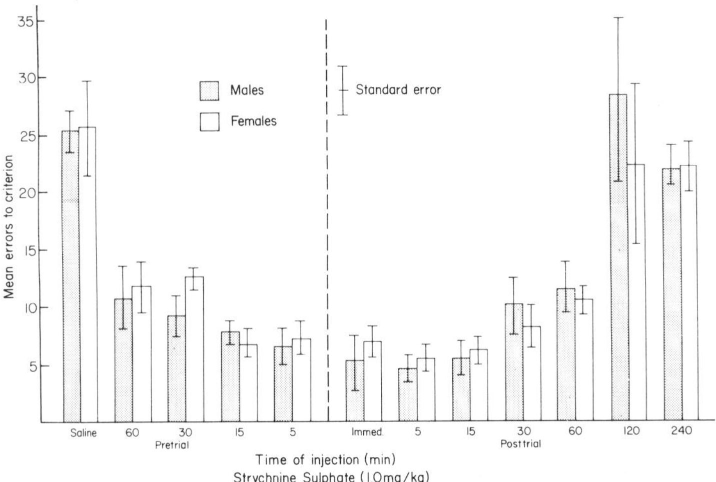

Figure 5. The effects of pre- and/or posttraining administration of strychnine sulphate (1.0 mg/kg I.P.) on visual discrimination learning in mice. The mice received three rewarded trials each day and were trained to a criterion of 9 out of 10 correct responses. The bars indicate the mean number of errors made by mice in the different groups. Significant facilitation was found in groups injected between 1 hour before and 1 hour after the daily training. (From McGaugh and Krivanek, 1970).

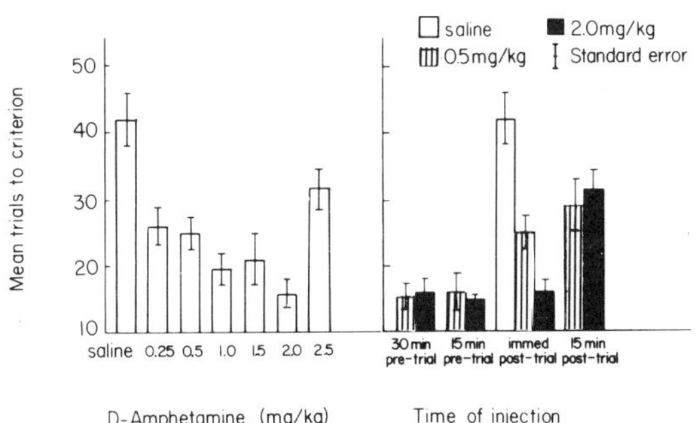

Figure 6. The effect of d-amphetamine on visual discrimination learning in mice (same task as that used for experiment summarized in Figure 5). Results on left show the effect of dose of posttrial d-amphetamine on learning (trials to criterion). Results on right show the effect of time of administration, pre- or posttraining, of 0.5 and 2.0 mg/kg (I.P.). (From Krivanek and McGaugh, 1969).

high (2.0 mg/kg) dose of d-amphetamine. With d-amphetamine as well as strychnine, the degree of facilitation decreased as the training-treatment interval was increased.

We interpret these findings as indicating that the drugs affect retention by modulating the posttraining neural processes involved in memory storage. The use of posttrial drug administration effectively eliminates any interpretation that the drugs affect learning by directly influencing sensory, motivational, or motor processes. As we suggested above, it may be that these treatments enhance memory storage because they produce brain states which are favorable for memory storage. That is, the facilitation may be due to widespread nonspecific influences on neural activity. Our findings, as well as those of other investigators (Doty and Doty, 1966; Breda, Carlini and Sader, 1969; Evangelista and Izquierdo, 1971), that learning is enhanced by posttrial administration of d-amphetamine suggest that the modulating influences may be due to influences on brain catecholamines. The effects of amphetamine on behavior, such as locomotor activity, appear to require catecholamine synthesis (Sulser and Sanders-Bush, 1971). In support of this interpretation, there is some evidence that the facilitating effects of pretrial administration d-amphetamine are blocked if the animals are pretreated with alpha-methyl-para-tyrosine (α-MT) which inhibits the synthesis of dopamine and norepinephrine by inhibiting tyrosine hydroxylase (Orsingher and Fulginiti, 1971; Knoll, in press). As yet, there have been to our knowledge no investigations to determine whether α-MT blocks the facilitating effects of posttrial d-amphetamine on learning. Further, there have been no studies investigating the effect of α-MT on the learning facilitation produced by other CNS stimulants. It is possible that the different drugs facilitate learning by different actions on the nervous system. However, the possibility that the drugs act through a common mechanism cannot be ruled out at this time.

There is other evidence suggesting that catecholamine biosynthesis is required for memory storage. Several studies have reported that retention is impaired by posttraining administration of diethyldithiocarbamate (DDC) which inhibits norepinephrine synthesis by inhibiting dopamine-beta-hydroxylase (DBH) (Dismukes and Rake, 1972; Randt et al., 1971; Osborne and Kerkut, 1972). We have confirmed this effect in recent experiments (Van Buskirk, Haycock and McGaugh, in preparation). Most of our experiments have examined the effects of DDC on the retention of an inhibitory avoidance response in mice. The mice are given a single training trial and a single retention test trial one week later. Figure 7 shows the dose-response effects when the injections are administered immediately after training. As can be seen the most effective dose was 900 mg/kg (I.P.). The effects of varying the time of administration are shown on the right. Impairment of retention is

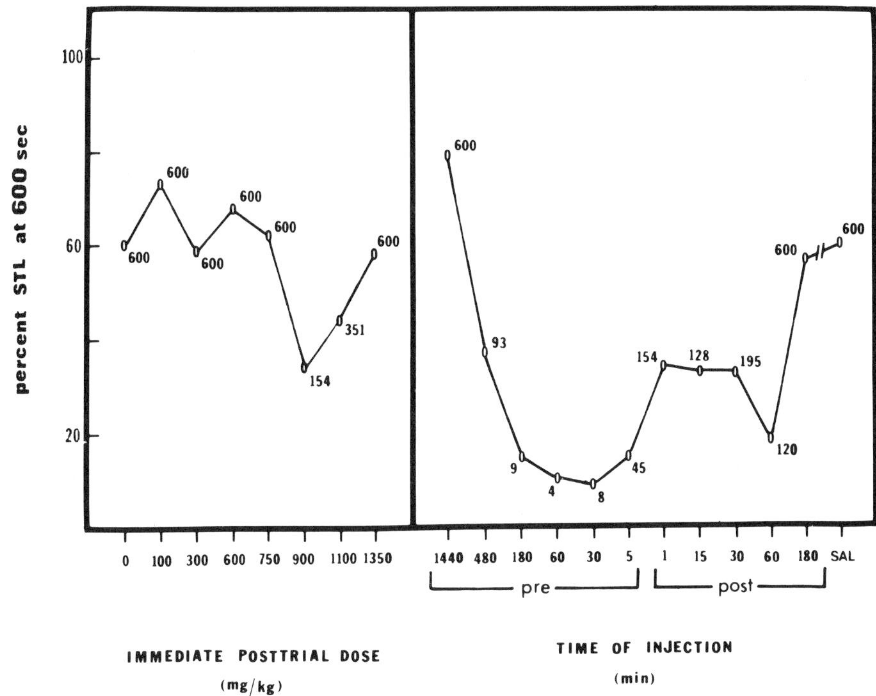

Figure 7. The effect of diethyldithiocarbamate (DDC) on retention of an inhibitory avoidance response in mice. The results on the left indicate the dose-response effect of DDC injections administered (I.P.) immediately after training. The animals were given a single training trial and a single retention test trial 1 week later. The percent of animals in each group with response latencies of 100 seconds is indicated. The numbers adjacent to each point are the median retention response latencies. The results on the right indicate the effects of time of administration, before or after training of a dose of 900 mg/kg (I.P.). Although minimum impairment of retention was produced by DDC administered within 3 hours before training, significant impairment was produced by injections administered up to 1 hour after training. (From Van Buskirk, Haycock and McGaugh, in preparation.)

produced by injections given within 3 hours before training or one hour after training. The findings that posttrial administration is less effective than pretrial is probably due to relatively slow onset of the inhibition of DBH by the drug. Thus, even though the drug is administered immediately after training it is probably not effective for half an hour (Randt et al., 1971). We have obtained comparable results in similar experiments using rats.

These findings, as well as those of other investigators, provide strong evidence that DDC impairs memory storage processes. Thus, the findings are consistent with the hypothesis that memory storage processes are modulated by newly synthesized catecholamines, in particular, norepinephrine. Kety (1970, 1972), for example, has suggested that biogenic amines released in affective states may modulate memory storage by regulating synaptic activity at recently activated synapses. Although our findings are consistent with this view, it is important to emphasize that DDC has a variety of effects on neural activity. We have found that, in mice, DDC produces brain seizure activity at doses which impair memory. Further, additional experiments are needed in order to determine whether memory storage is influenced by other drugs which influence catecholamine levels and biosynthesis. Should additional experiments provide further evidence that drugs which influence catecholamines also affect memory storage, it would be essential to attempt to determine whether such effects are due to influences on specific brain regions, or whether the effects are due to widespread neurochemical influences throughout the brain. Thus, it would be premature to conclude, on the basis of available evidence that catecholamines affect memory by modulating neural systems involved in memory. Obviously, the role, if any, of catecholamine systems in memory storage remains to be determined. It is also possible that memory storage processes will be modulated by any drug which influences transmitter functioning. And, several classes of drugs do, in fact, appear to influence memory storage processes.

CONCLUDING COMMENTS

Experiments of the kind reviewed here are, of course, not designed to discover the anatomical location or the chemical coding of specific memories. These experiments can only serve to provide information about changes in anatomical, physiological and chemical systems which are associated with or, hopefully, involved in the modulation of memory storage. It seems likely the anatomical location of cells involved in the memory of specific experiences such as those produced by a trip to an international conference will depend upon the neural systems activated by the experiences. The chemical coding of the memory will depend upon the transmitters used by the cells involved. Thus, it may well be that we will never be able to specify in detail the neurobiological basis of particular memories. It seems more likely that our understanding of memory will consist of understanding the general processes underlying the formation or utilization of memory systems. Understanding the way in which memory storage processes are modulated by electrical and chemical stimulation of the brain should contribute to this ultimate goal.

*FOOTNOTE

This paper is based on papers presented at the Fourth International Neurobiological Symposium in Magdeburg in May, 1973, and at the "Chemical Coding of Cerebral Function" symposium of the Science and Man Congress of the American Association for the Advancement of Science, in Mexico City, June, 1973.

ACKNOWLEDGEMENTS

Supported in part by Research Grant Mh 12526 from the National Institute of Mental Health, United States Public Health Service. We thank Roderick Van Buskirk and John Haycock for their contributions to some of the research reported here.

REFERENCES

Bloch, V. 1970. Facts and hypotheses concerning memory consolidation. Brain Research, 24, 561-575.

Breda, J. B., Carlini, E. Q. and Sader, N. F. A. 1969. Effects of chronic administration of (+)-amphetamine on maze performance of the rat. Brit. J. Pharmacol. 37, 79-86.

Bresnahan, E. and Routtenberg, A. 1972. Memory disruption by unilateral low level, sub-seizure stimulation of the medial amygdaloid nucleus. Physiol. Behav. 9, 513-525.

Cherkin, A. 1969. Kinetics of memory consolidation: Role of amnesic treatment parameters. Proc. Nat. Acad. Sci. 63, 1094-1101.

Chorover, S. L. and Schiller, P. H. 1965. Short-term retrograde amnesia in rats. J. Comp. Physiol. Psychol. 59, 73-78.

Denti, A., McGaugh, J. L., Landfield, P. W. and Shinkman, P. 1970. Effects of posttrial electrical stimulation of the mesencephalic reticular formation on avoidance learning in rats. Physiol. Behav. 5, 659-662.

Destrade, C., Soumireu-Mourat, B. and Cardo, B. 1973. Effects of posttrial hippocampal stimulation on acquisition. Behav. Biol., 1973, 9, in press.

Dismukes, R. K. and Rake, A. V. 1972. Involvement of biogenic amines in memory formation. Psychopharmacologia, 23, 17-25.

Doty, B. A. and Doty, L. A. 1966. Facilitative effects of amphetamine on avoidance conditioning in relation to age and problem difficulty. Psychopharmacologia, 9, 234-241.

Evangelista, A. M. and Izquierdo, I. 1971. The effect of pre- and posttrial amphetamine injections on avoidance responses of rats. Psychopharmacologia, 20, 42-47.

Glickman, S. E. 1961. Perseverative neural processes and consolidation of the memory trace. Psychol. Bull. 58, 218-233.

Gold, P. E., Bueno, O. F. and McGaugh, J. L. 1973. Training and task-related differences in retrograde amnesia thresholds determined by direct electrical stimulation of the cortex in rats. Physiol. Behav., in press.

Gold, P. E., Edwards, R. M. and McGaugh, J. L. In preparation.

Gold, P. E., Macri, J. and McGaugh, J. L. 1973. Retrograde amnesia gradients: Effects of direct cortical stimulation. Science, 179, 1343-1345.

Gold, P. E., Macri, J. and McGaugh, J. L. 1973. Amnesia produced by low-level subcortical stimulation. In preparation.

Gold, P. E., Macri, J. and McGaugh, J. L. 1973. Retrograde amnesia produced by subseizure stimulation of the amygdala. Behav. Biol., in press.

Gold, P. E., McDonald, R. and McGaugh, J. L. 1973. Direct cortical stimulation: A further study of treatment intensity effects on retrograde amnesia gradients. Behav. Biol. in press.

Gold, P. E. and McGaugh, J. L. 1973. Relationship between amnesia and brain seizure thresholds in rats. Physiol. Behav. 10, 41-46.

Gold, P. E., Zornetzer, S. F. and McGaugh, J. L. Electrical stimulation of the brain: Effects on memory storage. In: Newton, G. and Riesen, A. (Eds.), Advances in Psychobiology, Volume 2. New York, Wiley Interscience, in press.

Haycock, J. W., Deadwyler, S. A., Sideroff, S. I. and McGaugh, J. L. Retrograde amnesia and cholinergic systems in the caudate-putamen complex and dorsal hippocampus of the rat. Exper. Neurol., in press.

Kaada, B. R. 1972. Stimulation and regional ablation of the amygdaloid complex with reference to functional representation. In: Eleftheriou, B. E. (Ed.), The Neurobiology of the Amygdala. New York: Plenum Press, pp. 205-282.

Kety, S. S. 1970. The biogenic amines in the central nervous system: Their possible roles in arousal, emotion, and learning. In: Schmitt, F. O. (Ed.), The Neurosciences, New York: Rockefeller University Press, pp. 324-336.

Kety, S. S. 1972. Brain catecholamines, affective states and memory. In: McGaugh, J. L. (Ed.), The Chemistry of Mood, Motivation, and Memory. New York: Plenum Press, pp. 65-80.

Knoll, J. 1973. Modulations of learning and retention by amphetamines. Proc. Fifth Int. Cong. Pharm. Basel, Switzerland: S. Karger, in press.

Kopp, R., Bohdanecky, Z. and Jarvik, M. E. 1966. Long temporal gradient of retrograde amnesia for a well-discriminated stimulus. Science, 153, 1547-1549.

Krivanek, J. A. and McGaugh, J. L. 1969. Facilitating effects of pre- and posttrial amphetamine administration on discrimination learning in mice. Agents and Actions, 1, 36-42.

Landfield, P. W., Tusa, R. and McGaugh, J. L. 1973. Effects of posttrial hippocampal stimulation on memory storage and EEG activity. Behav. Biol., 8, 485-505.

McDonough, J. H., Jr. and Kesner, R. P. 1971. Amnesia produced by brief electrical stimulation of amygdala or dorsal hippocampus in cats. J. Comp. Physiol. Psychol. 77, 171-178.

McGaugh, J. L. 1966. Time-dependent processes in memory storage. Science, 153, 1351-1358.

McGaugh, J. L. 1968. Drug facilitation of memory and learning. In: Efron, D. H. et al. (Eds.), Psychopharmacology: A Review of Progress, 1957-1967, Washington, D. C.: U. S. Government Printing Office, PHS Publ. No. 1836, pp. 891-904.

McGaugh, J. L. 1973. Electroconvulsive shock: Effects on learning and memory in animals. In: Fink, M., Kety, S. S. and McGaugh, J. L. (Eds.), Psychobiology of Electroconvulsive Therapy. Washington, D.C.: V. H. Winston and Sons, Inc., in press.

McGaugh, J. L. 1973. Drug facilitation of learning and memory. Annual Review of Pharmacology, Volume 13, pp. 229-241.

McGaugh, J. L. and Dawson, R. G. 1971. Modification of memory storage processes. In Honig, W. K. and James, P. H. R. (Eds.), Animal Memory. New York: Academic Press, pp. 215-242.

McGaugh, J. L. and Herz, M. J. 1972. Memory Consolidation. San Francisco, California: Albion Publishing Company, 204 pps.

McGaugh, J. L. and Krivanek, J. 1970. Strychnine effects on discrimination learning in mice: Effects of dose and time of administration. Physiol. Behav., 5, 1437-1442.

McGaugh, J. L., Zornetzer, S. F., Gold, P. E. and Landfield, P. W. 1972. Modification of memory systems: Some neurobiological aspects. Q. Rev. Biophys., 5, 163-186.

Orsingher, O. A. and Fulginiti, S. 1971. Effects of alpha-methyl tyrosine and adrenergic blocking agents on the facilitatory action of amphetamine and nicotine on learning in rats. Psychopharmacologia, 19, 231-240.

Osborne, R. H. and Kerkut, G. A. 1972. Inhibition of noradrenalin biosynthesis and its effects on learning in rats. Comp. Gen. Pharmac., 3, 359-362.

Peeke, H. V. S. and Herz, M. J. 1971. Caudate nucleus stimulation retroactively impairs complex maze learning in the rat. Science, 173, 80-82.

Randt, C. T., Quartermain, D., Goldstein, M. and Anagnoste, B. 1971. Norepinephrine biosynthesis inhibition: Effects on memory in mice. Science, 172, 498-499.

Sulser, F. and Sanders-Bush, E. 1971. Effects of drugs on amines in the CNS. Ann. Rev. Pharmacol. 11, 209-230.

Van Buskirk, R. B., Haycock, J. W. and McGaugh, J. L. In preparation.

Wyers, E. J., Peeke, H. V. S., Williston, J. S. and Herz, M. J. 1968. Retroactive impairment of passive avoidance learning by stimulation of the caudate nucleus. Exp. Neurol. 22, 350-366.

Zornetzer, S. F., Chronister, R. B. and Ross, B. 1973. The hippocampus and retrograde amnesia: Localization of some positive and negative memory disruptive sites. Behav. Biol., 8, 507-518.

Zornetzer, S. and McGaugh, J. L. 1970. Effects of frontal brain electroshock stimulation on EEG activity and memory in rats: Relationship to ECS-produced retrograde amnesia. J. Neurobiol., 1, 379-394.

Zornetzer, S. and McGaugh, J. L. 1971. Retrograde amnesia and brain seizures in mice. Physiol. Behav., 7, 401-408.

Zornetzer, S. and McGaugh, J. L. 1972. Electrophysiological correlates of frontal cortex-induced retrograde amnesia in rats. Physiol. Behav. 8, 233-238.

ABSTRACT:

There is extensive evidence that memory storage processes are modified by treatments which affect neural functioning if the treatments are administered shortly after training. This paper summarizes recent findings of our research examining the effects, on memory, of 1) electrical stimulation of specific brain regions and 2) drugs which influence brain catecholamines. Our findings indicate that, in rats, retrograde amnesia is produced by direct electrical stimulation of the cortex, as well as other regions including the amygdala, hippocampus and caudate nucleus. In general, the degree of modification of memory varies with current intensity and duration. The elicitation of brain seizures is neither a necessary nor sufficient condition for producing retrograde amnesia. Facilitation of memory is produced by posttraining low intensity stimulation of the reticular formation and hippocampus. Facilitation of memory is also produced by posttraining administration of CNS stimulants including d-amphetamine. Drugs which inhibit norepinephrine biosynthesis, such as diethyldithiocarbamate, produce memory impairment comparable to that obtained with electrical stimulation of the brain. These findings are discussed in terms of a possible role of catecholamines in modulating memory storage processes.

NEURO-HUMORAL CODING OF SLEEP BY THE PHYSIOLOGICAL SLEEP FACTOR DELTA

M. Monnier* and G.A. Schoenenberger**

Physiological Institute*
Biol. Chem. Res. Div., Dept. of Surgery**
Univ. of Basel, Vesalgasse 1, CH-4051 Basel

Investigations on neuro-humoral coding of sleep were as yet chiefly focussed on amino-acids and biogenic amines such as serotonin. A possible involvement of peptides in the regulation of sleep was hardly ever taken into consideration. If a neuro-humoral coding of sleep exists, its first demonstration should be the neuro-humoral transmission of sleep.

The first experiments in that direction were carried out in Paris by Legendre and Piéron (1910; 1913). These investigators showed that cerebrospinal fluid of dogs, having fallen into deep sleep after prolonged sleep deprivation, induces sleep when injected into an alert dog They attributed this sleep to the depressant action of "hypnotoxins" on the waking mechanism.

The theory of sleep due to inhibition of a waking mechanism found a celebrated supporter in Pavlov (1923). He identified the mechanism of sleep with that of internal inhibition of conditioned reflexes by the cerebral cortex.

In 1925, the theory of sleep through inhibition of wakefulness found a new protagonist in W.R. Hess. He stimulated, in the cat, the medio-ventral thalamus and elicited, according to the stimulation parameters, either sleep or arousal (Hess, 1927; 1929; 1944). Stimulation with low pulse frequency (6/sec), weak

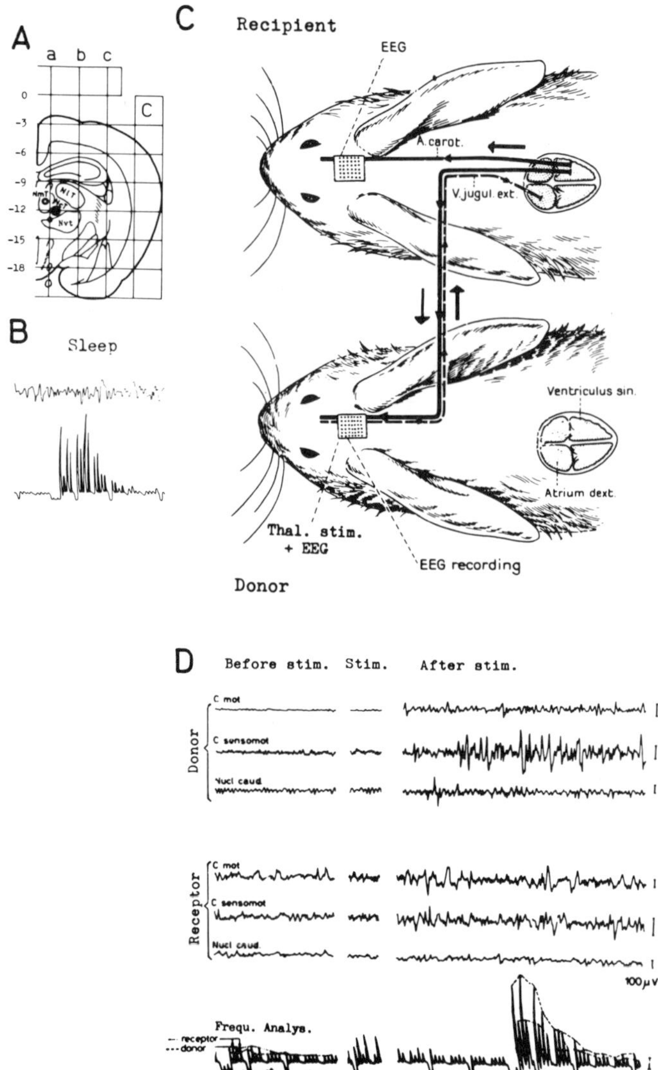

Fig. 1. A. Hypnogenic area in the medio-central intralaminary thalamus of the rabbit (Hösli and Monnier, 1962), localized with the stereotaxic atlas of Monnier and Gangloff (1961).

B. Electro-encephalogram (EEG) of the rabbit during "orthodox sleep": Increased synchronized delta activities. Below: Quantification with an automatic frequency analyzer.

C. Humoral transmission of sleep demonstrated by cross-circulation in rabbits.

intensity (0.7 V) and long pulse duration (12 ms) induces sleep. By contrast, stimulation of the same thalamic substrate at higher frequency and intensity produces an arousal reaction.

Sleep research was strongly promoted by electrophysiological methods, particularly by EEG recording, correlated with observation of the sleep behaviour in various animals.

EEG analysis of the thalamic sleep mechanism in cats (1950) and rabbits (1961) showed that the rabbit is a choice animal for the study of orthodox sleep. (Monnier, 1950; Monnier and Hösli, 1961). As a lagomorph, it mostly sleeps with half open eyes, always ready to wake up and escape. This medium deep sleep develops rapidly and may be quantified with great accuracy.

A. Method for inducing delta sleep in the rabbit

In order to study the mechanism of sleep on restrained or freely moving rabbits, a stereotaxic method, with a corresponding atlas, was worked out (Monnier and Gangloff, 1961). Thanks to a coordinates system, electrodes could be introduced into the different brain structures to be stimulated, with concurrent recording of the EEG (Fig.1A).

The optimal points and stimulation parameters for eliciting depressant effects and sleep from the thalamus had first to be defined. As in Hess' cats, the most sensitive points are located in the intralaminary thalamus, between the medial, lateral and ventral thalamic nuclei (Fig.1A; Hösli and Monnier, 1962). Stimulation of this substrate with low frequency stimuli (6/sec), weak voltage (0.7 V) and long pulse duration (10 ms) elicits a typical sleep in the rabbit.

In deep sleep, the animal lies down and exhibits marked hypotonia, almost complete closing of the lids and slower breathing. In the electroencephalogram, the delta

---

D. Sleep induced in the donor by stimulation of the thalamic sleep area (synchronized delta activities in the sensory motor cortex) is humorally transmitted to the recipient (Monnier, Koller and Graber, 1963).

activities become more abundant, slower (0.5-2.5 c.p.s.) (Hösli and Monnier, 1962) and increase in voltage. This synchronization of the delta activities is characteristic of the orthodox delta sleep.(Fig.1B)

By contrast, another type of very deep sleep with complete relaxation of the body muscles and rapid eye movements (REM) was called paradoxical sleep, because of the paradoxical desynchronization of the cortical EEG (Dement, 1958; Jouvet, 1965). This type of sleep did not develop in the rabbit under acute experimental conditions, in contrast to the orthodox delta sleep, with which we shall exclusively deal here.

As in the cat, the same thalamic substrate may elicit moderating or activating effects, i.e. sleep or arousal according to the stimulation frequency.

1) Stimulation of the medio-central intralaminary thalamus with low frequency (6/sec) produces via thalamo-cortical projections a synchronization of the cortical EEG. In other words, it induces an orthodox delta sleep (2-3 c.p.s.)(Fig.1B; Monnier et al., 1972).

2) Conversely, stimulation of the same medio-central intralaminary thalamus with higher frequency stimuli (75/sec) activates the cortex through the ascending projections of the reticular system. It desynchronizes the EEG of the neocortex, induces a theta rhythm (6-7 c.p.s.) in the archicortex (hippocampus) and elicits a waking reaction.

The alterations of the EEG can be quantified by automatic frequency analysis.

Sleep with increased slow delta activities and synchronization of the EEG is expressed by high deflections in the left low frequency band of the frequency spectrum (2-3 c/s).(Fig.1B)

Alertness with desynchronization of the EEG is characterized, on the other hand, by minimal deflections in the low frequency delta band of the frequency spectrum, and by deflections in the middle (6-7 c.p.s.) theta band of the frequency spectrum.

The reciprocal effects of both thalamo-cortical projection mechanisms are an important prerequisite for the rapid reversibility of neurally induced sleep. Should a humorally induced sleep exist, similar reciprocal chemical mechanisms would be expected in order to explain the same rapid reversibility of sleep.

B. Humoral transmission of sleep demonstrated on cross-circulation rabbits

Humoral transmission of cortical activation or alertness had already been demonstrated in cats (Ingvar, 1955; Purpura, 1956).

In the rabbit, cross-circulation experiments were carried out with the purpose of studying a possible humoral transmission of sleep (Monnier et al., 1963; Fig.1C). The jugular vein of the rabbit donor, in which sleep was elicited by electrical stimulation of the thalamic sleep area, was connected with that of the recipient. Thus, a sleep substance released into the cerebral venous blood of the donor could penetrate the brain of the recipient over its heart, lung and carotid artery.

Simultaneous recording of the EEG in donor and recipient showed an increase in delta activity of the sensory motor cortex, both in the stimulated donor and in its partner (Fig.1D). Automatic frequency analysis objectively confirmed the increase of delta activity in the donor (dotted line), and, to a lesser extent, in the recipient (points-line). Statistical evaluation of the results confirmed their significance. No such delta increase could be detected in the parafocally stimulated or non-stimulated control animals.

These experiments on the rabbit and those of Kornmüller et al. (1961) on the cat led to the conclusion that stimulation of the thalamic sleep area somehow triggers off the release in the blood of the donor of a sleep substance, capable of inducing sleep in the recipient by humoral transmission. These parabiosis experiments were repeated and confirmed many years later by Japanese workers (Matsumoto et al., 1972). They found that the amount of synchronized EEG activities during the sleep phase is higher in the parabiotic than in the control rat. This underlines the important part played by humoral factors in the maintenance of sleep.

C. Humoral transmission of sleep by hemodialysate from sleeping rabbits

Attempts at extracting the sleep substance from the cerebral venous blood of sleeping rabbits were made by extracorporal hemodialysis with the artificial kidney of Kuhn et al. (1957). The optimal site for obtaining cerebral venous blood is the cranial sinus (confluens sinuum) in which a cannula can easily be introduced (Koller et al., 1964). The cerebral venous blood of the rabbit donor is then conveyed to the dialyzing apparatus by means of a roller pump and therefrom back to the femoral or jugular vein (Monnier and Hösli, 1964; 1965; Fig.2A). In the artificial kidney, part of the smaller molecules with a MW below 40'000, diffuse from the blood through the pores of a cellophane membrane (25 Å) into the dialyzing fluid of an outer circuit.

After 30 min. predialysis with concurrent recording of the EEG, the thalamic sleep area is intermittently stimulated during 50 min. under further dialysis and EEG control. At the end of the experiment, about 35 ml sleep dialysate are obtained from the outer circuit and 35 ml from the dialyzer itself. As control animals, rabbits are used in which the thalamus was either stimulated outside the thalamic area, or not stimulated at all.

For quantification of sleep, the amount of delta activities in the donor is plotted on the ordinate, against the prestimulation and stimulation times on the abscissa (Fig.2B; Monnier and Hatt, 1971). The initial amount (100% = base line) is the mean delta amount of the prestimulation time, i.e. the average of 6 x five min. periods expressed in mm delta deflection or uV/min. For instance, an optimal donor has a mean initial delta value of about 200 mm computer deflection, i.e. about 2300 uV/5 min. The increased delta amount induced by thalamic stimulation is, in each case, determined with reference to the initial amount (=100%). Thus, in a series of 5 sleeping donors, the mean delta amount reached 136%, against 92% in 5 control animals. Concurrently, the significance of the differences in the delta amounts of sleeping donors and control donors was statistically analyzed. The corresponding p-values were expressed by symbols for each 5 min. period.

The hypnogenic effect of the sleeping donor dialysates

A

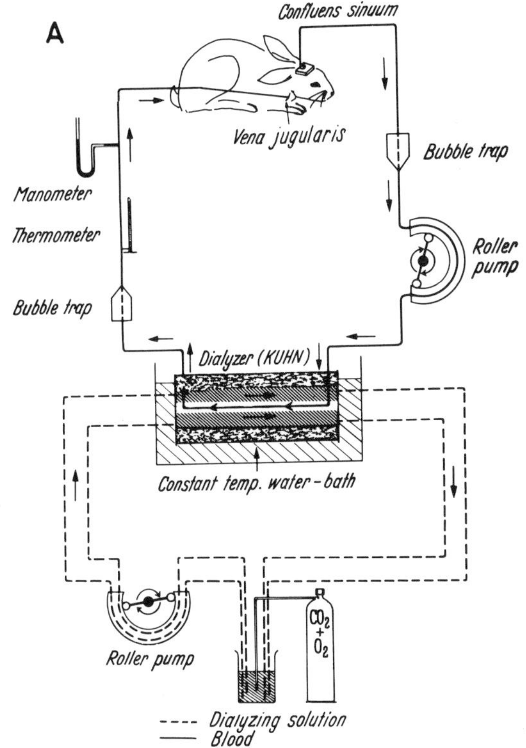

B SLEEP DONORS

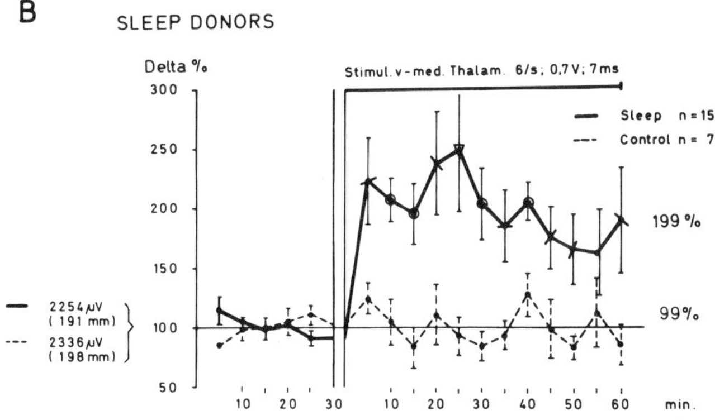

Fig. 2. A. Extracorporal hemodialysis in sleeping, alert and control rabbits (Monnier and Hösli, 1964).

B. Increased delta activity in sleeping donors during thalamic stimulation and hemodialysis (199%) as referred to the initial delta activity (100%) and to that of control donors (99%)(Monnier and Hatt, 1971).

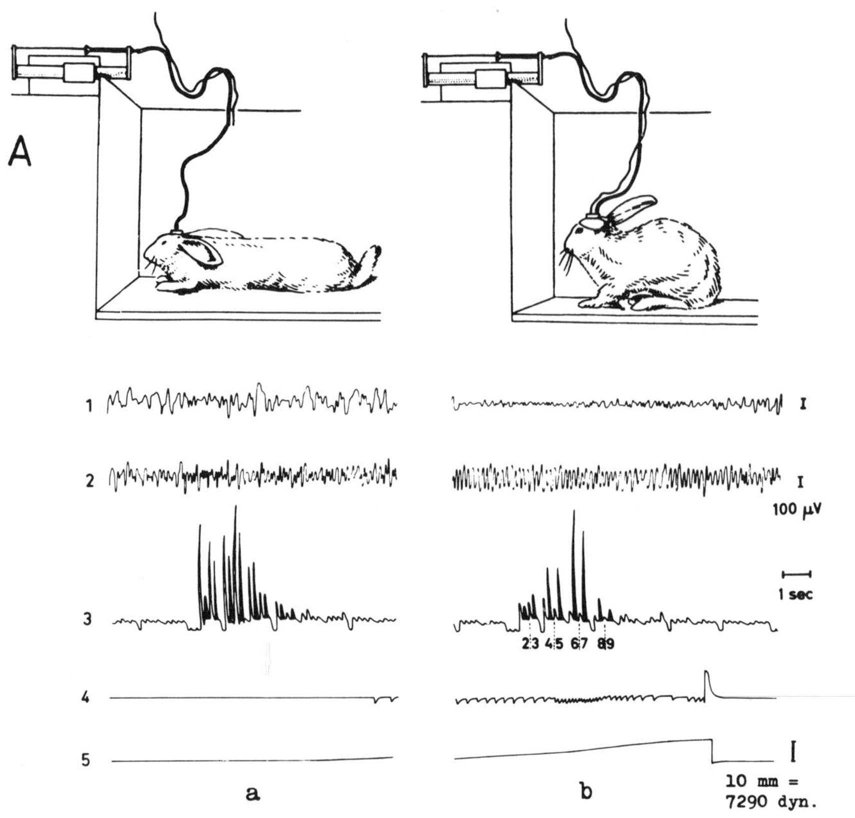
A
1
2
3
4
5
100 μV
1 sec
23 45 67 89
10 mm =
7290 dyn.
a
b

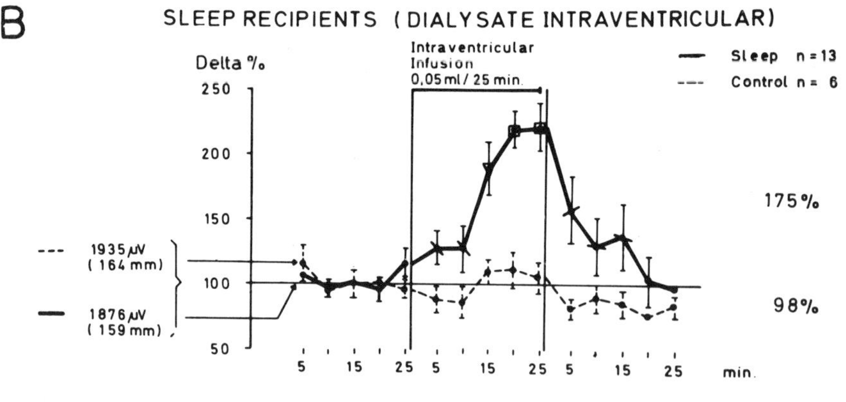
B
SLEEP RECIPIENTS (DIALYSATE INTRAVENTRICULAR)
Delta %
Intraventricular
Infusion
0,05 ml / 25 min.
Sleep n = 13
Control n = 6
250
200
150
100
50
1935 μV
(164 mm)
1876 μV
(159 mm)
175 %
98 %
5 15 25 5 15 25 5 15 25 min.

Figure 3

was originally assessed by intravenous injection into rabbit recipients. The criteria of sleep were, on the one hand, the increased delta amount in the EEG-test and, on the other hand, the decreased motor activity in the kinesigraphic behavioural test.

The most important improvement of the testing method consisted in the replacement of intravenous injection by intraventricular infusion to the mildly restrained or freely moving animal (Monnier and Hatt, 1970; Fig.3A). The sleep dialysate was infused into the mesencephalic ventricle by means of an implanted cannula (Fig.3B). Instead of 20 ml sleep dialysate for intravenous injection, 0.05 ml only were necessary for intraventricular infusion.

For assessment of the hypnogenic effect of the dialysate, 0.05 ml is infused intraventricularly with an automatic infusion pump within 25 min. The first sleep parameter remains the delta amount, recorded in the recipient with an electroencephalograph (Fig.3A, 1+2) and quantified with the automatic frequency analyzer (Unit: uV/min.)(Fig.3A,3). The second sleep parameter is the animal's behaviour, recorded with a kinesigraph (Unit: dyn/min(Fig.3A, 4+5). In the sleeping animal, during infusion of sleep dialysate, the kinesigram remains flat, the delta activity increases in the EEG (Fig.3Aa). By contrast, in the alert animal, during infusion of dialysate from an aroused donor, the activity increases in the kinesigram, whereas the delta rhythm disappears in the EEG (Fig.3Ab).

Fig. 3. Assessment of the hypnogenic activity of rabbits' hemodialysates.

A. Intraventricular infusion of hypnogenic (a) and waking (b) dialysate (Method of Monnier and Hatt, 1970). In the sleeping recipient (a), the EEG test shows increased, synchronized delta activities in the neocortex (a : 1), with increased deflections in the left low frequency band of the automatic frequency analyzer (3). The kinesigraphic test reveals no motor activity (a : 4+5), in contrast to the alert recipient (b : 4+5)(Monnier and Hatt, 1971; Monnier et al., 1972).

B. Increased delta activity (175%) in recipients during and after intraventricular infusion of sleep dialysate, in contrast to recipients of control dialysate (98%)(Monnier and Hatt, 1971).

In recipients of sleep dialysate under optimal conditions the initial average delta value for 5 min. is expressed by 154 mm computer deflection, corresponding to about 1800 uV/5 min. (Fig.3B; Monnier and Hatt, 1971). In this series of 13 animals during and after the infusion of sleep dialysate, the mean delta amount rises to 153%, against 98% for recipients of control dialysate. Concurrently to the EEG delta increase, the motor activity in the kinesigram decreases.

These experiments of sleep transmission by extracorporal sleep dialysate were not repeated in other laboratories. By contrast, the experiments of Legendre and Piéron (1910; 1913), reconsidered in 1939 by Schnedorf and Ivy, were successfully repeated by Pappenheimer et al. (1967). These investigators demonstrated on rats (by injection into the lateral ventricle) the sleep promoting effects of cerebrospinal fluid from sleep deprived goats. A new assay was developed by Myers (1967) for transfusion of chemical tissue factors from one monkey to the other.

In between, pharmacologists, histochemists and biochemists exhibited a growing interest for the problems of peripheral and central neurohumoral coding. It had been recognized that many activating mechanisms (among them those inducing alertness) are transmitted or modulated by certain amino-acids, biogenic amines (such as catecholamines) and by acetylcholine (Monnier and Herkert, 1972). On the other hand, it had been established that many inhibitory mechanisms (among them those inducing sleep) are also transmitted or modulated by certain amino-acids or biogenic amines like serotonin. The important works of Hernández-Peón et al. (1967), Jouvet (1969), Koella et al. (1968), Myers and Beleslin (1970), and Drucker-Colin *et al.* (1970) improved the climate for a better understanding of the neuro-humoral regulation of sleep. Concurrently, the chances for a successful cooperation of neurophysiologists and biochemists considerably increased.

D. Isolation of the hypnogenic fraction of sleep dialysate (sleep factor delta)

Extracorporal dialysis had permitted a first fractionation of the hypnogenic blood plasma by the pore size of the dialyzing membrane (20-30 Å). It was decided to carry further the fractionation of this sleep inducing hemodialysate.

Preliminary experiments showed that repeated freezing and thawing of the primary sleep dialysate (for lyophilization and purification purposes) reduced the hypnogenic activity; heating at 100°C for 30 min. or standing at extreme pH overnight even abolished it (Monnier and Schoenenberger, 1972). The results of these experiments on the lability of the hypnogenic dialysate and those of Pappenheimer (factors in the cerebrospinal fluid from goats affecting sleep and activity in rats) suggested that the hypnogenic factor found in both fluids - hemodialysate and cerebrospinal fluid - might be a compound of low molecular weight (Fencl et al., 1971; Monnier and Schoenenberger, 1972). The chemical nature of the hypnogenic substance remained, however, unknown.

Fractionation of the sleep dialysate by Gel filtration, (Sephadex G-10). The sleep dialysate was first purified by Gel filtration over a Sephadex G-10 column. The fractions thus obtained - one desalted fraction of higher molecular weight (MW) and a salt fraction of lower MW - were again tested on the recipient. Only the desalted fraction (Pool I) had hypnogenic effects, attributed to a 'sleep factor delta'; when infused to the recipient, in a series of 6 animals, it induced a delta increase of 204% as an average, against about 108 % in 11 control recipients which had received a control fraction (Fig.4A; Monnier et al., 1973).

Correspondingly, the motor activity in the kinesigram fell to 46%, against 103% in the control animals (Fig.4B). Therefore, there is complete agreement between the results of the EEG and behaviour tests. By contrast, the intraventricular infusion of the salt fraction (Pool II) of sleep dialysate did not increase the average delta amount. The value remained low (95%) as in the control animals (92%) (Monnier et al., 1972).

The effects of the hypnogenic desalted fraction from sleep hemodialysate on the chief visceral activities (blood circulation, respiration) were also tested in order to ascertain whether they are primary or secondary (side) effects. During the infusion and the increased delta activity, the arterial blood pressure recorded with a catheter in the femoral artery remained constant. The heart rate, on the ECG, showed a slight, but significant decrease similar to the bradycardia observed in physiological sleep; this phenomenon was not observed in the control animal. Finally, the respiration rate, recorded with a thermo-

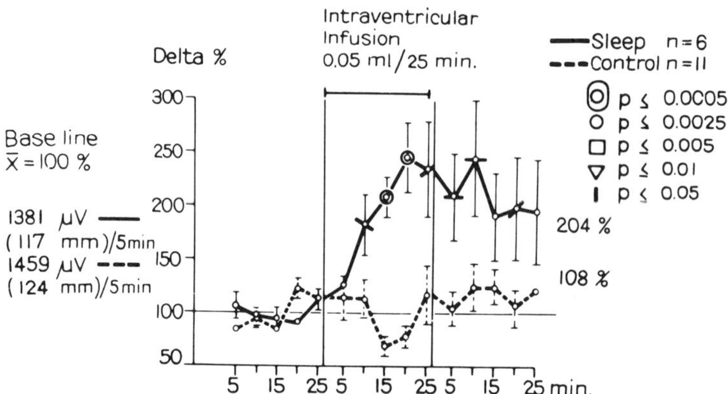

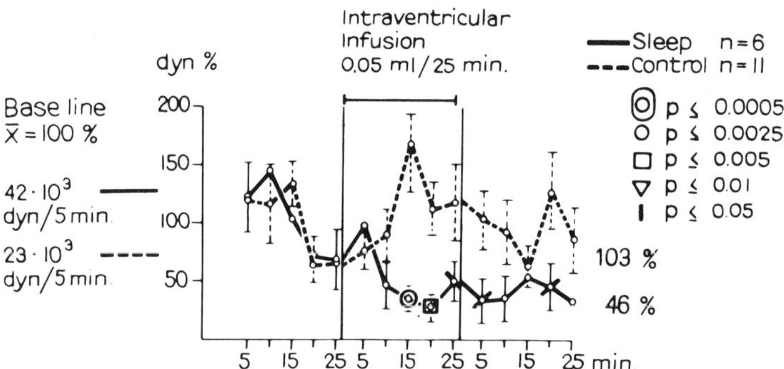

Fig. 4. Isolation of the sleep factor delta by Gel filtration of the primary sleep dialysate on Sephadex G-10

A. Electroencephalographic test: increased delta activity up to 204% during and after intraventricular infusion of the hypnogenic desalted fraction Pool I.

B. Kinesigraphic behavior test: decreased motor activity down to 46% due to the sleep factor delta (After Monnier et al., 1972; Monnier et al., 1973).

couple in front of the snout, also remained unaltered.

From these observations, it follows that the hypnogenic action of the desalted fraction of the primary sleep dialysate is not a secondary "side" effect of visceral changes. Sleep is a primary, specific effect of the dialysate, as is also the bradycardia, i.e. the slowing down of the heart rate, which always accompanies normal sleep.

We may therefore conclude that the desalted, higher MW fraction of the primary sleep dialysate contains a specific hypnogenic factor, which we called 'sleep factor delta'. In the behavioural tests, performed in freely moving recipients, the progressive development of sleep, its deepening and prolongation with decreased motor activity and slowing down of the heart rate, and its rapid reversibility under the influence of waking factors correspond to what may be expected of a physiological sleep substance.

E. Physical-chemical characterization of the sleep factor delta

It was then decided to follow up further fractionation by the Ninhydrin-reaction. The Ninhydrin-positive materials were tested in bio-assays in adequate amounts, calculated by weight in respect to the weight found to be active in the original dialysate. All work was done at 4°C. Metal-free distilled water and acid-clean glassware were used. Care was taken to ensure an adequate control by testing, at each isolation step, the solvents without material. All experiments were carried out under "double blind" conditions. Thus, the physiologists testing the incoming fractions during the isolation procedure did not know which material they analysed.

a. Fractionation by thin layer chromatography (TLC). The desalted material was separated by thin layer chromatography (TLC) on Silica Gel, developed with acetone/water 7 + 3. Six Ninhydrin-positive fractions with different Rf-values were separated (Fig.5A; Schoenenberger et al., 1972).

Only one fraction (Band 2) from the starting point, contained the hypnogenic activity, with an increased delta amount of 160%, in contrast to 75% in the control recipients of other fractions (Fig.5B). The analytical

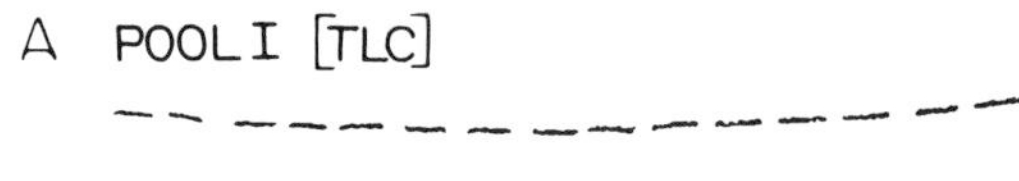

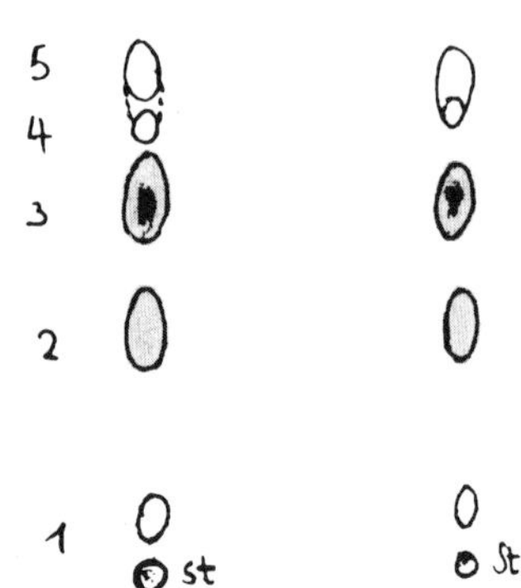

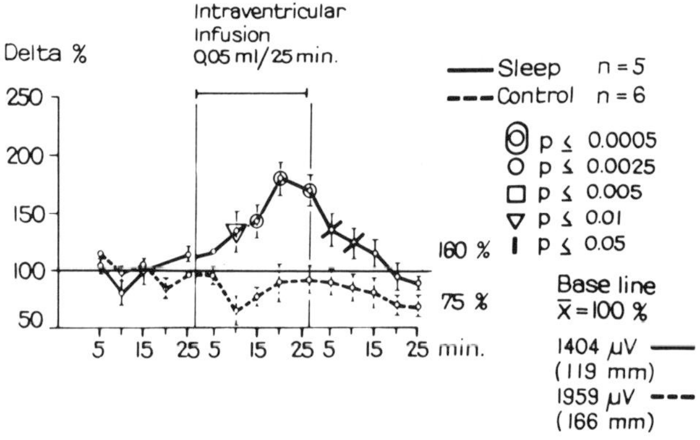

Fig. 5. Physical-chemical characterization of the sleep inducing factor delta (Schoenenberger, Cueni, Monnier and Hatt, 1972).

A. Thin layer chromatography (TLC) of the hypnogenic fraction Pool I (G-10). Bands 1-5. Running medium: Acetone:H_2O (7+3).

B. Sleep induced by infusion of the hypnogenic Band 2 eluate from Pool I. Increased delta activity (160%) in contrast to control eluates from other bands (75%).

high voltage electrophoresis of this fraction revealed the presence of 6 Ninhydrin-positive substances.

b. Fractionation by Gel filtration on Sephadex G-15. A further fractionation of Band 2 on Sephadex G-15 elicited 3 distinct Ninhydrin-positive peaks emerging from the column (Fig.6A). Only Peak 1, with the smallest V_e (solid line), which was not present in the TLC fractions 1, 3, 4, 5 and 6 of Fig.5A contained the whole hypnogenic activity. Infusion of this Peak 1 material to the recipients increased the delta amount up to 199% against 80% in the recipients of other fractions (peaks 2 and 3; Fig.6B). The relation V_o (Blue Dextran) $>$ V_e (Peak) suggested, by the characteristics of the Gel, that the hypnogenic peak has a molecular weight below 1500.

c. Fractionation by rechromatography on Sephadex G-15. A more precise molecular weight of the hypnogenic material was obtained by rechromatography of Peak 1 over a longer and pressure-packed G-15 column. Calibration and conditions of the column are indicated on Fig.7A (Schoenenberger et al., 1972). A peak with two shoulders (1a + 1b) was eluted. The first shoulder 1a, containing the hypnogenic material, was eluted with a V_e smaller than that of the peptide marker (MW 355). Its molecular weight might therefore be above 355. The increase in delta activity reached 196%, in contrast to the value in control recipients (98%; Fig.7B).

d. Fractionation by high voltage paper electrophoresis. As a quantitative separation of the two shoulders by the previous technique had proved unsuccessful, a preparative fractionation by high voltage electrophoresis was carried out on Whatman 2 MM paper and acid pH. An analytical example of this separation, later on used for preparative purposes, is shown on Fig.8A. Four electrophoretically different Ninhydrin-positive fractions were present. The hypnogenic activity resided only in fraction II from the starting point.

This fraction produced an increase of 170%, as compared to 71% for the controls (Fig.8B). In this graph, the electrophoretic fractions I, III and IV, also eluted quantitatively from the same paper, were taken as control.

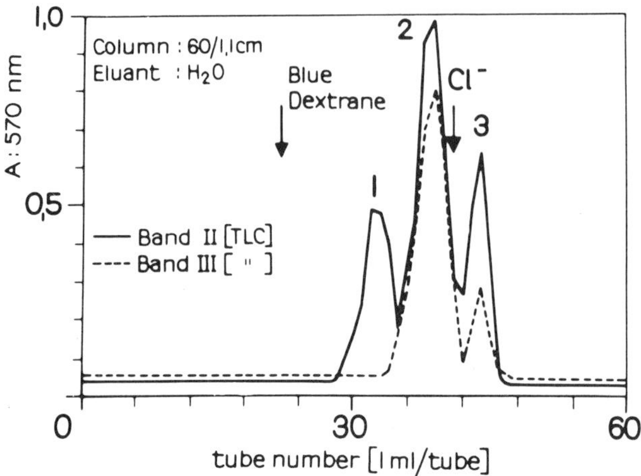

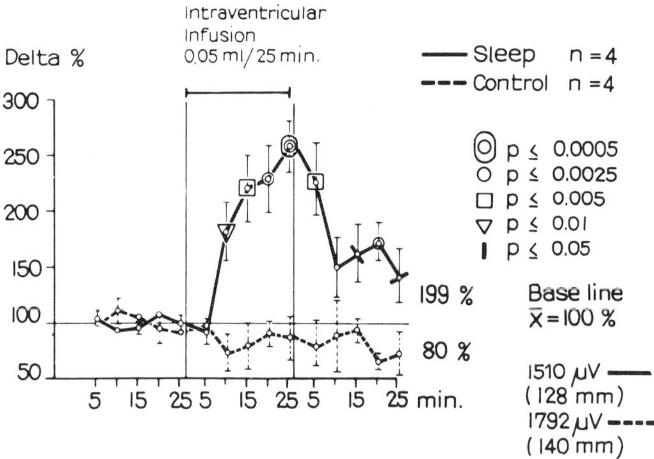

Fig. 6. Gel chromatography of Band II (——) and Band III (---) on Sephadex G-15 (Schoenenberger et al.,1972)

A. Separation of the hypnogenic Band II fraction into three peaks 1, 2, 3, with different V_e. The non hypnogenic fraction Band III contains only peaks 2 and 3.

B. Sleep induced by infusion of Peak 1 fraction (higher molecular weight) from band II. Increased delta activity (199%) in contrast to peaks 2 and 3 (low MW) control fractions (80%).

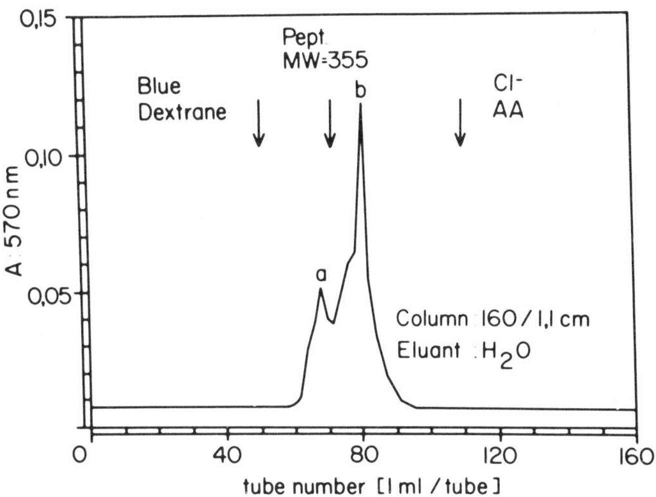

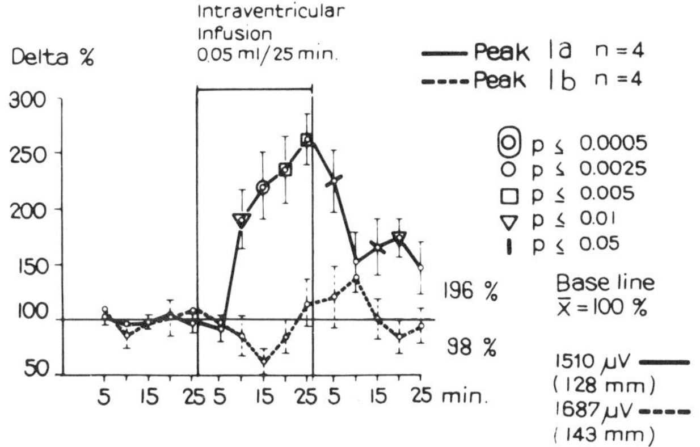

Fig. 7. Rechromatography of Peak 1 fraction on Sephadex G-15 (Schoenenberger et al., 1972)

A. Estimation of the molecular weight of the hypnogenic fraction peak 1 and separation into subfractions Peaks 1a and 1b. Reference substances: blue dextran ($MW > 2 \times 10^6$). Peptide marker (Gly-Leu-Met-NH_2 . HCl (MW 355), Alanine (MW 89).

B. Sleep induced by infusion of subfraction 1a (196% delta) in contrast to subfraction 1b (98%).

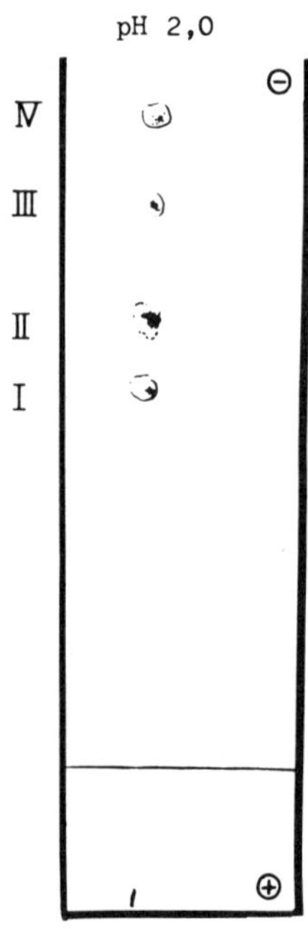

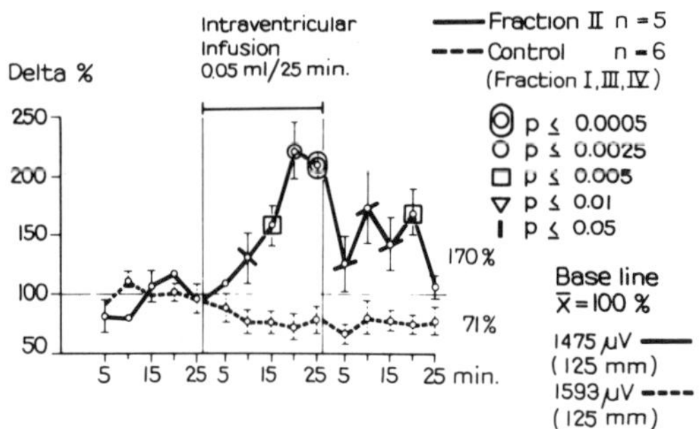

Fig. 8. A. Separation of Peak la material into 4 Ninhydrin-positive spots (I - IV) by high voltage paper electrophoresis: 2000 V/45 mA/75 min.

B. Sleep with increased delta amount (170%) induced only by infusion of the hypnogenic fraction II (——), in contrast to fractions I, III and IV used as controls (---).

Amino-acid analysis. The amino-acid analysis of the hypnogenic Ninhydrin-positive peak obtained from Sephadex G-15 revealed, after hydrolysis, seven additional amino-acid residues, as compared to the analysis of the same amount of the identical material before acid hydrolysis, namely alanine, aspartic acid, glutamic acid, glycin, leucine, threonine and serine.

After paper electrophoresis, the hypnogenic fraction II was dansylated (Gray, 1967) before hydrolysis and underwent a two-dimensional chromatography/electrophoresis. A single dansyl-positive spot, not identifiable as a free amino-acid, was eluted from the plate, hydrolyzed and redansylated, followed by a second electrophoretic/chromatographic separation. Here again, seven different DNS-amino-acid residues were liberated by the hydrolysis.

Thin layer chromatography controls showed that the active compound was not identical with -γ- OH-butyrate (γ- Hydroxybutyric-Natrium salt), -L-Glu (L-Glutamic acid), GABA (γ- Aminobutyric acid), 5HT-Hydrogenoxalate (Serotonin-Hydrogenoxalate), 5-HTP (DL-5-Hydroxytryptophan), L-DOPA (L-2-(3,4-Dihydroxyphenyl)-Alanine.

The molecular weight between 355 and 1500, most likely around 700, confirmed the previous assumption that the hypnogenic activity of the sleep dialysate must be related to a specific, labile organic compound of low molecular weight (Monnier and Schoenenberger, 1972).

The effective dose estimated on the base of a molecular weight of 700 was about 5.0 ng/kg for an intraventricular infusion over 25 minutes. These facts suggested the peptide nature of the highly active sleep factor delta, found in the extracorporal hemodialysate from sleeping rabbits.

The characteristics of the 'sleep factor delta', isolated from the hemodialysate of sleeping rabbits, differ slightly from those of the sleep promoting 'factor S' obtained by Pappenheimer from the cerebrospinal fluid of sleeping goats. This 'factor S' obtained after filtration through molecular sieves had a molecular weight below 500 (Fencl et al., 1971).

G. Summary and conclusions

Preliminary experiments were necessary for inducing sleep in rabbits by electrical stimulation of the thalamic sleep area of W.R. Hess and for quantifying this sleep by electroencephalographic and kinesigraphic (behaviour) tests. The rabbit was chosen as experimental animal because, belonging to the lagomorph species, it provides an optimal orthodox delta sleep.

Humoral transmission of sleep was first demonstrated in cross-circulation experiments in rabbits, then by intravenous injection of hemodialysate from sleeping rabbit donors to awakened rabbit recipients.

The pore size of the dialysing membrane used for extracorporal hemodialysis in sleeping rabbits permitted a first fractionation of the sleep substance from the blood by excluding disturbing blood components with molecular weights above 40,000.

In order to facilitate further isolation of the sleep substance, the technique for producing pure sleep dialysate, i.e. with a minimum of waking contaminants, was improved, as well as the technique for testing the hypnogenic activity of the sleep dialysate or of its subsequent fractions. The intraventricular infusion of this material in very small doses to rabbit recipients was the most reliable physiological test used for monitoring further fractionation.

Gel filtration of the primary sleep dialysate on Sephadex G-10 revealed that the hypnogenic activity is present only in the desalted (higher molecular weight) fraction. After analysis of its cerebral and visceral effects, this activity was attributed to a specific sleep factor delta.

The physical-chemical characterization of this factor by numerous techniques (thin layer chromatography, gel filtration on Sephadex G-15, rechromatography on Sephadex G-15, high voltage paper electrophoresis) showed that the hypnogenic activity was associated with a Ninhydrin-positive material throughout all fractionation steps. The molecular weight of the hypnogenic substance lies between 355 and 1500, most likely around 700. The last hypnogenic fractions submitted to amino-acid analysis and dansylation contained seven different amino-acid residues. No other fraction submitted to identical contact with solvents and preparation techniques showed any hypnogenic activity.

The effective dose, estimated on the base of a molecular weight of 700, was about 5.0 ng/kg for an intraventricular infusion over 25 min.

These properties argue in favor of the peptide nature of the highly active sleep factor delta, isolated from the extracorporal hemodialysate of sleeping rabbits.

The final characterization by structural analysis and synthesis of the physiological sleep substance will definitely enlarge our knowledge of the neuro-humoral coding of sleep.

Acknowledgement. We should like to thank Dr. R.O. Studer, Head of the Protein Research Group (Hoffmann-La Roche AG, Basel) and his co-workers for their technical assistance in carrying out the amino-acid analyses. We are also very grateful to Hoffmann-La Roche AG, the CIBA Foundation (Basel), the Sandoz AG (Basel) and the 'Fonds zur Förderung von Lehre und Forschung, Universität Basel' for their financial support, to W. Mehlhose and V. Zabotto for their technical help. The chemical investigations were partially carried out with financial support from the Swiss National Research Fund (Nr. 3.373.70).

REFERENCES

Dement, W., 1958, The occurrence of low voltage, fast electroencephalogram patterns during behavioral sleep in the cat, EEG clin. Neurophysiol. 10:291.

Drucker-Cólin, R.R., Rojas-Ramirez, J.A., Vera-Trueba, J., Monroy-Ayala, G., Hernández-Peón, R., 1970, Effects of crossed perfusion of the midbrain reticular formation upon sleep, Brain Res. 23:269.

Fencl, V., Koski, G., and Pappenheimer, J.R., 1971, Factors in cerebrospinal fluid from goats that affect sleep and activity in rats. J. Physiol. (London) 216: 565.

Gray, W.R., 1967, Dansyl chloride procedure, in 'Methods in Enzymology' (C.H.W. Hirs, ed.), pp.139-151, Academic Press, New York-London.

Hernández-Peón, R., O'Flaherty, J.J., Mazzuchelli-O'Flaherty, A.L., 1967, Sleep and other behavioural effects induced by acetylcholine stimulation of basal temporal cortex and striate structure, Brain Res. 4:243.

Hess, W.R., 1924/25, Ueber die Wechselbeziehungen zwischen psychischen und vegetativen Funktionen, Schweiz.Arch. Neurol.& Psychiat.15:250; 16:36+285.

Hess, W.R., 1927, Das Schlafzentrum, Verh.Schweiz.Naturforsch.Ges.247:889.

Hess, W.R., 1929, Lokalisatorische Ergebnisse der Hirnreizversuche mit Schlafeffekt, Psychiatr. Nervenkr. 88:813.

Hess, W.R., 1944, Das Schlafsyndrom als Folge diencephaler Reizung, Helv.Physiol.Pharmacol.Acta 2:305.

Hösli, L., and Monnier, M., 1962, Schlaf- und Weckwirkungen des intralaminären Thalamus, Pflügers Arch. 275:439.

Ingvar, D.H., 1955, Extraneuronal influences upon the electrical activity of isolated cortex following stimulation of the reticular activating system, Acta Physiol. Scand. 33:169.

Jouvet, M.,1965, Paradoxical sleep - a study of its nature and mechanisms, in 'Progress in Brain Research'(Akert, Bally and Schadé, ed.),pp.20-57, Elsevier, Amsterdam.

Jouvet, M., 1969, Biogenic amines and the states of sleep, Science, 163:32.

Koella, W.P., Feldstein, A. and Czicman, J.S., 1968, The effect of parachlorophenylamine on the sleep of cats, EEG clin. Neurophysiol. 25:481.

Koller, Th., Monnier, M. and Gamp, R., 1964, Technik der Kanülierung des Confluens sinuum beim Kaninchen für Versuche mit partiellem extra-korporellem Kreislauf, Experientia (Basel) 20:108.

Kornmüller, A.E., Lux, H.B., Winkel, K. and Klee, M.,1961, Neurohumoral ausgelöste Schlafzustände an Tieren mit gekreuztem Kreislauf unter der Kontrolle von EEG-Ableitungen, Naturwissenschaften 48:503.

Krupp, P., and Monnier, M., 1963, Inhibitory and excitatory action of the intralaminary thalamus on single cortical neurones, Exper. Neurol. 7:24.

Kuhn, W., Majer, H., Heusser, H., and Zen Ruffinen, B., 1957, Künstliche Niere mit Capillarsystem für den Stoffaustausch, Experientia (Basel) 13:469.

Legendre, R. and Piéron, H., 1910, Les résultats histophysiologiques de l'injection intra-occipito-atlantoïdienne des liquides insomniques, C.R.Soc.Biol. Paris 68:1108.

Legendre, R., and Piéron, H., 1913, Recherches sur le besoin de sommeil consécutif à une veille prolongée, Z.allg.Physiol. 14:235.

Matsumoto, J., Sogabe, K., and Hori Santiago, Y., 1972, Sleep in parabiosis, Experientia (Basel) 28:1043.

Monnier, M., 1950, Action de la stimulation électrique du centre somnogène sur l'électro-corticogramme chez le chat (Réactions hypniques et réactions d'éveil). Rev. Neurol. 83:561.

Monnier, M. and Gangloff, H., 1961, Atlas for stereotaxic brain research on the conscious rabbit, Elsevier, Amsterdam.

Monnier, M. and Hatt, A.M., 1970, Intraventricular infusions in acute and chronic rabbits. Application of the stereotaxic method of Monnier and Gangloff, Pflügers Arch. 317:268.

Monnier, M., and Hatt, A.M., 1971, Humoral transmission of sleep. V. New evidence from production of pure sleep hemodialysate, Pflügers Arch. 329:231.

Monnier, M., Hatt, A.M., Cueni, L., and Schoenenberger, G.A., 1972, Humoral transmission of sleep. VI. Purification and assessment of a hypnogenic fraction of 'sleep dialysate' (factor delta), Pflügers Arch. 331:257.

Monnier, M., Hatt, A.M., Dudler, L., Cueni, L.B., and Schoenenberger, G.A., 1973, Humoral transmission of sleep. I. Cerebral, behavioural and visceral effects of a purified sleep factor delta, lrst Europ. Congr. Sleep Res., Karger, Basel (in press).

Monnier, M., and Herkert, B., 1972, Concentration and seasonal variations of acetylcholine in sleep and waking dialysates, Neuropharmacol. 11:479.

Monnier, M. and Hösli, L., 1961, Les paramètres de stimulation des systèmes somnogènes synchronisants du thalamus médian et de l'appareil réticulaire, Excerpta Medica, Int. Congr. Series 37 (Reports and Free Communications, pp.58-59.

Monnier, M., and Hösli, L., 1964, Dialysis of sleep and waking factors in blood of the rabbit, Science 146:796.

Monnier, M., and Hösli, L., 1965, Humoral transmission of sleep and wakefulness. II. Hemodialysis of a sleep inducing humor. Pflügers Arch. 282:60.

Monnier, M., Koller, Th., and Graber, S., 1963, Humoral influences of induced sleep and arousal upon electrical brain activity of animals with crossed circulation. Exper. Neurol. 8:264.

Monnier, M., and Schoenenberger, G.A., 1972, Some physical-chemical properties of the rabbit's sleep hemodialysate, Experientia (Basel) 28:32.

Myers, R.D., 1967, Transfusion of cerebrospinal fluid and tissue bound chemical factors between the brains of conscious monkeys; a new neurobiological essay, Physiol. Behav. 2:373.

Myers, R.D., and Beleslin, D.B., 1970, The spontaneous release of 5-hydroxytryptamine and acetylcholine within the diencephalon of the unanaesthetized rhesus monkey, Exp. Brain Res. 11:539.

Pappenheimer, J.R., Miller, T.B., and Goodrich, C.A.,1967, Sleep promoting effects of cerebrospinal fluid from sleep deprived goats, Proc.nat.Acad.Sci.(Washington) 58:513.

Pavlov, J., 1923, Innere Hemmung der bedingten Reflexe und der Schlaf, ein und derselbe Prozess, Scand.Arch. Physiol. 54:42.

Purpura, D.P., 1956, A neurohumoral mechanism of reticulo-cortical activation, Amer. J. Physiol. 186:250.

Schnedorf, J.G., and Ivy, A.C., 1939, An examination of the hypnotoxin theory of sleep, Amer. J. Physiol. 125:491.

Schoenenberger, G.A., Cueni, L.B., Monnier, M., and Hatt, A.M., 1972, Humoral transmission of sleep. VI. Isolation and physical chemical characteristics of the sleep inducing factor delta, Pflügers Arch. 338:1.

Tissot, R., and Monnier, M., 1959, Dualité du système thalamique de projection diffuse, EEG clin. Neurophysiol. 11:675.

ABSTRACT:

Sleep dialysate was obtained by extracorporeal dialysis in rabbit donors kept asleep by stimulation of the thalamic sleep area. The hypnogenic effect of the dialysate and of its subsequent fractions was tested by infusion into the mesodiencephalic ventricular system of rabbit recipients submitted to a quantitative electroencephalographic analysis of the cortical delta activity and kinesigraphic analysis of the motor behavior.

Sleep dialysate was separated by Gel filtration on Sephadex G-10 into 2 fractions: a salt-free fraction of higher M.W. (Pool I) and a salt fraction of lower M.W. (Pool II). The effects of these fractions were compared to those of the fresh or of the lyophilized sleep dialysate. Infusion of Pool I induced a marked, significant increase of the slow delta activities, symptomatic of orthodox delta sleep and a corresponding sleep behavior with decreased motor activity. The specific hypnogenic activity of Pool I was higher than that of the sleep dialysate before fractionation. By contrast, infusion of Pool II neither increased the amount of delta activities, nor induced a sleep behavior. The hypnogenic effect of the desalted Pool I (30 µg/0.05 ml/25 min) suggests the existence of a "sleep factor delta" in this fraction. Its hypnogenic property and the concomitant decrease in heart rate are specific effects and not visceral side effects, since the pressure of the cerebrospinal fluid and of the arterial blood remained unchanged, as well as the respiration rate, during intraventricular infusion to recipient rabbits.

The hypnogenic fraction Pool I was further separated into 6 different Ninhydrin-positive fractions by thin layer chromatography. The tests in recipient animals showed that the hypnogenic activity resided in one of them only. After rechromatography on Sephadex G-15, this material emerged as 3 different Ninhydrin-positive peaks. The hypnogenic activity was found only in the material corresponding to the first peak, emerging with a Ve between Blue dextran and aminoacids. Rechromatography of this peak over an analytical G-15 column calibrated with peptides of different molecular weights suggested a molecular weight around 1000. High voltage paper electrophoresis revealed 4 different Ninhydrin-positive fractions in this active material. After elution from the paper, the hypnogenic activity was found only in one of these fractions. After acid hydrolysis, this active fraction contained at least seven different aminoacids. The peptide nature of the compound apparently associated with hypnogenic activity (sleep factor delta) is thus suggested.

THE POSSIBLE NATURE OF SLEEP INDUCING BRAIN PERFUSATES: THEIR RELATIONSHIP TO SEIZURE INHIBITION

R. R. Drucker-Colín*

Instituto Miles de Terapéutica Experimental

Apdo. Postal 22026, México 22, D.F.

INTRODUCTION

The search for a relationship between a particular neurotransmitter and a certain type of behavior has become somewhat of a scientific vogue. Yet, despite the vast literature on the subject and despite some direct demonstrations that neurohumors are released as a result of synaptic activity or nerve function (McLennan, 1964; Baldessarini and Kopin, 1966; Szerb, 1967; Yaksh and Myers, 1972), there is little evidence showing that a particular behavior or behavioral component is induced as a result of a neurohumor acting locally at a brain site. As Kety (1967) has suggested, when we come to behavioral-neurohumoral relationships, we find that most of the evidence depends on indirect pharmacological studies, whose drawbacks are many, with perhaps the most important one being related to the multiplicity of effects induced by any one drug.

*Present Address: Department of Psychobiology
University of California
Irvine, California

When we turn to the area of sleep research, we find a similar pattern. First of all, in the past decade or so, a great deal of the sleep researchers' time has been devoted to the search for the neurohumors associated with sleep. And secondly, despite the ingenious studies of Jouvet and many others, we still seem to be far from knowing which are the hypnogenic substances. Perhaps, here again, the neuropharmacological approach is too indirect to give us clear-cut answers, and it may be useful for us to turn to more direct approaches.

In the past few years the "push-pull" cannula technique, developed originally by Gaddum (1961), has found its way into some laboratories, where attempts have been made to extract "active" substances from particular brain sites, as they are presumably released into the extracellular fluid in connection with particular physiological states (Myers and Beleslin, 1971).

Since sleep has such well defined electrophysiological and behavioral characteristics, there is good reason to believe that the "push-pull" technique may be particularly fruitful for the study of the neurochemistry of sleep, more so if we seriously take into consideration Moruzzi's (1972) recent suggestion that the existence of a sleep-wakefulness cycle "is likely to be due to an accumulation and dissipation of chemical substances within well defined groups of neurons". This suggestion is really a revival of the old idea of Pieron (1913) that in wakefulness there accumulates a substance which is eliminated during sleep. Moreover, this idea was put to experimental testing by Legendre and Pieron (1911). They showed that cerebrospinal fluid (CSF) obtained from sleep-deprived dogs induced signs of drowsiness and sleep when injected into the cisterna magna of normal dogs. Since then a few investigators have replicated these studies with similar techniques. Schnedorf and Ivy (1939) confirmed Legendre and Pieron's observations in 9 out of 24 experimental dogs. More recently, Pappenheimer _et al._ (1967) obtained perfusates from sleep deprived goats whose ventricular systems were cannulated. When the perfusate was injected into the ventricles of cats or rats, it induced clinical signs of sleep (no EEG was taken) lasting sometimes up to 18 hours. They further reported the sleep inducing factor to be a dialyzable substance with a molecular weight of less than 500. Ringle and Herndon (1969), however, were unable to demonstrate a similar effect with CSF extracted from sleep deprived rabbits. Perhaps the most critical part of these studies is the lack of EEG recordings, which make these observations difficult to interpret, in view of the many postural states which, upon ocular observation, may resemble sleep.

A different approach has been attempted by Monnier and his collaborators. They have used a crossed circulation arrangement between two rabbits, and have shown that sleep induced by electrical stimulation of the medial thalamus of a donor rabbit releases a blood-borne factor capable of producing sleep in the recipient rabbit (Monnier and Hosli, 1964; Monnier and Hatt, 1971). They have further shown that the sleep factor is of a peptide nature whose molecular weight lies somewhere between 355 and 1500 (Schoenenberger <u>et al</u>. 1972). Monnier's results, however, are not consistent with a recent study of Lenard and Schulte (1972) who have shown that twins born extensively conjoined at the skulls and with a common circulation have independent sleep and waking cycles. These findings make the postulation of an extraneuronal sleep inducing factor difficult to interpret.

We have in our laboratory taken a different approach based on the assumption that the sleep-wakefulness cycle can be accounted for in terms of a reciprocal inhibition between two antagonistic systems. We suggest that either through inhibition, release of inhibition or an interplay of both, sleep ensues as a result of reticular deactivation, and that this effect is mediated by some neurotransmitter. Should this reasoning be correct, it should be possible to extract from the MRF a substance presumably released into the extracellular fluid during sleep, and test the effect of this substance on the same MRF neurons of an awake animal. Eventually, of course, extraction of this substance could lead to its identification.

In previous studies we perfused sleep-deprived cats with "push-pull" cannulae implanted in the MRF, and showed that the perfusates obtained from cats during rebound of sleep induced sleep upon awake recipient cats having "push-pull" cannulae in the homologous MRF (Drucker-Colîn <u>et al</u> 1970). The experiments to be reported below are an extension of these observations and include experimental evidence suggestive of a neurohumoral relationship between sleep and seizure inhibition.

CROSSED PERFUSION AND SLOW WAVE SLEEP (SWS)

With the aid of the stereotaxic technique, a specially devised "push-pull" cannula system was implanted in the MRF of cats. In addition, electrodes were implanted in the MRF of cats. In, addition electrodes were implanted, allowing electrophysiological recordings of sleep. For further methodological details see Drucker-Colín (1973). Animals were randomly assigned the role of either donors or recipients, donor cat (D) being one from which "push-pull" perfusates were obtained during either sleep or wakefulness, and a recipient cat (R) being one through whose "push-pull" cannula the donor's perfusate was tested.

Since cats are animals which seem to fall asleep very easily, the purpose of these experiments was to show that perfusates obtained from the MRF of sleeping donor cats do in fact accelerate the onset of slow wave sleep (SWS) and fast wave sleep (FWS) in awake recipients, when compared to experimental sessions in which these same recipients were perfused with either donor's awake perfusate, Ringer or not perfused at all.

Thus the EEG of recipient cats were recorded for 5 consecutive days, after which on the next two consecutive days they were perfused with either donor's sleep perfusate or awake perfusate. They were then recorded for another 5 consecutive control days, and finally on the following day they were perfused with Ringer. For each animal all recordings were done at the same time of the day, and total recording time was 60 min. Perfusions into recipients were always done for 15 min only, at a speed of 40-60 μl/min. Figure 1 illustrates the results of these experiments. The sleep perfusate significantly decreased the latency and increased the durations of SWS ($p < 0.001$), but had no effect upon FWS. The effects of the awake perfusate were sufficiently different from those of the sleep perfusate so that it maintained the animal at least as awake as controls. In addition in all 11 cats tested no FWS appeared for at least one hour (which represented the total recording session).

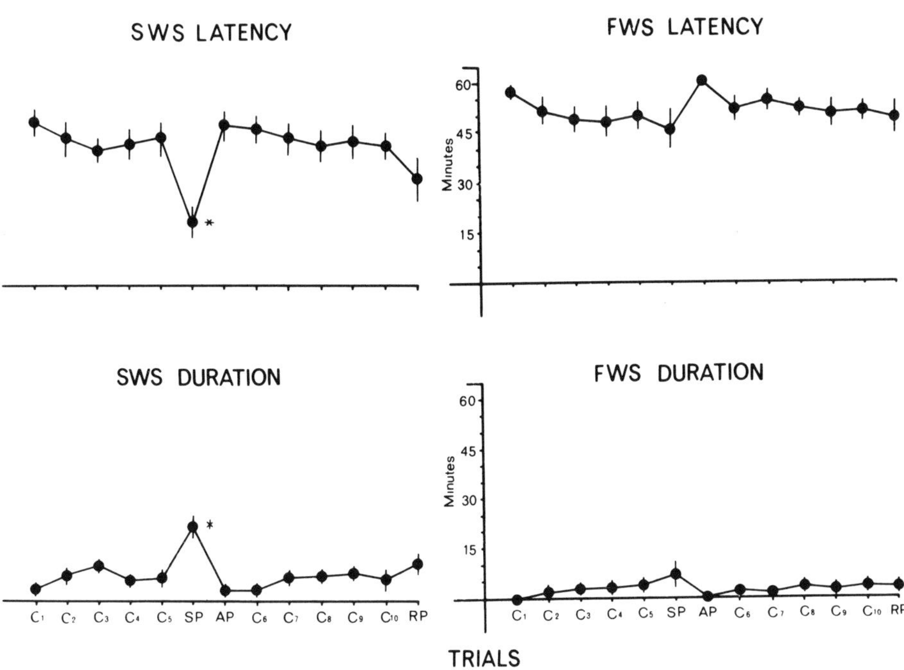

Fig. 1. Graph representing the means and standard errors of latencies and durations of SWS and FWS. Each point represents the same 11 cats during the various control and perfusion sessions *Significant ($p < .001$, one tail test). C: control; SP: sleep perfusate; AP: awake perfusate; RP: Ringer perfusion. (From Drucker-Colín, 1973).

These results are most likely due to a local effect exerted by the sleep perfusates upon the MRF, since histological material showed that after perfusions with Janus Green, which stains the tissue in dark blue, the stained area never exceeded 3 mm (for 2 hrs perfusion time), and was usually in the order of 1-2 mm. At any rate, even if the sleep perfusates are injected into the recipients rather than "push-pulled", thus allowing for diffusion, the cats still show a reduced sleep latency (Drucker-Colín, 1972, 1973).

From our experiments two things become apparent. First of all, although we would all agree that a cat's spontaneity to sleep is great, we believe that what is

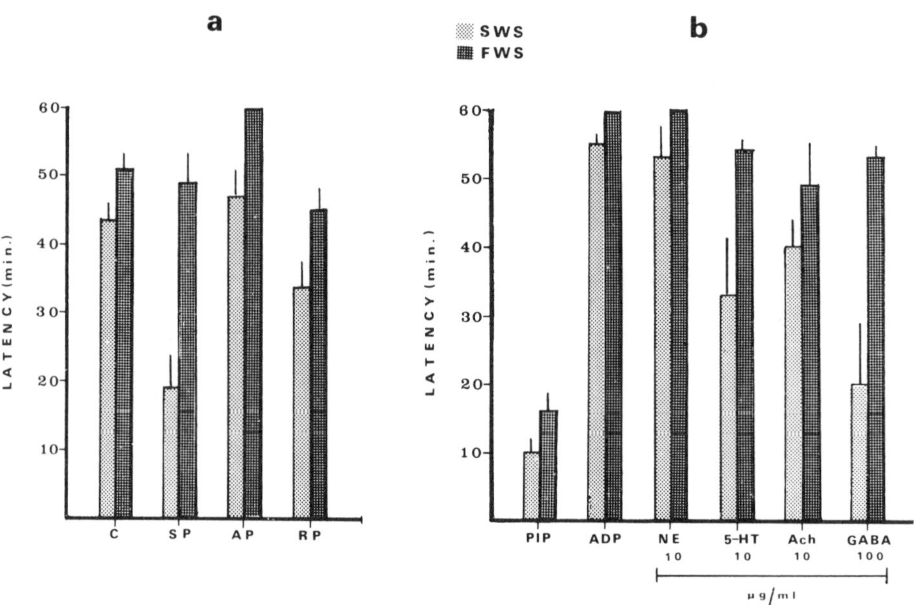

Fig. 2. Means and standard deviations of SWS and FWS latencies of recipient cats. a: represents comparison between SP, AP, RP perfusion sessions (n=11 for each) and all control sessions (n=110), regardless of the chronological order in which they were run. b: represents the effects of the post-ictal perfusates (PIP, n=6); after-discharge perfusates (ADP, n=6); 5-HT (n=9); ACh (n=8) and GABA (n=5). Note great latency decrease due to PIP.

usually meant by this is that if we record a cat during 24 hours it will sleep most of the time. However, if we measure the onset of sleep with EEG recordings beginning immediately after placing the animal in the recording cage, sleep is no longer so greatly spontaneous. We have calculated the mean and standard deviations of all control sessions (110) and compared it to sleep perfusate sessions (11), awake perfusate sessions (11) and Ringer sessions (11). These data, which are shown graphically in Figure 2a, mean that if we take a cat from its home cage, place it into the recording cage, immediately plug it to the EEG, and from there determine the time it will take it to fall asleep, the most likely figure will be somewhere around 40 min for SWS and 50 min for FWS. The remarkable stability of these control figures leaves no doubt that the sleep perfusates did in fact exert a sleep inducing effect.

The second observation which stands out is the fact that the sleep perfusates do nothing to the recipients' fast wave phase of sleep. This is true also even if perfusates are obtained during donor's FWS only (Drucker-Colín *et al.* 1970). On account of this we felt it was necessary to try on the MRF a different type of perfusate. We therefore developed another perfusion technique based on observations made in other laboratories, which have suggested that there is a close relationship between sleep and epilepsy.

SEIZURE INHIBITION AND FAST WAVE SLEEP (FWS)

According to Pompeiano (1969) generalized epileptic discharges are facilitated during SWS and blocked during FWS, while focal epileptic discharges are facilitated during both phases of sleep. Although the relationship between epileptic discharges and FWS seems contradictory, the differences may be reconciled if we assume that the facilitation of focal epilepsy during FWS is due to local disinhibition of interneurons in certain brain areas, resulting in focal non-spreading high frequency discharges (Pompeiano 1969), while the blocking of generalized epilepsy in FWS can be related to inhibition of the reticular activating system and invasion of such inhibition into the thalamus as suggested by Hernández-Peón (1969), thus depressing thalamo-cortical recruiting responses (Rossi *et al.* 1961; Yamaguchi *et al.* 1964). This hypothesis is supported by recent studies which have shown that thalamic lesions which block spindles also suppress seizures (Feeney and Gullotta, 1972; Kusske *et al.* 1972). To this we should add Berlucchi's findings (1965) showing that during FWS there is a decrease of the interhemispheric traffic of impulses coursing along the corpus callosum.

At any rate, whatever the explanation may be, it remains evident that a striking relationship exists between seizures and FWS.

Anyone who has worked with FWS deprived animals cannot but notice the sort of sub-convulsive state which appears during FWS recovery. Since convulsed animals seem to spend less time in FWS (Cohen and Dement, 1966), it could be possible that convulsions somehow substitute for FWS, from which it could follow that in the absence of FWS the brain would be in a subconvulsive state. In fact this is just what happens. Rats deprived of FWS show a clear drop in the threshold for electroconvulsive shock (Cohen and Dement, 1965), while entorhinal evoked potentials increase some 300% in FWS deprived cats. (Satinoff *et al.* 1971).

More interesting yet is the fact that when animals are deprived of FWS, then allowed to sleep, there is a FWS rebound. However, when ECS is given during sleep-deprivation, FWS rebound is prevented (Cohen et al 1967). Pompeiano (1969) has suggested that convulsions may therefore "use up" the substances responsible for FWS. This could well be, since it is interesting to note that during the first 10 days or so of the neonatal period of certain mammals, sleep is almost exclusively composed of FWS (Jouvet-Mounier *et al.* 1970), whereas seizures are very infrequent in newborn mammals, regardless of whether they are elicited by electrical means, by drugs, sound stimulation or fever (Lennox-Buchtal, 1973). Furthermore the FWS of newborns displays high frequency and intensity of phasic muscular contractions similar to that which occurs in adults during rebound after FWS deprivation (Ferguson and Dement, 1967).

On the other hand, since Fernândez-Guardiola and Ayala (1971) have given evidence that at the termination of generalized convulsions signs of FWS appear the alternative to Pompeiano's suggestion would be that a substance inhibiting convulsions could promote FWS. However, this is a contention difficult to hold in view of the fact that anticonvulsants such as diphenylhydantoin (DPH) significantly reduce FWS (Cohen *et al.* 1967).

However, since the termination of seizures has been suggested to be an inhibitory phenomenon (Gastaut and Fischer-Williams, 1959; Dow et al. 1962) most likely mediated by some neurotransmitters, maybe GABA (see Lovell, 1971), we have devised a technique for the study of the relationship between the inhibition of seizure and FWS, which we view in part as an inhibitory phenomenon.

Again we have divided our experimental animals into donors and recipients. However, whereas recipients had identical implants as the ones described above, donors were implanted with a modified "push-pull" system, which allows simultaneous EEG recordings and extraction or perfusion of substances in deep areas of the brain. Figure 3 illustrates our technique.

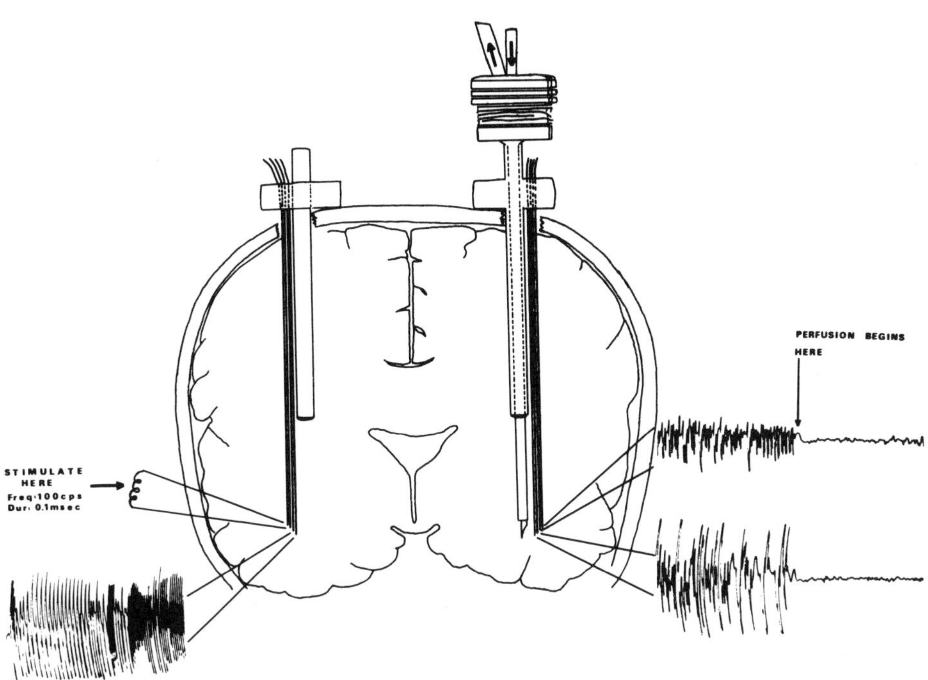

Fig. 3. Diagrammatic representation of the experimental arrangement allowing simultaneous stimulation, recording and perfusion of specific brain loci. PIP is obtained from the non-stimulated side the moment EEG silence appears. ADP can also be obtained if perfusion is done during the after-discharge period only.

Bilateral "push-pull" guide tubes with 4 pole electrodes are introduced stereotaxically into the brain (the hippocampus in our experiments) and fixed permanently. The electrodes are always located below the tip of the "push-pull" cannula guide. When perfusion is to be done the "push-pull" cannula is inserted through the guide tube, and lowered at the same level as the electrodes.

With this arrangement we are able to induce and record after-discharges (sub-clinical) from one side of the brain, while also record its spread to the other side. Should seizure extinction be due to the release of a neurohumoral agent, we should be able to collect such a substance through the "push-pull" cannula during the post-ictal period (i.e. after after-discharge termination). The important feature of this system is that perfusates are always obtained from the non-stimulated side, making the situation quite physiological. The problem is of course that post-ictal periods are very short; therefore many trials are needed in order to obtain a sufficient quantity of the perfusate.

At any rate, we obtained enough of such perfusates and tested them as to their effect when perfused for 15 min into the MRF of an awake recipient cat. To our delight they induced FWS in all cats tested. Figure 4 shows the EEG of one of the recipient cats, which manifested all the electrophysiological characteristics of REM sleep at min 6 after the beginning of perfusion of what we call the post-ictal perfusate (PIP). We tested six cats, and the mean latency for FWS sleep was 16.1 ± 2.52. This drop in FWS latency was significantly larger than any latency previously observed under any condition (see below) and was different from the effect induced by the after-discharge perfusates (ADP), which are perfusates obtained during the after-discharges period.

FWS duration was also changed significantly. Table I shows that the PIP perfusate nearly doubled the length of the first REM episode recorded, when compared to non-perfused or Ringer perfused cats. Our control figures for FWS duration in fact agree quite closely with those of Morgane and Stern (1973). In order to double check whether the drops in sleep latencies were not due to some artifact, we took our two best cats and recorded them for one hour each, once in the morning, once in afternoon, and once at night, for five consecutive days. Figure 5 shows the mean latency for their SWS and FWS: we can obviously see that REM latency does not nearly approximate our PIP value.

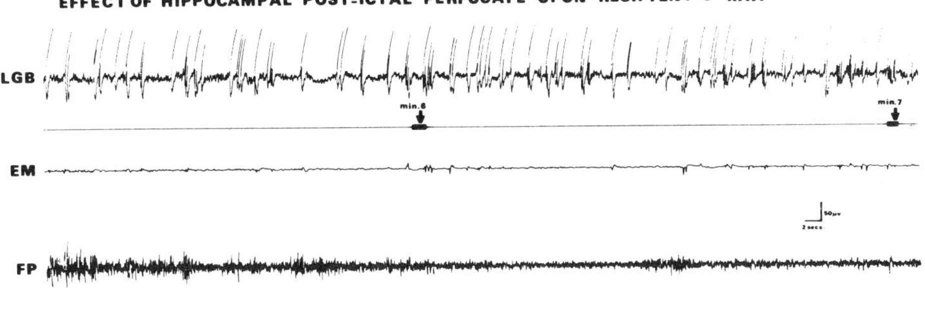

Fig. 4. Shows all the signs of FWS starting at minute 6 after beginning of PIP perfusion into the midbrain reticular formation of a recipient cat. LGB: lateral geniculate body; EM: eye movements; FP: fronto parietal cortex; EMG: electromyogram.

TABLE I: EFFECT OF POST-ICTAL PERFUSATE (PIP) UPON THE DURATION OF THE FIRST FWS EPISODE.

CAT No.	CONTROL (min)	RINGER (min)	PIP (min)
1	5.1	4.1	10.6
2	6.2	7.8	11.5
3	8.7	5.3	9.3
4	4.6	6.2	7.1
5	7.4	4.3	15.2
6	7.1	7.5	12.8
MEAN ± SD	6.5 ± 1.45	5.8 ± 1.47	11.0 ± 2.79*

* $p < 0.01$, two-tailed t test

We will recall that our initial assumption was that a substance having inhibitory properties upon convulsions could promote FWS. Since the results are highly suggestive of this assumption, the following experiments attempt to determine the possible nature of this substance. We perfused a number of known neurotransmitters into the donors, in order to determine their effect upon hippocampal after-discharges, and then perfused the same neurotransmitters into the MRF of recipient cats. Should a substance having inhibitory influences upon after-discharges induce REM sleep within periods of time as short as those induced by PIP, we could suggest that our PIP contained in part some of that substance, or a structural or functional analog of it.

We perfused the left hippocampus of our donor animals with varying doses in the microgram range of norepinephrine (NE), serotonin (5-HT), acetylcholine (ACh)

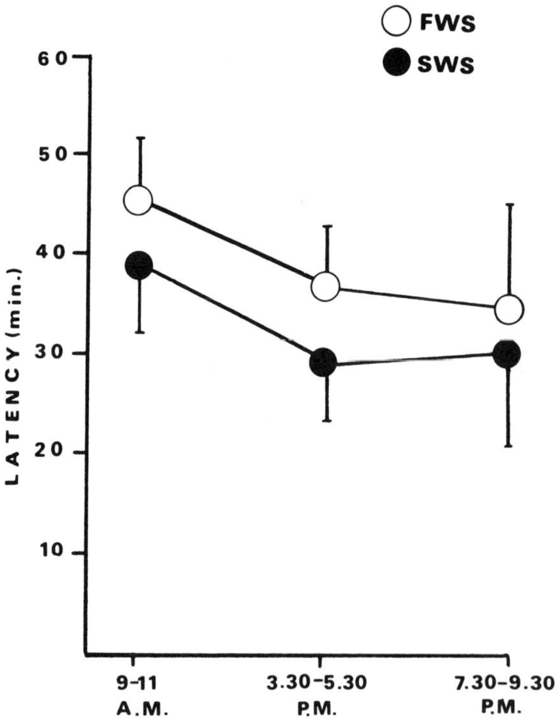

Fig. 5. Mean and standard deviations of SWS and FWS latencies of two cats at various times of the day. Each point represents the means for 5 days.

and gamma-aminobutyric acid (GABA) during 10 min, at which time after-discharges were induced from the right side. Figure 6 illustrates the effects of ACh while Figure 7 shows the results of these experiments. The graph represents the change from control in after-discharge duration after varying doses of NE, 5-HT, ACh and GABA. From it, we can see that only 5-HT and GABA inhibited after-discharges. Furthermore, all effects were dose dependent.

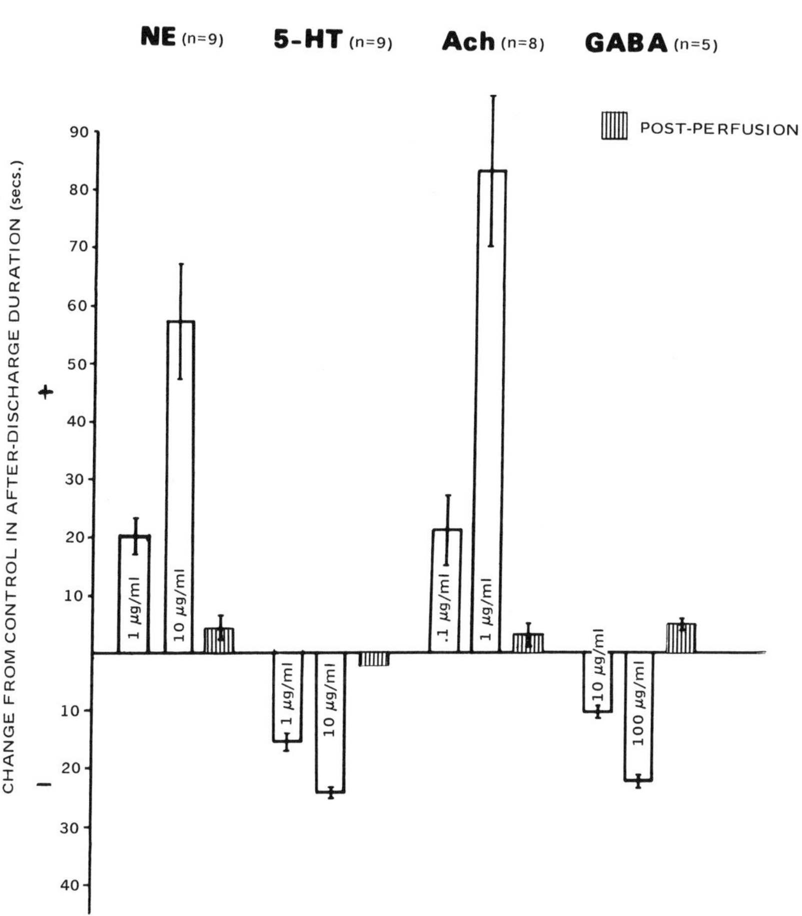

Fig. 6. Graph showing direction of change from control in hippocampal after-discharge duration as a result of perfusion into the non-stimulated hippocampus of varying doses of NE, 5-HT, ACh and GABA. Bars represent means, vertical lines represent standard deviations.

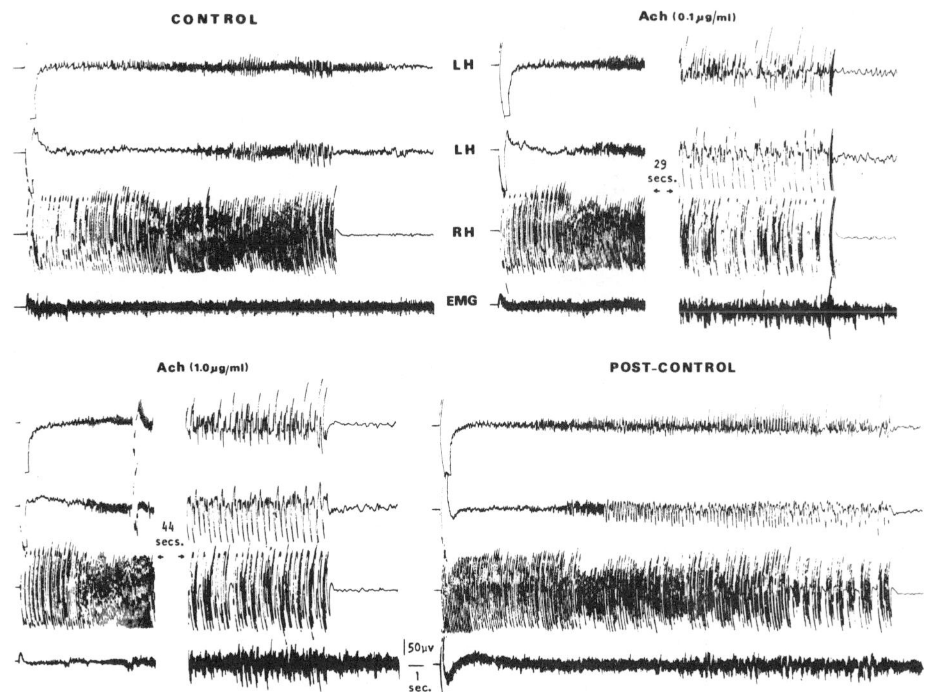

Fig. 7. Illustrates effects of varying doses of ACh upon hippocampal after-discharge duration. ACh was perfused during 10 min in the left hippocampus (LH), at which time the right hippocampus (RH) was stimulated (Freq: 100/sec; Dur: 0.1 msec; Intensity: 20 volts). Break in ACh recordings refers to time elapsed in seconds. Note increased duration and amplitude of left hippocampal after-discharges as a result of ACh perfusion, and absence of clinical convulsive signs (see EMG).

Though not shown in the graph, we did a few experiments with Ringer, with no change in after-discharge duration.

We then took our recipient animals and perfused through the reticular formation these same substances utilizing the highest dose used in the after-discharge experiments. From Fig. 2b we can see that none of these substances induced REM sleep. To our surprise, however, GABA induced a latency decrease in SWS, similar to the sleep perfusate, while serotonin had no effect. It is clear from these experiments that none of these substances are likely to represent a major component of the PIP perfusates, and we must therefore suggest that the substance(s) responsible for FWS are different from the ones so far suggested by other authors (Jouvet, 1972; Hartmann 1969; Wyatt, 1972).

GABA, PIP AND PGO WAVES

Finally we should like to direct our attention to an observation which may be of interest, though it was not systematically studied in view of its unexpected appearance. We found that as a result of perfusing either GABA or PIP into the MRF, PGO waves appeared very early after the beginning of the perfusions, when the animals were still awake. Normally PGO waves appear only during REM sleep and this has been suggested to be the result of a "gating effect" (Brooks and Gershon, 1971). However, PGO also appears during wakefulness in reserpine treated animals. Brooks and Gershon (1971) and Brooks *et al*. (1972) have therefore suggested that reserpine, by reducing serotonin, "opens the gate" of this "gating effect".

However, in view of our above mentioned observations we would like to suggest that in wakefulness PGO waves may be tonically inhibited by activation of the so-called MRF vigilance neurons. When GABA or PIP, which are assumed to possess an inhibitory action, are introduced into the MRF, the vigilance neurons become inhibited, thus releasing their inhibition upon the PGO waves. Eventually, continued perfusion of these substances leads to sleep, although their effects are different since GABA failed to induce FWS.

DISCUSSION

The main results of our experiments can be summarized as follows: we can extract from the MRF a substance presumably released during sleep, which can induce sleep when perfused into the homologous MRF of an awake animal . However this substance can only induce SWS. We can also obtain from the hippocampus a substance presumably released during the post-ictal period (i.e. seizure inhibition), which can induce FWS (REM sleep) when perfused into the MRF of an awake animal. We are unable to make a statement as to the characteristics of these perfusates; however, from our experiments it appears that none of the substances so far suggested by other authors as being responsible for sleep (see Jouvet 1972) form an important component of such perfusates. We therefore suggest that other neurohumors should begin to be explored systematically.

On the other hand, in view of the complicated machinery of sleep and wakefulness, it may be possible that in so far as the neurohumoral regulation of sleep is concerned, everyone is partly right. Thus, in keeping with the two antagonistic systems idea (Moruzzi, 1972) of sleep and wakefulness, SWS could occur as follows:

1) Ascending spinal cord influences (Pompeiano and Swett, 1962; Hodes, 1964), possibly of a cholinergic nature (Rojas-Ramírez and Drucker-Colín, in press), would excite SWS serotonergic ponto-bulbar neurons (Jouvet, 1972), which in turn would send inhibitory impulses towards the MRF.

2) Descending cholinergic inhibitory influences arising from cortical areas (Mazzuchelli-O'Flaherty et al. 1967) and preoptic regions (Sterman and Clemente, 1962; Hernández-Peón, 1965; Bremer, 1970 a,b) would both directly inhibit MRF neurons and activate ponto-bulbar neurons (Hernández-Peón and Drucker-Colín, 1970).

3) Fully developed signs of behavioral and EEG SWS occur when the ascending and descending influences work synergistically to deactivate MRF neurons through the release of an unknown humoral agent. Fluctuations between sleep and waking, as animals begin to sleep, could be due to the absence of one or more of these influences, with a concomitant reduction in the number of cells releasing the inhibitory substance. The latter could explain the varying degrees of effectiveness of the sleep perfusates in our experiments.

As for the transition from SWS to FWS, Jouvet (1972) has given ample evidence supporting his hypothesis that caudal raphé neurons prime dorso-lateral pontine tegmental FWS executive nuclei. For the sake of argument, it could be suggested that a sort of double feedback relation exists between MRF and pontine neurons, so that as the level of MRF deactivation becomes greater it releases inhibition upon pontine FWS neurons, which in turn, through some inhibitory substance releases the tonic and phasic events of FWS. This inhibitory substance may be related to our post-ictal perfusate (PIP) and may be similar to the substance contained within a new unidentified mesencephalic neuron system described by Bjorklund et al.(1971).

Unfortunately, such an hypothesis presupposes a certain corelation between amount of SWS and appearance of FWS. In fact, Jouvet (1969) has shown in his raphé lesioned cats that FWS occurs only if the amount of SWS exceeds a certain percentage of the total recording time. On the other hand, Rechtschaffen et al.(1972) have shown, in rats, that FWS rises gradually despite pronounced fluctuations in total sleep. This may suggest that FWS is triggered in terms of other brain needs.

Should it be true that FWS in neonates subserves maturation of the nervous system as suggested by Roffwarg et al. (1966), it may not be too far-fetched to suggest that in adults FWS regulates excitability.

Moreover, granting the widely accepted Haeckelian view that ontogeny is a shortened repetition of phylogeny, such a view does not seem to hold too well when speaking of FWS, since it has been argued that FWS is of a mammalian origin (Jouvet, 1967; Allison and Van Twyver, 1970), though this has been disputed (see Flanigan et al. 1973 for review), which could imply that FWS has appeared as some kind of mutation. The fact that lower animals do not have any FWS and have a very high threshold for seizures (Servit, 1962) could suggest that the development of the cortical mantle needed the parallel development of other functional systems, such as the one of FWS among others, as regulators of its excitability. This could explain the reason why a substance inhibiting seizures can promote FWS.

SUMMARY

Despite the fact that attempts to determine the nature of the hypnogenic substance(s) date back to the beginning of this century, today such substances are still unknown, although much progress has been made toward their clarification mostly, thanks to the work of Jouvet and his collaborators. One of the possible reasons for such difficulties may be due to the fact that most studies attempting to elucidate the neurohumoral basis of sleep utilize indirect neuropharmacological techniques. However, the utilization of the "push-pull" cannula technique, which presumably allows extraction of neurohumoral substances in specific brain sites during particular physiological conditions, may provide a means for determining the chain of transmitters responsible for the sleep-wakefulness cycle.

Studies in our laboratory have shown that it is possible to extract from the midbrain reticular formation (MRF) of a donor cat during physiological sleep a substance which, when "push-pull" perfused into the MRF of recipient cats, induces in the latter significant ($p < .001$) diminutions of slow wave sleep (SWS) latencies, when compared to latencies following perfusion with Ringer, with perfusates obtained from an awake donor cat, or those of non-perfused controls. One of the major puzzles in our studies is the inability of the perfusates in inducing significant effects upon the so-called REM phase of sleep, here to be called fast wave sleep (FWS).

In view of the fact that several studies suggest a close relationship between FWS and seizure extinction, i.e., FWS seems to exert an inhibitory influence upon seizure discharge, ECS during FWS deprivation reduces FWS rebound, immediately after seizure extinction signs of FWS appear and neonate's sleep is practically all REM while they have poorly developed tendency for seizures; we have developed a technique whereby through "push-pull" chemitrodes implanted bilaterally in the hippocampus, we have been able to extract, during the post-ictal after-discharge period, a substance which induces FWS when perfused into the MRF of a recipient cat.

In order to have an idea as to the transmitters involved in such effects, we have induced after-discharges in one hippocampus and perfused the contralateral side with 5-HT, NE, ACh and GABA and found that only 5-HT and GABA are capable of inhibiting the after-discharges. However, when these substances were perfused into the MRF, they did not induce FWS. The results suggest that the substance(s) involved in the inhibition of seizure may be the same as the one which induces FWS, but that none of the known transmitter substances seem to be involved in this effect.

The results are discussed in terms of the recently proposed neurochemical mechanisms of sleep and in terms of the ontogeny and phylogeny of sleep and epilepsy.

ACKNOWLEDGEMENT: The author wishes to express his gratitude to Mr. Ruben Corona for his skillful technical assistance, and to Dr. Roberto Cavanna for reviewing the manuscript.

REFERENCES

Allison T. and Van Twyver H. 1970, The evolution of sleep Natural History 79:56.

Baldessarini, R.J. and Kopin, I.J. 1966, Tritiated norepinephrine: release from brain slices by electrical stimulation, Science, 149:450.

Berlucchi, G. 1965, Callosal activity in unrestrained, unanesthetized cats, Arch. Ital. Biol. 103:623.

Bjorklund, A., Falck, B. and Stenevi, U. 1971, Classification of monoamine neurons in the rat mesencephalon: distribution of a new monoamine neurone system, Brain Res., 32:269.

Bremer, F. 1970a, Inhibitions intrathalamiques récurrentielles et physiologie du sommeil, EEG. Clin Neurophysiol., 28:1

Bremer, F. 1970b, Preoptic hypnogenic focus and mesencephalic reticular formation, Brain Res., 21:132.

Brooks, D.C. and Gershon, M.D. 1971, Eye movement potentials in the oculomotor and visual systems of the cat: a comparison of reserpine induced waves with those present during wakefulness and rapid eye movement sleep, Brain Res., 27:223.

Brooks, D.C., Gershon, M.D. and Simon, R.P. 1972, Brain stem serotonin depletion and ponto-geniculo-occipital wave activity in the cat treated with reserpine, Neuropharmacol., 11:511.

Cohen, H.B. and Dement, W.C. 1965, Sleep: changes in threshold to electroconvulsive shock in rats after deprivation of "paradoxical" phase, Science, 150:1318.

Cohen, H.B. and Dement, W.C. 1966, Sleep: suppression of rapid eye movement phase in the cat after electroconvulsive shock. Science 154:396.

Cohen, H.B., Duncan,R.F. and Dement, W.C. 1967, Sleep: the effect of electroconvulsive shock in cats deprived of REM sleep, Science 156:1646.

Cohen, H.B., Duncan, R.F. and Dement, W.C. 1968, The effect of diphenylhydantoin on sleep in the cat, EEG. Clin. Neurophysiol. 24:401.

Drucker-Colín, R.R. 1972, Transmisión neurohumoral en el sueño: efecto del líquido de perfusión obtenido de la formación reticular mesencefálica (FRM) XV Congreso Nacional de Ciencias Fisiológicas pp. 8.

Drucker-Colín, R.R. 1973, Crossed perfusion of a sleep inducing brain tissue substance in conscious cats, Brain Res., 56:123.

Drucker-Colín, R.R., Rojas-Ramírez, J.A., Vera-Trueba, J., Monroy-Ayala, G., Hernández-Peón, R. 1970, Effect of crossed perfusion of the mid-brain reticular formation upon sleep, Brain Res. 23:269.

Dow, R.S., Fernández-Guardiola, A. and Manni, E. 1962, The influence of the cerebellum on experimental epilepsy, EEG. Clin. Neurophysiol., 14:383.

Feeney, D.M. and Gullotta, F.P. 1972, Suppression of seizure discharges and sleep spindles by lesions of the rostral thalamus, Brain Res., 45:254.

Fencl, V., Koski, G. and Pappenheimer, J.R. 1971, Factors in cerebrospinal fluid from goats that affect sleep and activity in rats, J. Physiol (Lond.) 216:565.

Ferguson, J.M. and Dement, W.C. 1967, The effect of variations in total sleep time on the occurrence of rapid eye movement sleep in cats, EEG. Clin. Neurophysiol., 22:2

Fernândez-Guardiola, A.and Ayala, F. 1971, Red nucleus fast activity and signs of paradoxical sleep appearing during the extinction of experimental seizures. EEG. Clin. Neurophysiol. 30:547.

Flanigan, W.F., Wilcox, R.H. and Rechtschaffen, A. 1973, The EEG and behavioral continuum of the crocodilian, Caiman Sclerops. EEG.Clin. Neurophysiol., 34:521.

Gaddum, J.H. 1961, Push-Pull Cannulae, J. Physiol. (Lond.) 155:1.

Gastaut, H. and Fischer-Williams, M. 1959, The physiopathology of epileptic seizures, in "Handbook of Physiology (Field, J., Magoun, HW and Hall, V.E. eds. pp. 329-363, American Physiological Society, Washington.

Hartmann, E.L. 1969, The biochemistry and pharmacology of the D-state (dreaming sleep) Exptl. Med. Surg., 27:105

Hernândez-Peôn, R. 1965, A cholinergic hypnogenic limbic forebrain hindbrain circuit, in Aspects Anatomofonctionnels du Sommeil (Jouvet, M. ed.) pp. 63-88. Centre National de la Recherche Scientifique, Paris.

Hernândez-Peôn, R. and Drucker-Colîn, R.R. 1970, A neuronographic study of cortico-bulbar hypnogenic pathways. Physiol Beh. 5:721.

Hodes, R. 1964, Electrocortical desynchronization resulting from spinal block: evidence for synchronizing influences in the cervical cord, Arch. Ital. Biol., 102: 183.

Jouvet, M. 1967, The states of sleep, Scientific Amer., 216:62.

Jouvet, M. 1969, Biogenic amines and the states of sleep. Science, 163:32.

Jouvet, M. 1972, The role of monoamines and acetylcholine containing neurons in the regulation of the sleep-waking cycle., Ergbn. Physiol.64:166.

Jouvet-Mounier, D., Astic, L. Lacote, D. 1970, Ontogenesis of the states of sleep in rat, cat and guinea pig during the first postnatal month. Develop. Psychobiol. 2:216.

Kety, S.S. 1967, The central physiological and pharmacological effects of the biogenic amines and their correlations with behavior, in "The Neurosciences (Quarton, G.C., Melnechuk, T. and Schmitt, C. eds.) pp. 444-451, Rockefeller Univ. Press, N.Y.

Kusske, J., Ojemman, G.A. and Ward, A.H. 1972, Effects of lesions of the ventral anterior thalamus on experimental focal epilepsy. Exp. Neurol., 34:279.

Legendre, R. and Pieron, H. 1911, Du development, an cours de l'insonnie experimental, des propriétés hypnotoxiques des humeurs en relation avec le besoin croissant de sommeil, C.R. Soc. Biol. (Paris) 70:190.

Lenard, H.G. and Schulte, F.J. 1972, Polygraphic sleep study in craniopagus twins, J. Neurosurg. Psychiat., 35:756.

Lennox-Buchtal, M.A. 1973, Febrile Convulsions: a Reappraisal, EEG. Clin. Neurophysiol. Suppl. 32:1.

Lovell, R.A. 1971, Some neurochemical aspects of convulsions, in "Handbook of Neurochemistry" (Lajtha, A. ed.) pp. 63-102. Plenum Press, N.Y.

Mazzuchelli-O'Flaherty, H.L., O'Flaherty, J.J. and Hernández-Peón, R. 1967, Sleep and other behavioral responses induced by acetylcholinic stimulation of the frontal and mesial cortex, Brain Res., 4:268.

McLennan, H. 1964, The release of acetylcholine and of 3-hydroxytyramine from the caudate nucleus, J. Physiol. (London) 174:152.

Monnier, M. and Hatt, A.M. 1971, Humoral transmission in sleep, V. New evidence from production of pure sleep hemodialysate, Pflugers. Arch. ges. Physiol.329:231.

Monnier, M. and Hosli, L. 1964, Dialysis of sleep and waking factors in blood of the rabbit, Science 146: 796.

Morgane, P.J. and Stern, W.C. 1973, Effects of serotonin metabolites on sleep-waking activity in cats, Brain Res., 50:205.

Moruzzi, G. 1972, The sleep-waking cycle, Ergebn. Physiol. 64:1.

Myers, R.D. and Beleslin, A.B. 1971, Changes in serotonin release in hypothalamus during cooling and warming in the monkey, Amer. J. Physiol., 220:1746.

Pappenheimer, J.R., Miller, T.B. and Goodrich, C.A. 1967, Sleep promoting effects of cerebrospinal fluid from sleep deprived goats, Proc. int. Acad. Sci. (Wash.) 58:543.

Pieron, H. 1913, Le problême physiologique du sommeil. Masson, Paris pp. 520.

Pompeiano, O. 1969, Sleep Mechanisms in "The Basic Mechanisms of the Epilepsies" (Jasper, H.H., Ward, H.A. and Pope A. eds.) pp. 453-473. Little Brown Co. Boston

Pompeiano, O. and Swett, J.E. 1962, EEG and behavioral manifestations of sleep induced by cutaneous nerve stimulation in normal cats, Arch. Ital. Biol., 100:343

Rechtschaffen, A.,Lovell, R.H., Freedman, D.X., Whitehead, W.E. and Aldrich, M. 1972, The effect of parachlorophenylalanine on sleep in the rat: some implications for the serotonin-sleep hypothesis. Conference on Serotonin and Behavior, Stanford University.

Ringle, D.A. and Herndon, B.C. 1969, Effects on rats of CSF from sleep-deprived rabbits, Pflugers Arch. ges. Physiol., 306:320.

Roffwarg, H. P., Muzio, J.N. and Dement, W.C. 1966, Ontogenetic development of the human sleep-dream cycle. Science, 152:604.

Rojas-Ramîrez, J.A. and Drucker-Colîn, R.R. 1973, Sleep induced by spinal cord cholinergic stimulation, Int. J. Neurosciences, in Press.

Rossi, G.F., Favale, E., Hara, T., Guisani, A. and Sacco, G. 1961, Researches on the neurons mechanisms underlying deep sleep in the cat, Arch. Ital. Biol.,99:270.

Satinoff, E., Drucker-Colîn, R.R. and Hernândez-Peôn, R. 1971, Paleocortical excitability and sensory filtering during REM sleep deprivation, Physiol. Beh. 7:103.

Schnedorf, J.G. and Ivy, A.C. 1939, On examination of the hypnotoxin theory of sleep, Amer. J. Physiol. 125:491.

Schoenenberger, G.A., Cueni, L.B., Hatt, A.M. and Monnier M. 1972, Isolation and physical-chemical characterization of a humoral sleep inducing substance in rabbits (Factor Delta), Experientia, 28:919.

Servit, Z. 1962, Phylogenesis and ontogenesis of the epileptic seizure, World Neurol. 3:259.

Sterman, M.B. and Clemente, C.D. 1962, Forebrain inhibition mechanisms: sleep patterns induced by basal forebrain stimulation in the behaving cat, Exp. Neurol., 6:103.

Szerb, J.C. 1967, Cortical acetylcholine release and electroencephalographic arousal. J. Physiol. (London) 192:329.

Yaksh, T.L. and Myers, R.D. 1972, Neurohumoral substances released from hypothalamus of the monkey during hunger and satiety, Amer. J. Physiol. 222:503.

Yamaguchi, N., Ling, G.M. and Marczynski,T.J. 1964, Recruiting responses observed during wakefulness and sleep in unanesthetized chronic cats, EEG. Clin. Neurophysiol., 17:246.

Wyatt, R.J. 1972, The serotonin-catecholamine dream bicycle. A clinical study, Biol. Psychiat., 5:73.

5-HYDROXYINDOLEACETIC ACID IN CEREBROSPINAL FLUID DURING WAKEFULNESS, SLEEP AND AFTER ELECTRICAL STIMULATION OF SPECIFIC BRAIN STRUCTURES

Miodrag Radulovački

Department of Pharmacology, College of Medicine
University of Illinois
Chicago, Illinois

Jouvet has implicated 5-hydroxytryptamine (5-HT, serotonin) in the regulation of slow-wave sleep (SWS). He and his co-workers showed that subtotal destruction of the raphe nuclei decreased the level of 5-HT in the brain and led to a state of severe insomnia (Jouvet et al., 1966) and that the injection of p-chlorophenylalanine (pCPA), which inhibits 5-HT synthesis, produced similar results (Delorme et al., 1966). This effect was reversed by administration of 5-hydroxytryptophan (5-HTP), a precursor of 5-HT, which restored SWS and paradoxical sleep (PS).

Available evidence suggests that alterations of brain 5-HT metabolism can be detected in cerebrospinal fluid (CSF) (Bowers, 1970; Curzon et al., 1971). Although 5-HT is not detectable in CSF (Coppen, 1971; Isaac and Radulovački), its metabolite 5-hydroxyindoleacetic acid (5-HIAA) has been measured in CSF of man and animals and its concentration has been taken as an index of 5-HT metabolism in the brain (Bowers, 1970; Eccleston et al., 1970; Moir et al., 1970; Roos et al., 1964; Roos and Sjostrom, 1969). Since there is no brain-CSF barrier (Davson, 1967), 5-HIAA diffuses from the brain to CSF from which it is transported to blood by an active process (Neff et al., 1967). It has been estimated that about 10-20% of 5-HIAA from the brain reaches CSF (Meek and Neff, 1973; Bulat). The remaining 5-HIAA in the brain is transported directly to blood by an active process (Neff et al., 1967). Although the percentage of brain 5-HIAA reaching CSF is relatively small, it is of central origin (Moir et al., 1970; Weir et al., 1973; Bulat and Zivkovic, 1973) and is present in measurable quantities. These and other conditions [i.e., rapid equilibration of 5-HIAA in the brain with CSF

(Anderson and Roos, 1968) and the possibility of repeated CSF sampling in the same animal] make determination of monoamine metabolites in CSF a useful research tool in human and animal studies.

We have attempted to test Jouvet's theory of sleep on the assumptions that metabolism of 5-HT in the brain follows the function of serotonergic neurons and that metabolic changes in the brain are reflected in CSF composition. We measured 5-HIAA in CSF according to the method of Korf and Valkenburgh-Sikkema (1969). With our improved cannulation technique, we tapped CSF from the cisterna magna and correlated concentration of 5-HIAA with wakefulness, SWS, EEG and behavioral states. These states were manipulated by application of drugs, deprivation of PS and stimulation of specific brain structures.

CANNULATION PROCEDURE IN THE CAT

The cannulation procedure described here (Fig. 1) is an improvement of our previous method (Radulovački and Girgis, 1968). A midline incision is made through the skin the full length of the top of the skull from a point

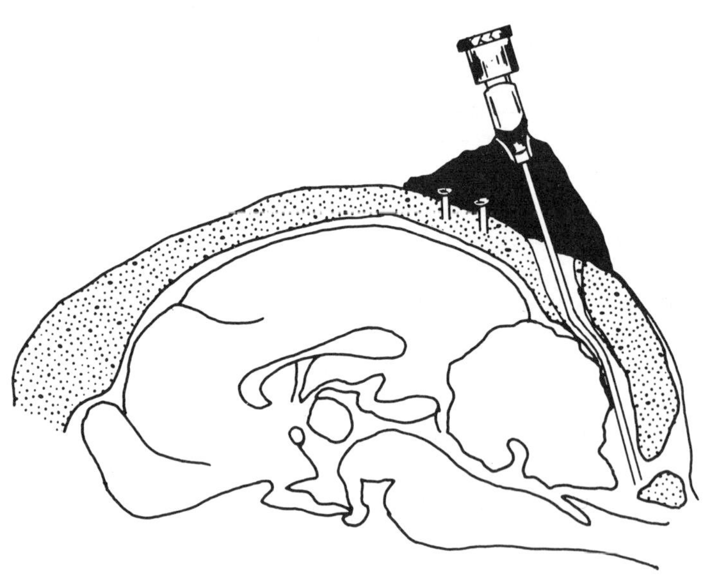

Figure 1. Diagram showing the implanted cannula in position.

level with frontal sinuses. Ligaments, muscles and periosteum are stripped from the bone with a dental spatula. A low-speed dental drill is used for making a hole (4 mm in diameter) on the left parietal bone 3 mm lateral to the sagittal suture and 4 mm cranial to the lambdoid suture. The caudal wall of the hole descends vertically upon the dura while the frontal wall meets the dura at a 45° angle. This is an important part of the procedure since a 45° angle is necessary to direct the cannula to the cisterna magna. Four stainless steel screws are fixed into the parietal bones to support the adhesion of dental acrylic to the skull. The cannula is made from No. 20 B-D Special Needle (Becton – Dickinson, Rutherford, N. Y.) cut to a shaft length of 10 mm. The upper tip of a 35-mm-long Teflon tube (20 gauge, Small Parts, Miami, Fla.) is fixed to the shaft of the needle and a hole at the side of the lower tip of the tube is made. The cannula is placed horizontally on the left parietal bone, inserted caudally under the cut dura and directed downward and medially between the dura and cerebellum. The cannula is always pressed against the dura so that it does not penetrate the cerebellum on its way down to the cisterna. CSF is withdrawn preferably by a 1-cc syringe and the hole is covered with gelfoam. Thin acrylic cement (Kold-weld resin, Precision Dental Manufacturing, Chicago) is poured over the parietal bones and gelfoam to fix the cannula. A B-D male plug for female luer-lok needles (Becton—Dickinson, Rutherford, N. Y.) is screwed to the top of the cannula to prevent leakage of CSF. Periodic flushing of the cannula with heparin solution keeps it patent up to 60 days.

5-HIAA IN CSF DURING WAKEFULNESS AND SWS

This study was done to determine a relationship between the 5-HIAA concentration in CSF and the states of wakefulness and SWS. We expected that an increase of activity in 5-HT neurons during SWS might be followed by an increased 5-HIAA concentration in the CSF. For this purpose, cats were implanted with a cannula in the cisterna and with appropriate electrodes to record cortical electroencephalogram (EEG), electromyogram (EMG) and electrooculogram (EOG). After the operation, the animals were kept in an isolated room where EEG activity was recorded intermittently for several days with a Grass VIII channel electroencephalograph. CSF samples were taken from cats during wakefulness and during SWS after 1 h of monitoring.

In 107 CSF samples taken from 12 cats during wakefulness and SWS, there was almost no difference in the mean values of 5-HIAA in CSF of awake and sleeping cats (Table 1). The Mann-Whitney U-test showed no significant difference between wakefulness and SWS ($P = 0.4$). However, there were exceptions in 2 cats in which the test of significance between the mean values of wakefulness and SWS showed (a) a significant difference in cat 4 (with higher 5-HIAA concentration in wakefulness), and (b) a significant difference in cat 12 (with elevated 5-HIAA concentration in SWS).

TABLE I

5-HIAA ng/ml CSF DURING WAKEFULNESS AND SWS

n: number of determinations (1 ml samples CSF).

Cats	Wakefulness			SWS		
	$\bar{X}$	S.D. ±	*n*	$\bar{X}$	S.D. ±	*n*
1	142	4	2	108	26	2
2	120	11	4	103	26	4
3	243	37	3	182	25	2
4	130	15	3	84	11	2
5	73	40	4	112	12	3
6	65	—	1	99	25	7
7	83	28	8	63	11	4
8	58	32	7	80	30	7
9	100	14	4	103	20	6
10	60	24	2	60	38	2
11	86	15	5	83	10	6
12	107	12	11	126	25	8
Mean	101 ± 46			99 ± 32		

5-HIAA IN CSF AFTER ADMINISTRATION OF RESERPINE AND TRYPTOPHAN

Matsumoto and Jouvet (1964) reported that administration of reserpine (0.5 mg/kg) delayed the occurrence of the first SWS episode by 12 h and PS first appeared after 24 h. This finding is in accordance with the postulation that 5-HT is involved in sleep since suppression of sleep by reserpine coincided with the depletion of brain monoamine stores. We have compared the time course of the action of reserpine on brain 5-HT metabolism with its effects on sleep during the same period. Cats were implanted with a cannula in the cisterna magna and electrodes for EEG, EMG and EOG recording. The animals were given reserpine (0.5 mg/kg, i.p.) and their EEG was continuously recorded for 24 h. CSF samples taken immediately prior to drug administration served as controls; the experimental sample was withdrawn 1 h after drug administration and every 4 h subsequently.

Figure 2 shows the time course of the concentration of 5-HIAA in the CSF of 3 cats after reserpine. After 24 h, the concentration was still above the pre-drug level. The effect on sleep was similar to the one observed by Matsumoto and Jouvet: the occurrence of first SWS episode was delayed by 8-11 h and the first PS period appeared after 22-24 h. It is not easy to interpret the increased 5-HIAA concentration during the period of suppression of SWS. However, it could be that the 5-HT released from granules was metabolized prior to its interaction with receptors.

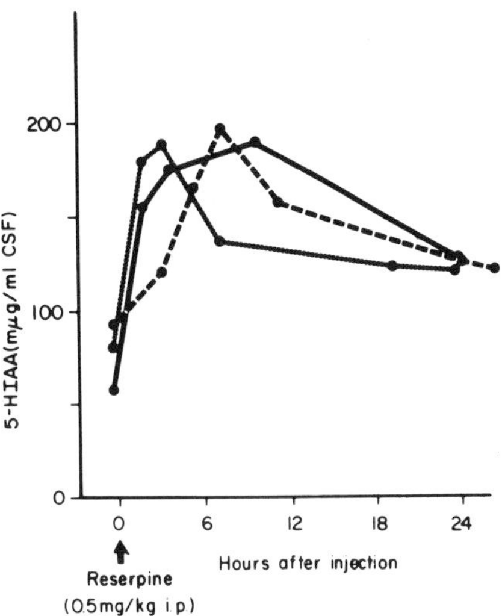

Figure 2. Effect of reserpine on 5-HIAA concentration in CSF in 3 cats.

We have used L-tryptophan, a 5-HT precursor, because it is converted to 5-HT only in 5-HT neurons due to specific location of tryptophan hydroxylase (Hartman et al., 1971) and to study the relationship between this effect and its reported sleep-inducing effect. It has been reported that administration of tryptophan to insomniac patients increased the duration of sleep, reduced sleep latency and reduced the number of awakenings (Hartman et al., 1971). Tryptophan increased SWS and decreased the length of SWS-PS cycle in normal subjects (Williams et al., 1968). Also, there is a report that tryptophan decreased the latency of PS onset in a few patients (Oswald et al., 1966). The experimental procedure in cats was similar to that after reserpine. L-tryptophan was administered to cats in a single oral dose of 135 mg/kg. SWS ensued 15-20 min after administration and the first PS episode appeared after 30 min. The duration of the induced sleep was for an hour. The 5-HIAA concentration in CSF rose significantly (Fig. 3) and reached the peak between 3 and 4.5 h after tryptophan administration. This effect was similar to that of Eccleston et al. (1970) who reported marked increases in brain 5-HT and 5-HIAA and in CSF 5-HIAA following administration of L-tryptophan to rats and dogs. Thus, episodes of SWS and PS followed the administration of tryptophan; periods of wakefulness, SWS and PS appeared sequentially, irrespective

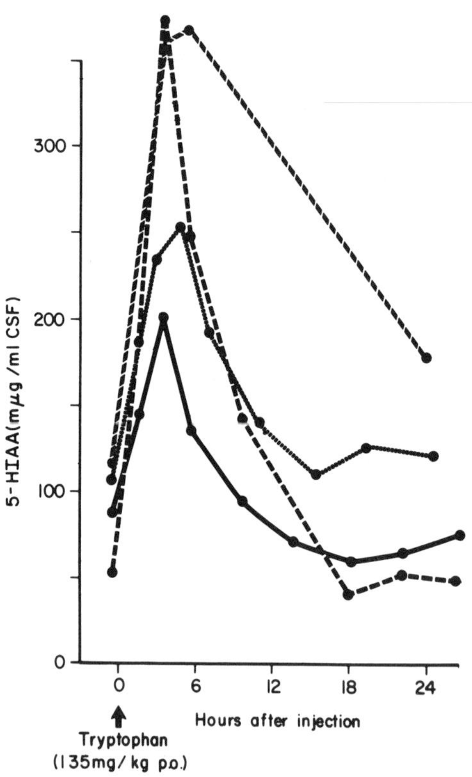

Figure 3. Effect of oral administration of L-tryptophan on 5-HIAA concentration in CSF in 4 cats.

of the concentration of 5-HIAA in CSF. It was of interest to note that within the first 5-h interval after drug administration (a period in which the highest concentration of 5-HIAA was found in CSF) there were episodes of complete wakefulness. Therefore, if the initial sleep-inducing effect of tryptophan was via increased 5-HT activity, one would anticipate a correlation between increased 5-HT metabolism and sleep, but we found no support for this (Fig. 3). Also, if natural SWS were due to increased 5-HT metabolism in the brain, increased 5-HIAA might be found during SWS, but again there was no evidence for this (Table 1).

CSF 5-HIAA CONCENTRATION DURING DEPRIVATION OF PS AND IMMOBILIZATION

An increase of brain 5-HIAA and a decrease of brain 5-HT following deprivation of PS in cats was reported by Weiss et al. (1968). Bliss et al. (1972) found an increase of brain 5-HIAA in rats deprived of PS. On the other hand, Stern et al. (1971) reported no change in endogenous 5-HT and norepinephrine (NE) levels in cerebral hemispheres, diencephalon and brain stem in rats deprived of PS when compared to stress controls. The standard technique for deprivation of PS (Jouvet et al., 1964) immobilizes the animals on an island surrounded by water. This technique, in addition to producing sleep-depriving effect, may elicit immobilization stress (Curzon and Green, 1969).

Since deprivation of PS and immobilization stress (Corrodi et al., 1968) have both been reported to cause a decrease in brain 5-HT, our intention was to determine what effect deprivation of PS had on 5-HT metabolism (as reflected in the concentration of 5-HIAA in CSF) and to distinguish between the effects caused by sleep loss and those induced by immobilization alone. To test this, we have implanted cats with a cannula in the cisterna magna and divided them in two groups: Animals in the first group were deprived of PS (DPS) by placement on square supports surrounded by water which were small enough to prevent relaxation without falling off. The cats remained on supports for 6 days and nights. Animals in the second group were immobilized by the same procedure except that the supports were large enough to permit them to relax, to lie down and sleep. These cats remained on supports for 4 days and nights. In both groups of cats, CSF samples were taken once a day at the same time of day for 3 days before, during and 3 days after immobilization.

During 144 h of DPS in 6 cats, there was a significant increase of cisternal 5-HIAA concentration (Fig. 4). In 4 cats, the 5-HIAA increased after 24 h of DPS, and in 2 cats, the increase was observed only after 72 h of

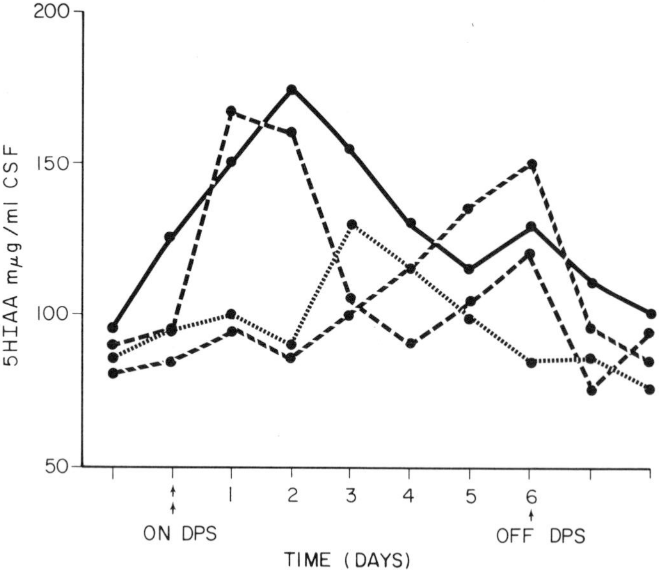

Figure 4. 5-HIAA mμg/ml CSF in 4 cats during DPS.

DPS. Also, the concentration of 5-HIAA peaked at widely different times in the various animals: in one cat, the peak occurred after 24 h of DPS, and in another was attained only after 144 h of DPS. Thus, no clear correlation was observed between the time course of the rise in 5-HIAA concentration and the duration of DPS. 5-HIAA concentration was back to normal values after a recovery period of 24-48 h following DPS.

In the animals of the second group, which were immobilized but could sleep, 5-HIAA also increased in the CSF (Table 2). A significant increase was obtained by the third day of immobilization. These findings are in accordance with the reports that immobilization alone affects brain 5-HT metabolism (Curzon and Green, 1969).

Table II

EFFECT OF IMMOBILIZATION ON 5-HIAA CONCENTRATION IN CSF

	5-HIAA Concentration (mμg/ml CSF)
Control	97 ± 52 (15)
Days on Immobilization	
1	90 ± 50 (5)
2	150 ± 73 (4)
3	157 ± 47 (5)*
4	129 ± 40 (5)
Days Following Immobilization	
1	114 ± 31 (5)
2	109 ± 42 (5)
3	92 ± 44 (5)

Number of CSF samples from 5 cats are shown in brackets. Results are given as mean ± SD. Control represents 5-HIAA concentration in CSF samples from 5 cats taken during 3 days. *Different from control where p is at least $<.05$.

CSF 5-HIAA AFTER ELECTRICAL STIMULATION OF THE PREOPTIC AREA, ANTERIOR HYPOTHALAMUS AND DORSAL RAPHE NUCLEUS*

Assuming that activation of 5-HT neurons would elicit SWS, electrophysiological studies have attempted to induce sleep using electrical stimulation of the raphe system, but have produced contradictory results. Electrical stimulation of pontine and magnus raphe nuclei in cats (2 and 10 c/sec, 1.0 m sec, 1-6 V) produced no change in behavior while stimulation of the dorsal raphe nucleus resulted in mild arousal (Jacobs et al., 1973). Kuroki (1958) stimulated the mesencephalic raphe in cats and reported no behavioral effects following low-frequency stimulation (10 c/sec). We have measured 5-HIAA concentration in CSF following electrical stimulation of the dorsal raphe nucleus (containing 5-HT neurons), anterior hypothalamus (containing 5-HT terminals) and lateral preoptic nucleus (without 5-HT neurons or terminals) (Fuxe, 1965). During the electrical stimulation, we observed the behavior of the animal.

Cats were stereotaxically implanted with stimulating electrodes in dorsal raphe nucleus, anterior hypothalamus and lateral optic nucleus.

*This study was done in association with Dr. G. Karmos.

Recording electrodes were placed in the dorsal hippocampus and frontal, parietal and occipital cortex, for EMG and EOG, with a cannula in cisterna magna. We stimulated the lateral preoptic nucleus (5 V, 0.3 m sec, 10 c/sec, and 100 c/sec) and recorded EEG in 4 cats. Stimulation of the anterior hypothalamus (5 V, 0.3 m sec, 100 c/sec) was done in 4 cats of the second group and the dorsal raphe nucleus was stimulated (5 V, 0.3 m sec, 10 c/sec) in 4 cats of the third group. The duration of electrical stimulation in all cats was 30 minutes during which 30-sec stimulation periods alternated with a 30-sec stimulus-free period. Samples of CSF (1 ml) were taken before, immediately after stimulation of 30 minutes and 2 h after 30 minutes of stimulation.

Sleep and synchronization in EEG followed electrical stimulation of the lateral preoptic nucleus (Fig. 5), but no significant increase of 5-HIAA followed stimulation (Table 3). Since there are no 5-HT neurons or terminals in the lateral preoptic nucleus, we did not expect to see the concentration of CSF 5-HIAA altered. However, since stimulation of the lateral preoptic area produced SWS, and if 5-HT is involved in SWS, one might have expected

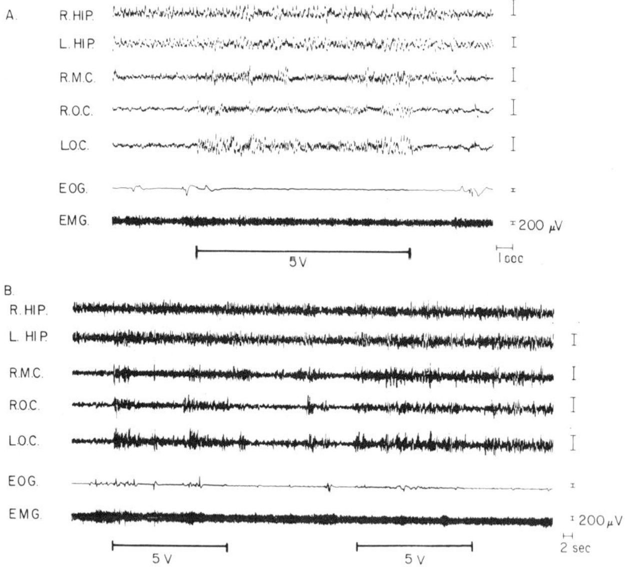

Figure 5. Electrical stimulation of the preoptic area produced synchronization in cortical EEG (A). Repeated 30-sec stimulation in the same experiment induced synchronization in cortical EEG and behavioral sleep in duration of 1 hour (B).

Table III

EFFECT OF ELECTRICAL STIMULATION OF SPECIFIC BRAIN STRUCTURES ON 5-HIAA CONCENTRATION IN CSF

	5-HIAA Concentration (mμg/ml CSF)		
	Before Stim.	Immediately after Stim. of 30 min.	2 h after Stim. of 30 min.
Preoptic Area	76 ± 14 (9)	78 ± 6 (9)	85 ± 11 (9)
Anterior Hypothalamus	54 ± 7 (11)	66 ± 14 (11)*	73 ± 20 (11)**
Raphe System	69 ± 8 (13)	85 ± 24 (13)*	101 ± 30 (13)***

Number of CSF samples from 4 cats in each group are shown in brackets. Results are given as mean ± SD. *Different from control where p is at least $\leq .05$; **$p < .02$; ***$p < .005$.

preoptic area stimulation to increase CSF 5-HIAA.

Excitation and strong emotional reaction followed the stimulation of anterior hypothalamus. Similar observations were made by Wasman and Flynn (1962) and Grastyan *et al.* (1966). 5-HIAA concentration was significantly increased following stimulation of this site (Table 3).

We have not observed any change in behavior during electrical stimulation of the dorsal raphe nucleus. There was only occasional twitching of the ears which coincided with the onset of the 30-sec stimulation period and was absent during the 30-sec stimulus-free period. However, this stimulation sufficiently increased the 5-HIAA levels in the CSF (Table 3).

CONCLUSION

In conclusion, our experiments failed to show a correlation between the concentration of 5-HIAA, a metabolite of 5-HT, in the cisternal CSF as a function of sleep and waking states. Administration of tryptophan produced a transient sleep-inducing effect and a long-lasting increase of 5-HIAA, but was not followed by continuous sleep. Administration of reserpine produced a rise in 5-HIAA for 24 h and a suppression of SWS which lasted only 11 h.

5-HIAA was affected by immobilization alone and by the loss of PS caused by the immobilization technique. This suggests that changes in 5-HT metabolism which occur following use of the island technique are secondary to immobilization stress rather than deprivation of paradoxical sleep.

Electrical stimulation of the brain structures containing 5-HT neurons or terminals was followed by an increase of 5-HIAA in cisternal CSF. However, the behavior elicited was very different. Thus, stimulation of anterior hypothalamus was accompanied by excitation and a strong emotional reaction, while stimulation of the dorsal raphe nucleus did not change behavior. Stimulation of brain areas without 5-HT neurons or terminals (lateral preoptic nucleus) was not followed by an increase in 5-HIAA even though SWS was produced. Our findings do not support the view concerning implication of 5-HT in sleep in that we were unable to correlate brain 5-HT neuronal activity, as reflected by 5-HIAA concentration in CSF, with natural or induced sleep states.

REFERENCES

Andersson, H., and Roos, B., 1968, 5-Hydroxyindoleacetic acid in cerebrospinal fluid after administration of 5-hydroxytryptophan, Acta pharmacol. (Kbh.) 26:293.

Bliss, E. L., Thatcher, W., and Ailion, J., 1972, Relationship of stress to brain serotonin and 5-hydroxyindoleacetic acid, J. psychiat. Res. 9:71.

Bowers, M. B., Jr., 1970, 5-Hydroxyindoleacetic acid in the brain and cerebrospinal fluid of the rabbit following administration of drugs affecting 5-hydroxytryptamine, J. Neurochem. 17:827.

Bulat, M., Personal communication.

Bulat, M., and Zivkovic, B., 1973, Penetration of 5-hydroxyindoleacetic acid across the blood-cerebrospinal fluid barrier, J. Pharm. Pharmacol. 25:178.

Coppen, A., 1971, Biogenic amines and affective disorders, in "Brain Chemistry and Mental Disease," (B. T. Ho, and W. M. McIsaac, eds.), pp. 123-133, Plenum Press, New York.

Corrodi, H., Fuxe, K., and Hokfelt, T., 1968, The effect of immobilization stress on the activity of central monoamine neurons, Life Sci. 7:107.

Curzon, G., and Green, A. R., 1969, Effects of immobilization on rat liver tryptophan pyrrolase and brain 5-hydroxytryptamine metabolism, *Br. J. Pharmacol*. 37:689.

Curzon, G., Gumpert, E. J. W., and Sharpe, D. M., 1971, Amine metabolites in the lumbar cerebrospinal fluid of humans with restricted flow of cerebrospinal fluid, *Nature New Biol*. 231:189.

Curzon, G., Joseph, M. H., and Knott, P. J., 1972, Effect of immobilization on food deprivation on rat brain tryptophan metabolism, *J. Neurochem*. 19:1967.

Davson, H., 1967, "Physiology of the Cerebrospinal Fluid," J. and A. Churchill, London.

Delorme, F., Froment, J. L., et Jouvet, M., 1966, Suppression du sommeil par la parachlorometamphetamine et la parachlorophenylalanine, *C. R. Soc. Biol. (Paris)* 160:2347.

Eccleston, D., Ashcroft, G. W., Moir, A. T. B., Parker-Rhodes, A., Lutz, W., and O'Mahoney, D. P., 1968, A comparison of 5-hydroxyindoles in various regions of dog brain and cerebrospinal fluid, *J. Neurochem*. 15:947.

Eccleston, D., Ashcroft, G. W., Crawford, T. B., Stanton, J. B., Wood, D., and McTurk, P. H. T., 1970, Effect of tryptophan administration on 5-HIAA in cerebrospinal fluid in man, *J. Neurol. Neurosurg. Psychiat*. 33:269.

Fuxe, K., 1965, The distribution of monoamine terminals in the central nervous system, *Acta Physiol. Scand*. 64, Suppl. 247.

Grastyan, E., Karmos, G., Vereczkey, L., and Kellenyi, L., 1966, The hippocampal electrical correlates of the homeostatic regulation of motivation, *Electroenceph. clin. Neurophysiol*. 21:34.

Hartman, E., Chung, R., and Chien, C., 1971, L-tryptophane and sleep, *Psychopharmacologia (Berl.)* 19:114.

Isaac, L., and Radulovački, M., Unpublished observation.

Jacobs, B. L., Asher, R., Henriksen, S. J., and Dement, W. C., 1972, Electroencephalographic and behavioral effects of stimulation of the raphe nuclei in cats, in "Sleep Research," (M. H. Chase, W. C. Stern, and P. L. Walter, eds.), p. 23, BIS-BRI, UCLA, Los Angeles.

Jouvet, D., Vimont, P., Delorme, J. F., and Jouvet, M., 1964, Etude de la privation de phase paradoxale du sommeil chez le chat, C. R. Soc. Biol. (Paris) 158:756.

Jouvet, M., Bobillier, P., Pujol, J. F., et Renault, J., 1966, Effets des lesions du systeme du raphe sur le sommeil et la serotonine cerebrale, C. R. Soc. Biol. (Paris) 160:2343.

Karmos, G., and Radulovački, M., 5-Hydroxyindoleacetic acid in cerebrospinal fluid after electrical stimulation of the dorsal raphe nucleus, anterior hypothalamus and preoptic area, (in press).

Korf, J., and Valkenburgh-Sikkema, T., 1969, Fluorimetric determination of 5-hydroxyindoleacetic acid in human urine and cerebrospinal fluid, Clin. chim. Acta 26:301.

Kuroki, T., 1958, Arrest reaction elicited from the brain stem, Folia psychiat. neurol. jap. 12:170.

Matsumoto, J., et Jouvet, M., 1964, Effets de reserpine, DOPA et 5-HTP sur les deux etats de sommeil, C. R. Soc. Biol. (Paris) 158:2137.

Meek, J. L., and Neff, N. H., 1973, Is cerebrospinal fluid the major avenue for the removal of 5-hydroxyindoleacetic acid from the brain?, Neuropharmacol. 12:497.

Moir, A. T. B., Ashcroft, G. W., Crawford, T. B. B., Eccleston, D., and Guldberg, H. C., 1970, Cerebral metabolites in cerebrospinal fluid as a biochemical approach to the brain, Brain 93:357.

Neff, N. H., Tozer, T. N., and Brodie, B. B., 1967, Application of steady-state kinetics to studies of the transfer of 5-hydroxyindoleacetic acid from brain to plasma, J. Pharmacol. exp. Ther. 158:214.

Oswald, I., Ashcroft, G. W., Berger, R. J., Eccleston, D., Evans, J., and Thacore, V. R., 1966, Some experiments in the chemistry of normal sleep, Brit. J. Psychiat. 112:273.

Radulovački, M., and Girgis, M., 1968, The permanent cannula to cisterna magna in cats, Sudan med. J. 6:170.

Roos, B. E., and Sjostrom, R., 1969, 5-Hydroxyindoleacetic acid (and homovanillic acid) levels in the cerebrospinal fluid after probenecid application in patients with manic-depressive psychosis, Pharmacol. Clin. 1:153.

Roos, B. E., Anden, N.-E., and Werdinius, B., 1964, Effect of drugs on the levels of indole and phenolic acids in the central nervous system, Int. J. Neuropharmacol. 3:117.

Stern, W. C., Miller, F. P., Cox, R. H., and Maickel, R. P., 1971, Brain norepinephrine and serotonin levels following REM sleep deprivation in the rat, Psychopharmacologia (Berl.) 22:50.

Wasman, M., and Flynn, J. P., 1962, Directed attack elicited from hypothalamus, Arch. Neurol. (Chic.) 6:220.

Weir, R. L., Chase, T. N., NG, L. K. Y., and Kopin, I. J., 1973, 5-Hydroxyindoleacetic acid in spinal fluid: relative contribution from brain and spinal cord, Brain Res. 52:409.

Weiss, E., Bordwell, B., Seeger, M., Lee, J., Dement, W., and Barchas, J., 1968, Changes in brain serotonin (5-HT) and 5-hydroxyindole 3-acetic acid (5-HIAA) in REM sleep deprived rats, Psychophysiology 5:209.

Williams, H. L., Lester, B. K., and Coulter, J. D., 1968, Monoamines and the EEG stages of sleep, Psychophysiology 5:210.

MODULATION OF SUBCORTICAL INHIBITORY MECHANISMS BY MELATONIN

Augusto Fernández-Guardiola
and Fernando Antón-Tay

Instituto de Investigaciones Biomédicas
Ciudad Universitaria, UNAM, México 20, D.F.

The existence of an active arresting subcortical mechanism in the last state of convulsive activity has been suggested by several authors (Jung, 1949; Gastaut and Fisher-Williams, 1959; Fernández Guardiola <u>et al</u>., 1961; Kreindler, 1965). This assumption is based on experimental evidence and the existence of an inhibitory mechanism which integrates descending and ascending influences. Anatomically, this mechanism is located in the cerebellum, the ventral ponto-bulbar reticular formation (Magoun and Rhines, 1946), the caudate nucleus and the diffuse thalamic system. This inhibitory organization might be acting through various kinds of physiological inhibitions: Purpura and Housepian (1961) have shown direct cortical inhibition with prolonged IPSP (more than 200 msec), while Giaquinto <u>et al</u>. (1964) have reported a pre-synaptic inhibition of subcortical origin on spinal reflexes during natural sleep.

In previous works we reported the possible inhibitory role of the cerebellum and red nucleus on convulsive activity. Stimulation experiments have shown that the activation of both structures inhibits cortical postdischarges of cobalt experimental epilepsy as well as those elicited by electroshock. (Fernández-Guardiola and Ayala, 1971). Conversely, lesion procedures lead to the enhancement of the electrical signs of convulsive activity. The recording of cerebellum and red nucleus with both micro and macroelectrodes shows a frequency acceleration that increases progressively during the last stages of seizure, concurrently with interclonic intervals, and is prominent during the post seizure electrical "silence" (Fig. 1).

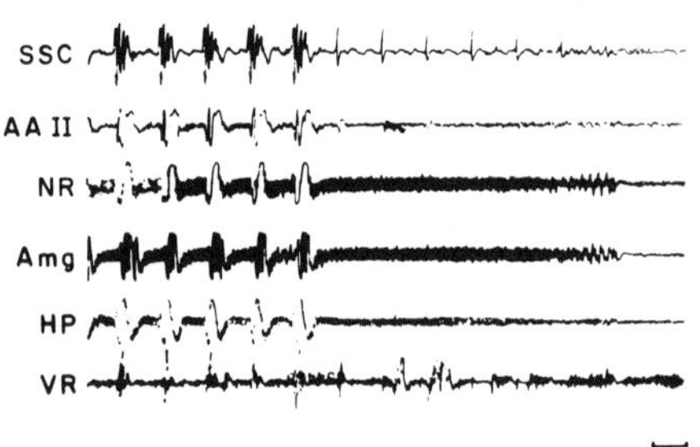

Figure 1. Final stages of an electroshock induced tonic-clonic seizure in the cat.
SSC, sensory motor cortex; AAII, secondary auditory cortical area; NR, Nucleus Ruber; Amg, basal amygdala; HP, hippocampus; VR, spinal ventral root.
The dorsal spinal root is electrically stimulated during and after the seizure (frequency of stimulation 0.5/second). The sensory evoked potentials are clearly seen during the cortical post-critical "silence".
Notice the fast high voltage activity present in NR and Amg during the interclonic intervals and seizure arrest.

Several signs of the REM phase of sleep are present during the arrest of convulsive activity, namely, inhibition of neck muscle tonus, rapid eye movements, fast multineuronal subcortical activity and hippocampal theta activity (Fernández-Guardiola *et al.*, 1968).

During this period, there is a marked and prolonged inhibition of the spinal monosynaptic reflex together with the signs of REM (Fernández-Guardiola and Ayala, 1971) (Fig. 2 and 3). Thus, it is very likely that the same mechanisms might be acting both on the arrest of seizures and on the REM phase of sleep.

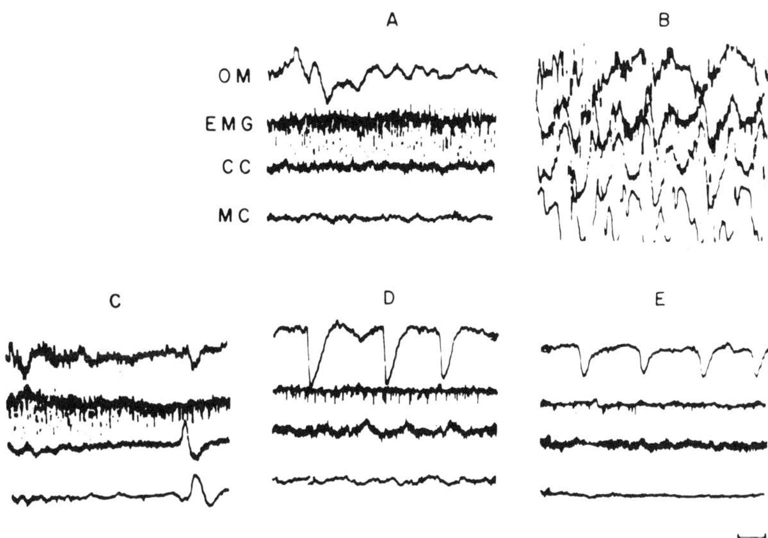

Figure 2. Signs of REM sleep appearing during seizure arrest in the cat.
OM, recording of eye movements; EMG, Electromyogram of neck muscles; CC, cerebellar cortex; MC, Motor cortex. A, control; B, clonic phase of electroshock induced seizure; C, D and E, seizure arrest. In the last seizure periods REMs and a progressive EMG depression take place.

The results reported by Jouvet (1965) on the enhancement of seizure susceptibility produced by artificial REM deprivation in cats gives further support to this hypothesis. Moreover, the relationship between seizure arrest and REM may help to understand, by means of a neurochemical approach the process of seizure extinction, since the role of neurotransmitters in the integration of the REM phase of sleep is already better known.

There is evidence that subcortical inhibitory mechanisms are involved in the effects of melatonin (5-methoxy-N-Acetyl tryptamine). This methoxy-indole is a hormone produced mainly by the pineal gland (Axelrod and Weissbach, 1961). Early evidence showed that

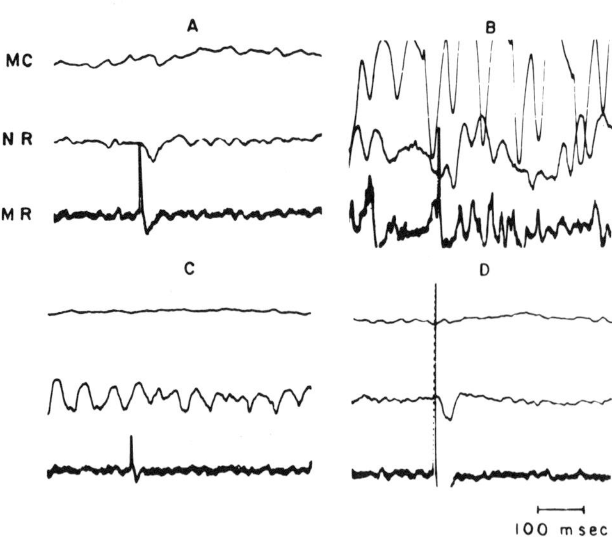

Figure 3. Spinal monosynaptic reflex depression during seizure arrest in the cat.
MC, motor cortex; NR, Nucleus Ruber; MR, Spinal monosynaptic relfex recorded in the ventral root. A, control; B, clonic phase; C, seizure spontaneous arrest; D, recovery.
The spinal monosynaptic reflex is markedly depressed, coinciding with rapid sinusoidal activity in red nucleus.

melatonin exerted inhibitory effects on the pituitary-ovary axis and, considering that melatonin implants in the midbrain or in the median eminence modify gonadal function, it has been suggested that they are mediated by CNS structures (Martini, Fraschini and Motta, 1968).

In recent years, metabolic, electrophysiological and clinical effects of melatonin have been described; melatonin administration is followed by sleep induction and electroencephalographic changes in the cat (Marczynski et al., 1962) and young chicken (Hishikawa et al., 1969) as well as by the potentiation of hexobarbital sleeping time in mice (Barchas et al. 1967). Intraperitoneal administration of melatonin increases brain GABA (Antón-Tay, 1971) and 5HT (Antón-Tay et al., 1968) concentrations; the loci at which melatonin produces the greatest increase of 5HT (the midbrain and the hypothalamus) are also the sites at which circulating H^3-melatonin is most highly concentrated (Antón-Tay and Wurtman, 1969). Thus both the neural and endocrine effects of administered melatonin have been related to the changes that this compound produces in the functional activity of the central serotonin containing neurons. It has also been reported that intraperitoneal administration of melatonin increases the brain pyridoxal phosphokinase activity (Antón-Tay et al., 1970), an enzyme that catalyzes the formation of pyridoxal phosphate. Pyridoxal phosphate is a co-factor for brain decarboxylases and transaminases. Thus the increase in brain 5HT and GABA concentrations after melatonin administration are probably mediated by increases in the activities of 5-hydroxytryptophan and glutamic decarboxylases, respectively, suggesting that melatonin modifies the metabolic activity of the brain by increasing the availability of pyridoxal phosphate.

Recently, evidence indicating that melatonin modifies also the activity of adenyl-cyclase was obtained (Ortega et al., 1973). Brain 3'-5'-adenosine monophosphate (cAMP) levels were assayed by radioimmunoassay after the intraperitoneal administration of melatonin to rats. The effect of melatonin on cAMP levels varied according to the region studied. The midbrain showed a 31% fall in contrast to the cerebellum which showed a 57% increase, while the levels in cerebral cortex remained unchanged.

The changes in brain cAMP observed after melatonin administration suggest that the hormone acts at least on two different kinds of receptors which are located in the midbrain and cerebellum. The midbrain receptor responds to melatonin by a fall in the cAMP level, while the receptors in the cerebellum produce an increase of the cyclic nucleotide. These results point out that melatonin effects are mediated through specific receptors.

Current electrophysiological evidence also supports the action of melatonin on inhibitory subcortical mechanisms. Besides the above mentioned behavioural and electroencephalographic effects of melatonin administration, we are studing the effects of this hormone on the neural activity of specific nuclei. In experiments with unanesthetized curarized cats we observed that, concurrently with EEG changes, there is an increase of red nucleus activity. In these experiments, multineuronal unit activity (MUA) of the red nucleus was bilaterally recorded. A microelectrode was placed in the magnocellular portion of the left side and a macroelectrode in the corresponding area of the right side. The motor cortex gross activity was also constantly monitored. The definite position of the stereotaxically oriented microelectrode in the red nucleus was established by the presence of a clear response to tactile stimuli applied to the distal portions of the hindlegs and forelegs.

The red nucleus MUA was processed by means of a comparator circuit eliminating low voltage activity. The high voltage spikes were measured second by second with an electronic counter. A one step generator calibrated to reset after 25 pulses and/or 1 second was coupled to the recording system to visualize the frequency changes. Frequency distribution histograms were computed.

After recording the control resting MUA activity, 10 mg/kg. of melatonin were administered I.V. The results showed an initial enhancement of discharge frequency reaching its maximum about 15 minutes after the melatonin injection, followed by a noticeable depression of MUA activity below the control frequency in 75 minutes. Control frequency figures recovered 120 minutes after the injection (Fig. 4).

The gross activity in red nucleus paralleled these changes, i.e. it showed an increased frequency at the beginning, followed by amplitude depression. The motor cortex displayed low amplitude high frequency gross activity as a result of melatonin administration.

In human subjects melatonin also elicited EEG and behavioural changes. Thus, the administration of 1.25 mg/Kg of melatonin to healthy volunteers is followed by slight EEG deactivation and sleep in the first 20 minutes (Antón-Tay et al., 1971). EEG monitoring during the following 2 hours showed an increase in alpha rhythms which appeared to have more amplitude. In the interview after these experiments, most subjects referred to a sensation of well being, comfort and elation.

On administering the hormone I.V. to epileptic patients wearing bilateral deep temporal lobe electrodes, we found that a single dose of melatonin ranging from 0.25 to 1.25 mg/kilogram was followed by a

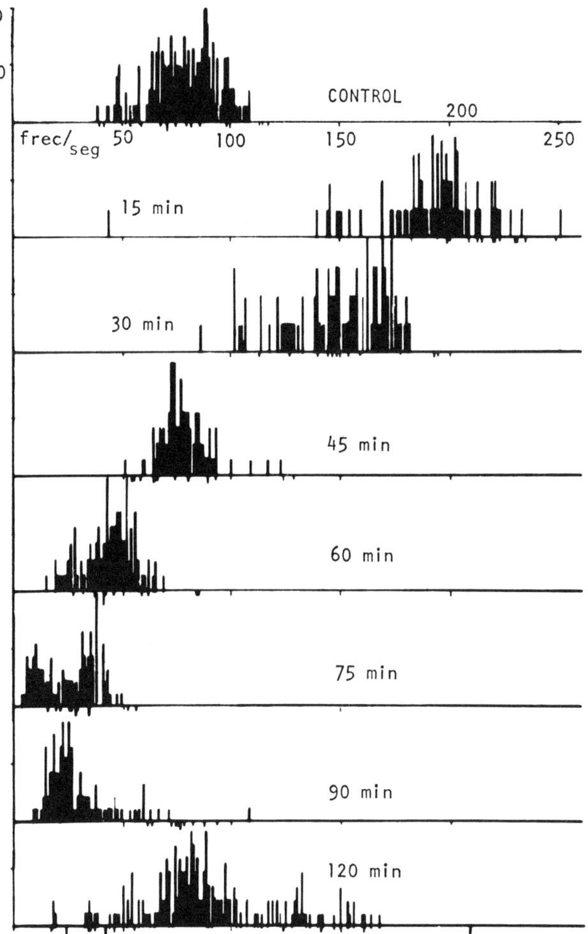

Figure 4. Frequency distribution histrograms of cat's nucleus ruber MUA activity. Effect of 10 mg/Kg of melatonin (I.V.). Fifteen minutes after the injection of melatonin the frequency increases up to 250 spikes per second. Later, there is a clear depression down to 50 per second.

generalized deactivation of brain electrophysiological patterns and depression of the paroxysmal localized activity. Also there was a remarkable inhibition of the paroxistic response to hyperventilation.

When administering melatonin to epileptic and Parkinsonian patients, it was noticed also that the subjects had a general feeling of well-being and euphoria.

Because of the above-mentioned data, the quantitative study of the effects on the normal human sleep pattern of chronic melatonin administration was considered useful. The experiments were performed on 6 university students (3 males and 3 females) whose ages averaged 25 years; a double-blind design was employed. After several habituation nights, control recordings were taken. The subjects received 250 mg oral melatonin, 4 times daily, for 6 days. The hormone was administered in a Carbowax 400 solution (250 mg/ml) and the same vehicle was used as placebo. The resting period between drug and placebo trials was 7 days.

The polygraphic recording consisted of 4 channels for occipital and central EEG, two for eye movements, one for electromyogram, one for the Galvanic Skin Response (GSR) and one for Heart Rate (HR). The Rechtschaffen and Kales (1968) technique was used to evaluate the distribution of the different sleep phases. A hypnogram was elaborated for each individual in every placebo, melatonin and control condition. The hypnograms were averaged and plotted as distribution histograms of phase duration. The spontaneous occurence of GSR was analyzed by counting the number of responses every 20 minutes. In a similar way HR was also analyzed.

From a clinical point of view the results were mild; no somnolence was reported except in one case. The subjects did not report any significant subjective change in the quality of their sleep. The time of onset of sleep was slightly increased by melatonin and so was the first REM latency, although these increases were not significant. The number of awakenings during sleep was significantly increased ($P<0.01$). In subjects under control conditions the average number of awakenings was 0.5 times per night, while in those receiving placebo, it was 2.08 times and in those under melatonin, 4.16 times.

Figure 5 shows the distribution histogram of sleep phases duration in control, placebo and melatonin situations. Under melatonin treatment there is a noticeable decrease of phase 4 of sleep, and also a significant increase of phase II duration. The Galvanic Skin Response was also increased by melatonin (Table I) and the most significant difference was found in the inital periods of sleep. Figure 6 shows a typical hypnogram of an average subject. The amount or density of rapid eye movements was also increased by melatonin, and this increase of REMS coincided with the GSR increase, and with a remarkable slowing down of heart rate.

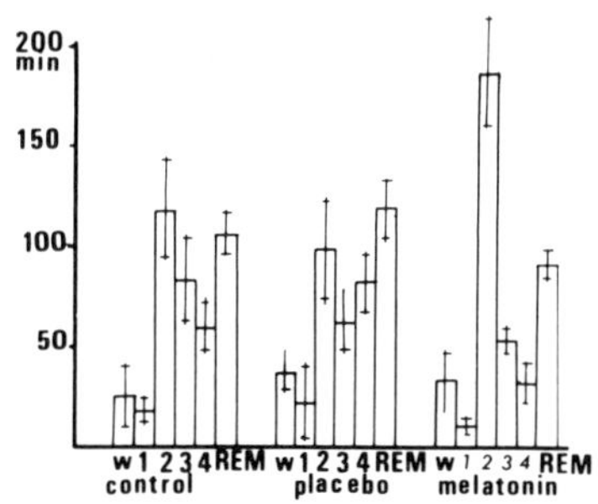

Figure 5. Sleep phase distribution histograms. The bars are the average of 6 all night recordings in control, placebo and melatonin conditions. W, periods of awakening; 1,2,3,4 and REM, sleep phases. There is a significant phase IV reduction and phase II enhancement induced by melatonin.

TABLE I.

CONTROL	PLACEBO	MELATONIN
37.147	34.670	61.350
±3.638	±4.920	±3.932

P < .001*

*Student test

The number of spontaneous GSR during the total sleeping time were counted and averaged for 20 min epochs. Under melatonin administration GSR were significantly increased in average.

These results seem to indicate that the pharmacological effects of melatonin on human sleep are related mainly to cholinergic mechanisms. GSR, bradycardia and rapid eye movements are all sleep phenomena blocked by atropine, (Jouvet, 1965). Thus, melatonin must rank between the substances that did not modify substantially the REM phase of sleep, though it reduces phase 4 and enhances phase 2.

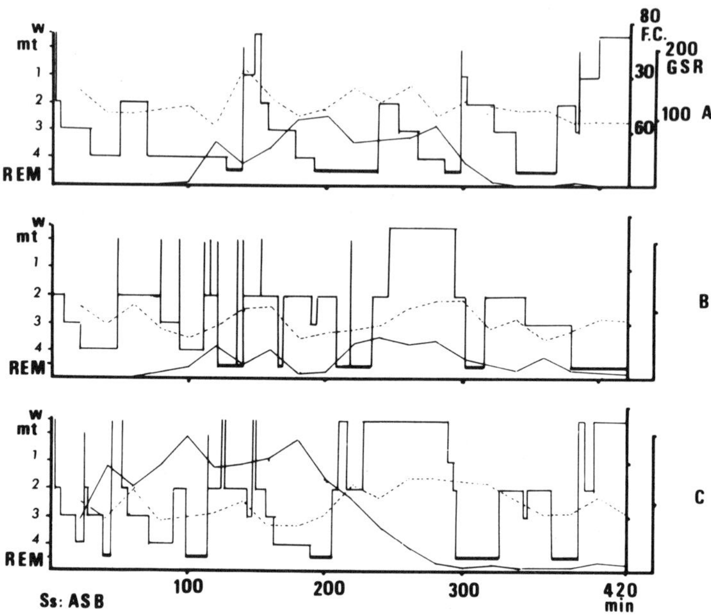

Figure 6. Hypnogram of a subject showing the sequential sleep phase distribution and the evolution of galvanic skin responses (GSR) (continuous line) and Heart Rate (dotted line). A, control; B, placebo; C, Melatonin. For description, see text.

As shown in Figure 7 the effects on sleep of melatonin and imipramine are similar. Considering the psychotropic effect of melatonin, this similarity with tricyclic derivatives suggests a possible antidepressant action of the pineal hormone that must be explored in the future.

The differential actions of melatonin on brain cAMP suggest that the effects of the hormone depend on its interaction with specific structures. Thus several effects several min. after its administration could be explained. Since serotoninergic neurones of the midbrain project their axons to the hypothalamus and to the limbic system through the medial forebrain bundle, the neuroendocrine and behavioural effects of melatonin are probably mediated by the activation of these serotonergic pathways. Similarly, the changes in the sleep pattern and convulsive threshold observed after melatonin administration may be related to the activation of the neural pathways that leave the cerebellum to enter the red nucleus and the reticular formation. Since melatonin and diphenylhydantoin (DPH) have a similar effect on MUA of the red nucleus and Julien et al. (1972) have shown the DPH enhances the spike discharge frequency of the cerebellar Purkinje cells, it is likely that the anticonvulsant properties of melatonin are related to an enhancement of the activity of the cerebellum-rubro-olivo-cerebellar network.

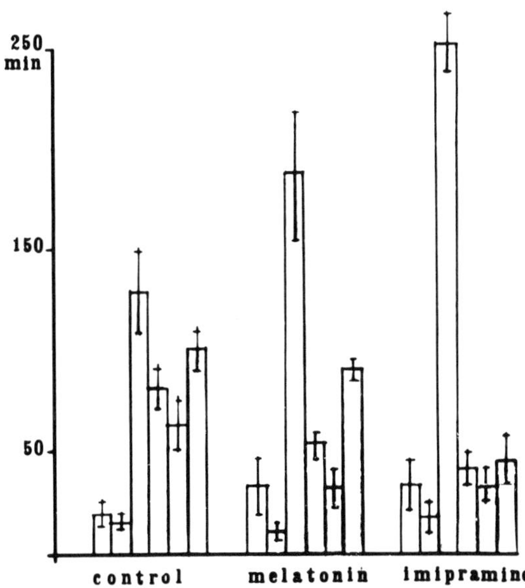

Figure 7. Sleep phases distribution histograms. Averages of 6 subjects in control, melatonin and imipramine (75 mg daily) conditions.
Imipramine and melatonin have similar effects on phase IV and phase II of sleep of normal volunteers. Imipramine decreases significantly REM duration. The order of bars is as in figure 5.

REFERENCES

Antón-Tay, F., 1971, Pineal-brain relationships. In "The Pineal Gland", (G.E.W. Wolstenholme, J. Knight, ed.), pp. 213-227, Churchill, London.

Antón-Tay, F., Chou, C., Antón, S., and Wurtman, R.J., 1968, Brain serotonin concentration: Elevation following intraperitoneal administration of melatonin, Science, 162: 277-278.

Antón-Tay, F., Díaz, J.L., and Fernández-Guardiola, A., 1971, On the effect of melatonin upon human brain. Its possible therapeutic implications, Life Science, 10: 841-850.

Antón-Tay, F., Sepúlveda, J., and González, S., 1970, Increase of brain pyridoxal phosphokinase activity following melatonin administration, Life Science, 9: 1283-1288.

Antón-Tay, F., and Wurtman, R.J., 1969, Regional uptake of H^3-melatonin from blood or cerebrospinal fluid by rat brain, Nature, 221: 474-475.

Axelrod, J., and Weissbach, H., 1961, Purification and properties of hydroxyindole-O-methyltransferase, J. Biol. Chem, 236: 211-213.

Barchas, J., Da Costa, F., and Spector, S., 1967, Acute pharmacology of melatonin, Nature, 214: 919-920.

Fernández-Guardiola, A., Alcaraz, M., and Guzmán-Flores, C., 1956, Modificación de la descarga convulsiva cortical por estimulación mesencefálica, Boletín del Instituto de Estudios Médicos y Biológicos (México) 14: 15-21.

Fernández-Guardiola, A., and Ayala, F., 1971, Red nucleus fast activity and signs of paradoxical sleep appearing during the extinction of experimental seizures, Electroenceph. Clin. Neurophysiol, 30: 547-555.

Fernández-Guardiola, A., Okujava, V.M., and Guma, E., 1968, Peripheral and central phenomena of post-epileptic extinction, Epilepsia (Amst), 9: 303-310.

Martini, L., Fraschini, F., and Motta, M., 1968, Neural control of anterior pituitary functions. In Recent Prog. Hormone Res. (E.B. Astwood, ed.), pp. 439-496. Academic Press. New York and London.

Gastaut, H., and Fischer-Williams, M., 1959, The physiopathology of epileptic seizures. In "Handbook of Physiology" (J. Field ed.), Amer. Physiol. Soc., Washington, pp. 1-329.

Giaquinto, S., Pompeiano, O., and Somogyi, I., 1964, Descending inhibitory influences on spinal reflexes during natural sleep. Arch. Ital. Biol., 102: 282-307.

Hishikawa, Y., Cramer, H., and Kuhlo, W., 1969, Natural and melatonin-induced sleep in young chickens, a behavioural and electrographic study, Experimental Brain Research, 7: 84-94.

Jouvet, M., 1965, Paradoxical sleep: a study of its nature and mechanism, In "Progress in brain research", (Akert ed.), Vol. 18, Elsevier, Amsterdam. pp. 20-57

Jung, R., 1949, Hirnelektrische Untersuchungen über den Elektrokrampf; die Erregungsablaufe in corticalen Hirnregionen bei Katze und Hund, Arch. Psychol. Nervenkr, 183: 206-244.

Kreindler, A., 1955, Epilepsia. Bucaresti, Acad. R.P.R.

Magoun, H.W., and Rhines, R., 1946, An inhibitory mechanism in the bulbar reticular formation, J. Neurophysiol., 9: 165-171.

Marczynski, T.J., Yamaguchi, N., Ling, G.M., and Grodzinska, L., 1964, Sleep induced by the administration of melatonin (5-methoxy-N-acetyl-tryptamine) to the hypothalamus in unrestrained cats, Experientia, 20: 435-436.

Ortega, B.G., Antón-Tay, F., Esparza, N., and Cancino, F.M., 1973, Melatonin: Effect on cAMP concentration in the brain. In Proceeding of the Fourth International Meeting of the International Society for Neurochemistry. Tokyo, Japan. In Press.

Purpura, D.P., and Housepian, E.M., 1961, Alterations in corticospinal neuron activity associated with thalamocortical recruiting responses, Electroencephalog. Clin. Neurophysiol., 13: 365-381.

Rechtschaffen, A., and Kales, A., 1968, A manual of standardized terminology, techniques and scoring system for sleep stages of human subjects, Public Health Service, U. S. Government Printing Office, Washington, D. C. pp. 1-12.

ACKNOWLEDGEMENTS

The investigations reported in this paper were supported by research grants from the Ford Foundation 710-0099 and Patronato Lucio Laguette.

ABSTRACT:

Growing evidence supports the idea that, in mammals, the brain is the target organ for melatonin (N-acetyl-5-methoxytryptamine). This methoxyindol is a hormone synthesized mainly by the pineal gland. Recent evidence from our laboratory suggests that its mechanisms of action on the brain are likely related to a modulatory effect on the subcortical and mesencephalic inhibitory networks. In the present paper, we will review these data and we will try to integrate them into a general hypothesis.

After a single dose of melatonin to cats, rats, or chickens, the more frequent changes are EEG synchronization, alpha rhythm enhancement and sleep induction. In cats, increasing firing of neurons of the red nucleus has also been recorded. In humans, acute and chronic melatonin administration is also followed by EEG synchronization and changes in the sleep patterns. All night sleep studies have shown that melatonin shortens sleep onset latency, increases phase II and reduces REM phase.

The administration of the hormone for larger periods of time, as has been done on Parkinsonians and epileptic patients, produces, besides the above mentioned changes and the amelioration of their clinical pictures, a notorious improvement of mood.

Hormonal changes due to melatonin administration had been observed. Most of them are also inhibitory in nature (i.e., blockade of LH and prolactin secretion). The effects mentioned are mainly integrated in brain structures in which radioactive melatonin is taken up (the brain stem and hypothalamus) and also the areas in which the hormone produces striking metabolic changes (i.e., increased 5-HT and GABA = decreased levels).

Thus, it is suggested that the hormone exerts its action upon subcortical mesencephalic structures which are related to the control of sleep cycles, integration of generalized seizures and neuroendocrine functions.

INTERACTION OF AMINE SYSTEMS IN THE CENTRAL NERVOUS SYSTEM IN THE REGULATION OF THE STATES OF VIGILANCE

P.J. Morgane and W.C. Stern

Laboratory of Neurophysiology
The Worcester Foundation for Experimental Biology
Shrewsbury, Massachusetts

One of the unique features of the nervous system is its tremendous organizational complexity. It is this complexity, well adapted to subserve function, which makes the particular <u>site</u> of a chemical reaction in the brain a parameter of essentially equal importance to substrate and enzyme activity in determining outcome. In the nervous system, perhaps more than in any other system in the body, biochemistry recognizes its dependence on structure. It is emphasized at the outset that a knowledge of the anatomical organization of the neurons specialized in neurohumoral function is a prerequisite for any approach to elucidating their roles in behavior. Although more is known about the chemical anatomy of the biogenic amine and cholinergic systems than other chemical pathways it should be kept in mind that such chemical neuron tracts as the histamine, glycine and GABA systems also play essential roles in behavior. To date, however, little is known of their topographical organization and chemical geometry.

Since neuronal actions are mediated by highly localized neuronal tracts containing neurotransmitter substances, one of our overall specific objectives has been to further characterize and topographically isolate these elements in order to manipulate them individually and in various combinations and thereby gain insight into their possible interactions. One of our aims has been to construct chemo-morphological models of neural systems using, as indices of their function, specific biologically constant neurophysiological indicators (such as ponto-geniculo-occipital (PGO) spikes and sleep-waking profiles) in conjunction with studies of amine alterations regionally in the brain. It is essential to localize transmitter systems in the brain before endeavoring to understand such complex but important phenomena as sleep, arousal, sexual activity, aggression, etc., all of which are dependent on synaptic linkages whose transmitters can be studied by histochemical, pharmacological, and physiological means. The development of transmitter histochemical methodology has led to the new field of chemical neuroanatomy in which chemical pathways are no longer viewed as mere threads tying together various parts of the brain, but rather as dynamic transmitter systems in which the pathway actually becomes the chemical message. Through the application of histofluorescence mapping a new view of the brain has emerged in which systems of neurons with specific neurotransmitter functions have been differentiated and show a specific chemical identity (Dahlström and Fuxe, 1964, 1965; Fuxe et al., 1968; Morgane and Stern, 1973). With this approach, for example, previously unrecognized neuronal systems have been identified in the reticular formation and this characterization of amine-specific pathways permits, for the first time, a promising functional view of the organization of that complex "system of systems". Since many of these chemical systems are located in places where lesions have been shown to impair sleep or waking, it is now possible to approach the chemical organization of the vigilance mechanisms by systematically studying these chemical systems through lesions, stimulations, recordings, neuropharmacological alterations and biochemical analysis. The identification of the neurons responsible for the biosynthesis and storage of monoamines is clearly essential to an understanding of the functional role of these biologically active substances in the central nervous system. One aspect of this problem is the appreciation of the role of certain fiber systems in the regulation of monoamine biosynthesis across neuronal systems in the brain. There is much evidence that this synthesis is dependent, in large part, on nerve impulse flow in these pathways. The technique of selective destruction of discrete groups of central neurons by the use of chemical lesions has provided a valuable tool in the identification of neuronal systems in the brains whose integrity is essential for the maintenance of monoamines in the central nervous system.

Both the indole and catecholamines have been implicated in various aspects of the regulation of the states of sleep and wakefulness. In particular, the studies of Jouvet (1969, 1972a,b) have demonstrated the complex interactions of chemical circuitry in the brain in regulating the oscillations between slow-wave sleep, REM sleep, and waking. To date, however, approaches to the elucidation of the relative roles of the various amine systems in generating and maintaining the states of vigilance, using electrolytic and psychopharmacological techniques, have produced ambivalent and conflicting results. Yet, it seems obvious that the main thrust in determining the role of chemical systems in the brain in various aspects of the vigilance states and other behaviors must be based on the chemical properties of these neurons.

To this end we have carried out chemical lesioning studies and unit recording in interrelated amine areas with the object of determining the nature of interactions in these chemical pathways.

In one group of studies in cats we have employed chemical lesioning with 6-hydroxydopamine, an agent which purportedly produces selective degeneration of catecholamine-containing neurons (Bloom et al., 1969; Uretsky and Iversen, 1969). In our studies (also see Panksepp et al., 1973; Zolovick et al., 1973) 6-hydroxydopamine was injected bilaterally through chronically implanted cannulae either into the locus coeruleus (N = 6, 80 μgm as salt per side) or into the area of the ventral noradrenergic pathway (N = 4, 12 μgm as salt per side) of adult cats. In this series of experiments the animals were wired for chronic recordings of sleep profiles with the usual array of electrodes, i.e., cortical, eye, neck and lateral geniculate nucleus. Sleep-waking patterns were recorded for 7 hours per day for the following 4 week period after which the animals were sacrificed for measurement of regional brain amines and histological analysis. The neurochemical patterns of serotonin (5-HT) and norepinephrine (NE) depletions (Table I) in 9 regions of the brain were almost identical in the two groups (locus coeruleus and noradrenergic pathway) in each brain area examined. Although for both 6-hydroxydopamine injection groups the absolute amine values differed from region to region the largest changes in norepinephrine occurred in the basal forebrain area and neocortex, with the least changes occurring in the brainstem. The correlation between the amount of amine changes between the 9 brain regions for the two 6-hydroxydopamine groups was +.54 for norepinephrine ($p < .02$) and +.63 for serotonin ($p < .01$). This indicates that a significantly similar pattern (across brain regions) of monoamine decreases occurred in cats given 6-hydroxydopamine into the locus coeruleus or the ventral noradrenergic pathway. Thus, in contrast to the rat (Bloom et al., 1969; Breese and Traylor, 1970), 6-hydroxydopamine does not appear to exert effects specific to catecholamine systems in the cat. Rather, a substantial degeneration of serotonin-containing neurons is also produced. Our findings agree with the recent report by Petitjean et al. (1972) that high doses (5.0 mg) of 6-hydroxydopamine given into the lateral ventricle of the cat lowered brain serotonin by an average of 33%.

In the chronic recording period, starting one week after 6-hydroxydopamine administration, sleep-waking patterns showed one important difference between the locus coeruleus and noradrenergic pathway groups (Figure 1). While both injection groups exhibited a tendency toward decreased waking and increased slow-wave sleep, the locus coeruleus group showed a 50% decrease in REM time while the ventral noradrenergic pathway group showed a 25 to 50% elevation of REM. We have considered as a possible explanation for these

TABLE I

COMPARISONS OF THE EFFECTS OF 6-HYDROXYDOPAMINE GIVEN INTO THE LOCUS COERULEUS (LC) OR THE VENTRAL NORADRENERGIC PATHWAY (VP) ON REGIONAL BRAIN LEVELS OF NOREPINEPHRINE (NE) AND SEROTONIN (5-HT) IN CAT

Percent of Normal Amine Concentrations[1]

	NE		5-HT	
	VP	LC	VP	LC
Occipital Cortex	42*	28*	54*	60*
Temporal Cortex	38*	48*	69*	76*
Ant. Pyriform Lobe	52*	62*	55*	72*
Post. Pyriform Lobe	74*	55*	43*	53*
Basal Forebrain Area	24*	38*	110	100
Striatum	64*	45*	90	71*
Hypothalamus	47*	49*	91	63*
Cerebellum	67*	49*	65*	81
Pons and Medulla	58*	63*	64*	55*
Mean	52%	49%	71%	70%

*$p < .05$ (2-tailed t-tests) comparing post-6-hydroxydopamine values (25-30 days after injection) to a pool of 16 normal cats.

[1]Normal amine values for these brain regions can be found in Panksepp et al. (1973).

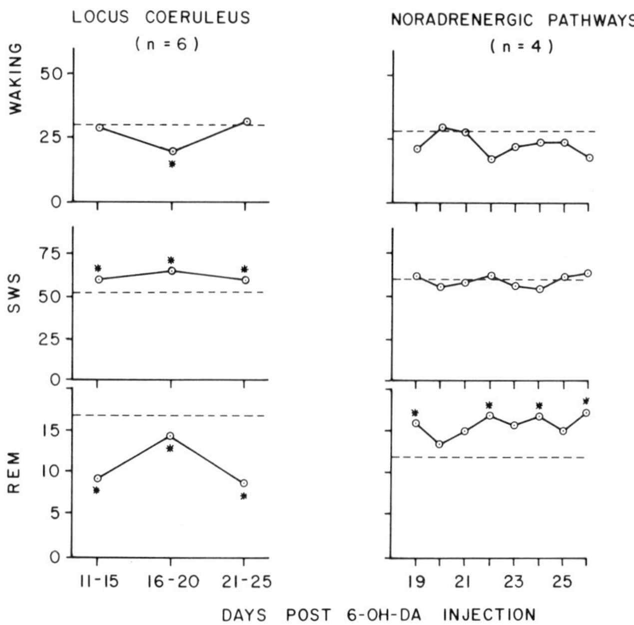

Figure 1. Graph showing effects on sleep-waking activity of 6-hydroxydopamine (6-OH-DA) administered directly into the locus coeruleus or ventral noradrenergic pathway. Ordinates are percent of total recording time. Dotted lines indicate baseline values for waking, slow-wave sleep (SWS) and REM. Asterisks indicate significant differences from baseline.

findings that, despite a very similar pattern of monoamine depletion in most brain regions examined, injections of 6-hydroxydopamine into the locus coeruleus produced non-specific neural damage in the dorso-lateral pontine reticular formation which may have also disrupted cholinergic mechanisms that have been reported to be involved in aspects of REM state (Hernández-Peón _et al._, 1962, 1963; Cordeau _et al._, 1963; George _et al._, 1964). There is some recent evidence that cholinergic mechanisms are involved in PGO spike generation (Magherini _et al._, 1971; Jacobs _et al._, 1972; Henriksen _et al._, 1972). Further, acetylcholinesterase staining techniques have demonstrated possible cholinergic elements in the region of the locus coeruleus (Shute and Lewis, 1963, 1967). Since there is considerable evidence of rather diffuse cholinergic projections ascending from this dorso-lateral pontine region it is possible these may have been lesioned by 6-hydroxydopamine injected into the region of the locus coeruleus. Evidence that 6-hydroxydopamine can abolish acetylcholinesterase staining was shown in the pineal body of rats by Eränkö and Eränkö (1971). In the ventral pathway group the cholinergic tracts might have been spared after 6-hydroxydopamine (since the injection site is several mm removed from the locus coeruleus) and REM time would then show the same elevation that we (Stern and Morgane, 1973) and others (King and Jewett, 1971) have reported to occur in the cat after injection of the catecholamine antagonist, alpha-methyltyrosine. On the whole this group of studies indicates that, contrary to one commonly accepted view of the chemistry of REM, this state does not appear to be wholly dependent on catecholamine systems since REM time actually increased after disruption of catecholamine-containing neurons (ventral pathway group). Figure 2 shows the brainstem areas involved in the two types of 6-hydroxydopamine lesions. It is apparent from this figure that the lesion in the area of the locus coeruleus (Figure 2, left) involves regions in which cholinergic circuitry projects whereas the lesion shown in Figure 2, right, involves an area in which this circuitry is sparse. Admittedly, the cholinergic circuits in the brain are more diffuse and do not tend to compact into discrete bundles. However, from what we know from acetylcholinesterase mapping, the more laterodorsal lesion would more likely involve the heaviest cholinergic projections.

During the past year we have also been studying the effects of 5,6 and 5,7-dihydroxytryptamine (DHT) on regional brain chemistry in the rat in order to determine the specificity of these agents in destroying serotonin systems. These two serotonin analogues have been reported in a series of experiments by Baumgarten _et al._ (1971, 1972, 1973) to exert prolonged and specific destructive effects on serotonergic systems. As a basis for carrying out physiological studies of effects of producing chemical imbalances in the brain, we have attempted to determine the specificity of the lesion effect on biogenic amines in the acute and chronic period and the regional effects on biogenic amines of injection of 5,6-DHT and its isomer 5,7-DHT in rats. In these studies both agents were injected into the lateral ventricle in doses of 50 or 100 μgm in 25 μl volumes. The regional effects on serotonin and norepinephrine levels were determined on post-injection days 2 and 10 (N = 6 per group) in the telencephalon, diencephalon and lower brainstem (remainder) and in a peripheral tissue, the stomach. As shown in Table II, 5,6-DHT significantly lowered serotonin levels to 50 to 65% of normal values on days 2 and 10 in all three brain regions examined, while only transiently lowering norepinephrine in one brain area (remainder) on day 2. The degree of depression in serotonin after 5,6-DHT injection was as great or greater

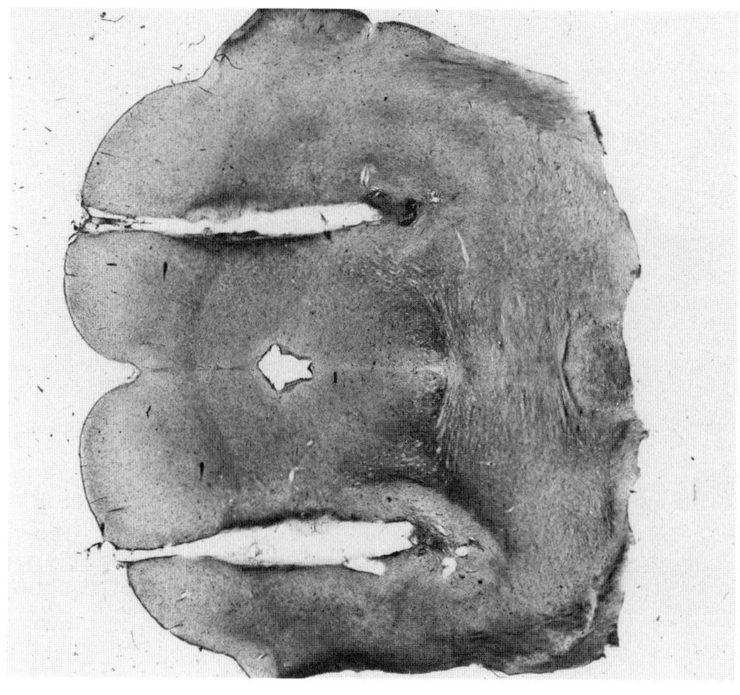

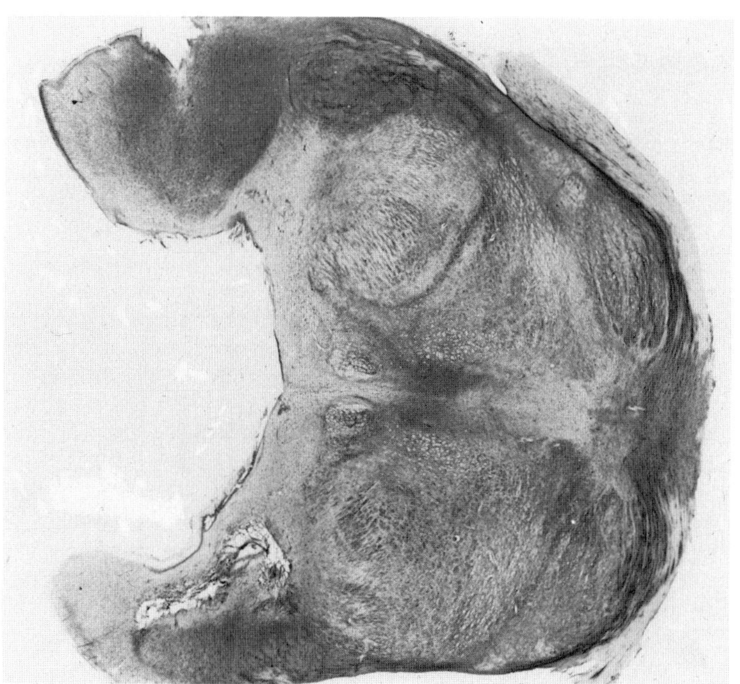

Figure 2. Photograph of 6-hydroxydopamine lesions in the locus coeruleus (left) and ventral noradrenergic pathways (right) of the cat. Klüver stain, approximately 8.5 X.

TABLE II

MONOAMINE LEVELS IN RAT AFTER INTRAVENTRICULAR INJECTION OF 5,6-DIHYDROXYTRYPTAMINE

Brain Region	Treatment	Mean ± S.E. ng/gm					
		SEROTONIN		5-HIAA		NOREPINEPHRINE	
		Day 2	Day 10	Day 2	Day 10	Day 2	Day 10
Telen.	Normals	514 ± 19	514 ± 19	299 ± 21	299 ± 21	255 ± 27	255 ± 27
	veh.	427 ± 32*	413 ± 28*	254 ± 32	274 ± 21	239 ± 19	224 ± 5
	50μg	324 ± 13*	319 ± 21*	178 ± 22*	260 ± 42	221 ± 15	235 ± 12
	100μg	301 ± 14*	303 ± 31*	183 ± 32*	227 ± 17*	278 ± 13	215 ± 12
Dien.	Normals	848 ± 28	848 ± 28	862 ± 107	862 ± 107	631 ± 31	631 ± 31
	veh.	810 ± 46	893 ± 45	582 ± 59*	558 ± 87*	553 ± 34	659 ± 36
	50μg	595 ± 30*	563 ± 40*	396 ± 65*	443 ± 69*	608 ± 40	595 ± 58
	100μg	567 ± 43*	574 ± 52*	469 ± 108*	539 ± 110*	510 ± 34*	630 ± 31
Remainder	Normals	614 ± 42	614 ± 42	309 ± 18	309 ± 18	350 ± 16	350 ± 16
	veh.	469 ± 19*	663 ± 49	329 ± 14	457 ± 62*	321 ± 11	387 ± 33
	50μg	434 ± 29*	359 ± 18*	300 ± 45	230 ± 22*	267 ± 12*	334 ± 16
	100μg	409 ± 36*	296 ± 52*	242 ± 28	283 ± 33	301 ± 13*	325 ± 31
Stomach	Normals	3345 ± 1091	3345 ± 1091	903 ± 522	903 ± 522	477 ± 239	477 ± 239
	100μg	3655 ± 791	4118 ± 1218	1945 ± 903*	910 ± 471	367 ± 55	486 ± 136

Abbrev: Veh.-vehicle control (saline and ascorbic acid); Telen.- telencephalon; Dien.-diencephalon; Remainder-cerebellum, midbrain, pons and medulla. Days 2 and 10 are the days of sacrifice following intraventricular injection (n=6 per group).
* p<.05 for comparisons of injected groups to normals, t-tests, 2-tailed.

on day 10 as on day 2 in all three brain regions. As seen in Table II, there is generally a dose/response type depletion of serotonin and 5-HIAA in the brain regions examined over the 2 doses. Also, serotonin was not depleted in the peripheral organ examined, i.e., the stomach, on either day 2 or day 10. Table III shows the effects of intraventricular administration of 5,7-DHT. On day 2 and day 10 in the telencephalon and remainder of the brain both serotonin and norepinephrine were significantly depressed. In the diencephalon serotonin was not depressed on day 2 but was significantly depleted on day 10, whereas norepinephrine was not significantly depressed in this region on either day. It is interesting that serotonin was markedly depleted in the stomach by 5,7-DHT on both days 2 and 10, which is in marked contrast to the effects of 5,6-DHT. Thus, with 5,7-DHT both serotonin and norepinephrine were lowered in the telencephalon and remainder of the brain to 55-80% of normal on day 2 and these effects persisted on day 10. As shown in Table III, 5,7-DHT also significantly decreased 5-HIAA in the telencephalon and diencephalon. Vehicle injections produced few long-lasting (day 10) effects on any chemical measure. These results indicate that, while 5,6-DHT produces relatively specific depleting effects on brain serotonin when injected intraventricularly, 5,7-DHT produces less marked and essentially non-specific monoamine depletions in brain tissue in both the short and long term situation.

There are few physiological studies of the effects of 5,6-DHT. In one preliminary study Froment et al. (1973) have shown that 5,6-DHT injected directly into the anterior raphé of cats does not affect sleep states or amine levels in the brain except for some decrease of serotonin in the cerebellum. They interpreted these results as indicating that 5,6-DHT does not alter the perikarya of serotonergic neurons. On the other hand, after intraventricular injection of 5,6-DHT they found a significant decrease of both slow-wave and REM sleep which lasted for 2 weeks. They also observed a permanent discharge of PGO waves during the first 3 days in the cats which showed the highest degree of insomnia. In biochemical studies they observed no alterations of either norepinephrine or dopamine but did find a significant decrease of serotonin and 5-HIAA in most cortical and subcortical structures. As noted, in our studies of chemical profiles in the rat we also observed (on both days 2 and 10) significant decreases in 5-HIAA in the telencephalon and diencephalon (Tables II and III) in the several brain regions examined following 5,6 and 5,7-DHT administration. Further studies are now in progress in our laboratory to determine the effects of intraventricularly injected 5,6 and 5,7-DHT on sleep profiles in rats and cats.

As a preliminary to describing some electrophysiological studies concerning interactions between the locus coeruleus complex and the raphé, we will briefly summarize some of the presumed

TABLE III

MONOAMINE LEVELS IN RAT AFTER INTRAVENTRICULAR INJECTION OF 5,7-DIHYDROXYTRYPTAMINE

Brain Region	Treatment	SEROTONIN Day 2	SEROTONIN Day 10	5-HIAA Day 2	5-HIAA Day 10	NOREPINEPHRINE Day 2	NOREPINEPHRINE Day 10
		Mean ± S.E. ng/gm					
Telen.	Normals	395 ± 14	395 ± 14	264 ± 30	264 ± 30	228 ± 19	228 ± 19
	veh.	385 ± 24	388 ± 33	229 ± 37	225 ± 32	174 ± 15	223 ± 11
	50μg	430 ± 30	256 ± 20*	147 ± 16*	152 ± 15*	154 ± 12*	202 ± 13
	100μg	254 ± 17*	210 ± 24*	283 ± 19	165 ± 26*	169 ± 11*	159 ± 5*
Dien.	Normals	685 ± 69	685 ± 69	734 ± 92	734 ± 92	719 ± 120	719 ± 120
	veh.	786 ± 68	712 ± 47	706 ± 83	833 ± 173	614 ± 59	759 ± 42
	50μg	655 ± 83	652 ± 119	311 ± 44*	601 ± 108	586 ± 28	687 ± 43
	100μg	660 ± 67	490 ± 51*	410 ± 64*	596 ± 72	601 ± 35	658 ± 62
Remainder	Normals	434 ± 14	434 ± 14	270 ± 16	270 ± 16	292 ± 15	292 ± 15
	veh.	547 ± 42*	399 ± 31	274 ± 18	288 ± 32	374 ± 27*	268 ± 20
	50μg	309 ± 18*	318 ± 20*	221 ± 30	264 ± 72	242 ± 13*	227 ± 13*
	100μg	361 ± 28*	301 ± 43*	239 ± 13	193 ± 29*	236 ± 17*	224 ± 21*
Stomach	Normals	3345 ± 1091	3345 ± 1091	903 ± 522	903 ± 522	477 ± 229	477 ± 229
	100μg	1104 ± 201*	1378 ± 380*	334 ± 122*	524 ± 178	402 ± 110	329 ± 91

Abbreviations and other information are given in the bottom of Table II.

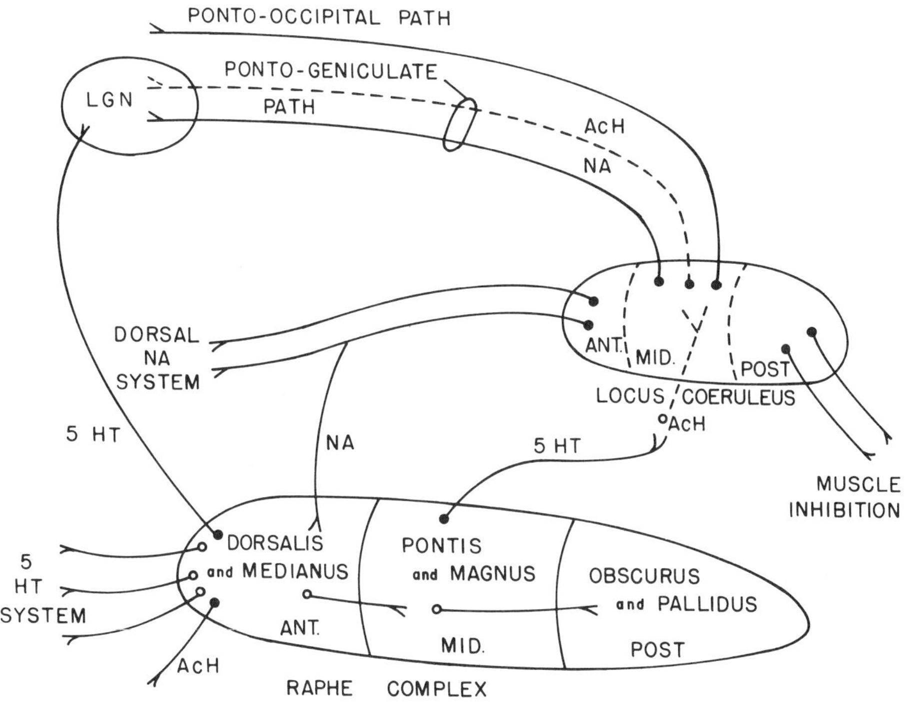

Figure 3. Schematic illustration of chemical connectivities between components of the locus coeruleus and raphé complex. Also shown are pathways to the lateral geniculate nucleus (LGN). Note the short serotonergic (5-HT) neurons linking the anterior (ant.), middle (mid.) and posterior (post.) raphé nuclei. 5-HT neurons from the middle raphé are shown synapsing on possible cholinergic (ACh) neurons, which, in turn, synapse on noradrenergic (NA) neurons in the locus coeruleus.

links between these two areas. Some of these are illustrated schematically in Figure 3. As shown in this figure, it is quite possible that there are 5-HT/5-HT interactions occurring in the raphé system itself, for example between the anterior and posterior raphé. It is likely that short 5-HT axons might effect some interaction between the caudal and rostral raphé systems. This serotonergic neuron in turn, projects to the locus coeruleus, possibly via a cholinergic interneuron. A variety of data indicates that there is a two-way interaction between the raphé system and the locus coeruleus with direct fibers passing from the raphé towards the locus coeruleus. Jouvet (1972b) has emphasized that serotonergic "priming" mechanisms are necessary to trigger the locus coeruleus. This latter complex, in turn, affects serotonergic neurons, presumably via noradrenergic neurons projecting to the anterior raphé. Since the PGO activity which presages and accompanies REM may be associated with a decrease of serotonin release in the terminals of raphé neurons (due to decreased raphé firing, McGinty *et al.*, 1973), it is possible that the fibers from the locus coeruleus may act upon serotonergic cell bodies thus inhibiting raphé firing and thereby suppressing the release of serotonin in the terminals of raphé neurons. Then, the fast cortical activity which is observed during REM or waking might be due to the combined effect of enhanced activity of ascending catecholamine and/or cholinergic mechanisms and a concomitant phasic decrease of firing of serotonin-containing neurons. In this regard there is now histofluorescence mapping evidence of noradrenergic pathways extending from the area of the locus coeruleus into the anterior raphé system (Loizou, 1969). Jouvet (1972b) has postulated that these noradrenergic elements might inhibit the firing of serotonin cells of the raphé, thus suppressing the slow-wave sleep mechanisms and leading to the generation of the REM state.

Thus, it appears more and more likely that complex chemical interactions occur between the components of the raphé and the locus coeruleus region and, in a recent study, we have attempted to determine the nature of some of these relationships. We examined the possible influence of locus coeruleus projections upon raphé unit activity by recording (extracellularly) single units from the nucleus raphé dorsalis following electrical stimulation of the locus coeruleus (Berman, 1973). In this series of studies unit activity of the nucleus raphé dorsalis was recorded with glass microelectrodes from rats under chloral hydrate anesthesia. Baseline firing rates of the raphé units were obtained for 2 minutes prior to presentation of unilateral single pulse, monophasic stimulation (0.5 msec., 20 μA) of the locus coeruleus. These pulses were presented once every 30 seconds via bipolar concentric needle electrodes. To date we have recorded a total of 14 raphé units in 14 different rats (1 unit per rat) and 8 of these units showed inhibition following locus coeruleus stimulation while 6 showed no effect. The mean level of inhibition for the first 2 seconds after the onset of the stimulation in the 8 cells affected was 30% (baseline spike rate = 1.1 spikes /sec.; post-stimulation rate = 0.74 spikes /sec.). In most instances, a small inhibitory effect was still noted 15-30 seconds after the stimulation. In another set of rats (N = 12) multipulse stimulation of the locus coeruleus was given 80-200 μA, 0.5 msec, 60 pulses /sec. for 1.0 sec. The results showed a marked inhibition of raphé firings (often 100% inhibition) for up to 3 seconds after cessation of stimulation. A sample of this inhibition is

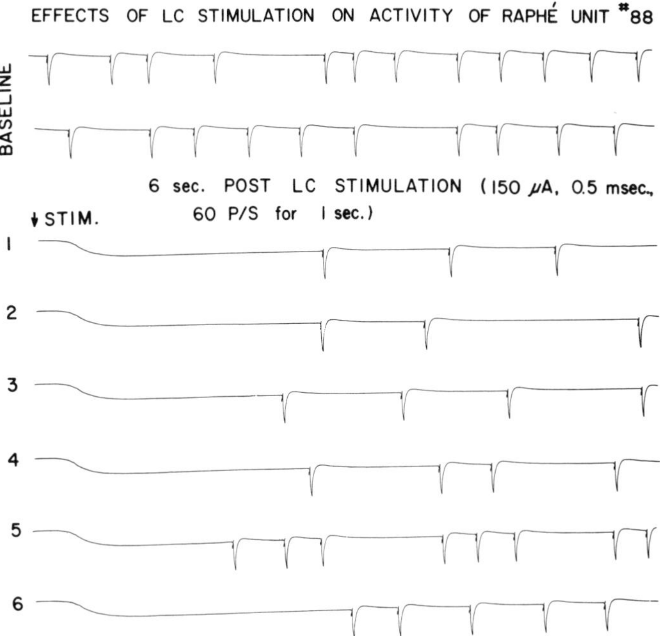

Figure 4. Polygraph recording of Schmitt trigger spikes of a single raphé unit illustrating the baseline firing rate (top 2 tracings) and firing rate (bottom 6 tracings) following stimulation of the locus coeruleus (6 separate stimulations). Each tracing is 6 seconds in duration. Abbreviations: LC - locus coeruleus; P/S - pulses per second.

illustrated in Figure 4. To date 9 out of 12 raphé units have shown statistically significant inhibition to locus coeruleus multipulse stimulation while none showed excitation. Prior to sacrifice the activity of raphé units were generally recorded following an intraperitoneal injection of LSD (250 μg/kg). All units tested with LSD (N = 18) showed cessation of firing, thereby providing indirect evidence of raphé localization. In terms of anatomical specificity it was observed that electrical stimulation of regions outside of the locus coeruleus did not inhibit raphé neuron firing (N = 4) and, similarly, locus coeruleus stimulation did not affect firing of cells approximately 1 mm lateral to the nucleus raphé dorsalis (N = 3). Figure 5 shows, respectively, a sample of electrode placement in the locus coeruleus (left) in which stimulation inhibited raphé unit firing. This same figure shows the microelectrode placement in the nucleus raphé dorsalis (right) in which area marked inhibition was produced following stimulation of the locus coeruleus.

In summary, these data indicate that the locus coeruleus can inhibit the activity of anterior raphé serotonergic units and suggest that one effect of activation of the locus coeruleus may be to decrease serotonin release in the terminals of anterior raphé projections. It is obvious that we need to know a great deal more about the interactions between the locus coeruleus and various other components of the raphé complex, including the posterior raphé, the latter possibly being a trigger for the "executive" mechanisms for REM sleep in the locus coeruleus. It is possible that influences from the anterior raphé to the posterior triggering mechanism may be the critical link between the anterior raphé and locus coeruleus. It is also possible that the anterior raphé may interact with the locus coeruleus by direct pathways not synapsing in the posterior raphé area. These possibilities are now being investigated by evoked response techniques in our laboratory.

These studies as a whole emphasize that various approaches to chemo-anatomical systems may be used profitably to study the nature of interrelations and create imbalances in them to alter aspects of the vigilance states. One approach employed is based on the chemical anatomy of these systems and the chemical properties of various analogues of serotonin or norepinephrine that allow them to somewhat selectively be incorporated into specific neuron systems. In chemical lesion studies we have attempted to disrupt function of specific amine pathways thereby creating chemical imbalances that may be reflected in sensitive behavioral and electrographic changes. In general, we have found that 6-hydroxydopamine does not produce specific monoamine depletions in cat brain and that the effect of this agent on sleep depends upon the site of injection. The serotonin analogue 5,6-DHT appears to be a more specific depletor of brain serotonin than 5,7-DHT, and, therefore, merits future physiological studies of its effects. The demonstration of powerful inhibitory effects of the locus coeruleus on raphé firing might be looked on as a first step toward understanding relations between these areas which would ultimately lead to attempts to correlate these alterations with chemical changes at the terminals of raphé units in the basal forebrain area. Such determinations may be carried out in the chronic period by use of types of _in vivo_ devices such as the push-pull cannula system or _in vivo_ biopsy-type devices. We are presently carrying out feasibility studies for

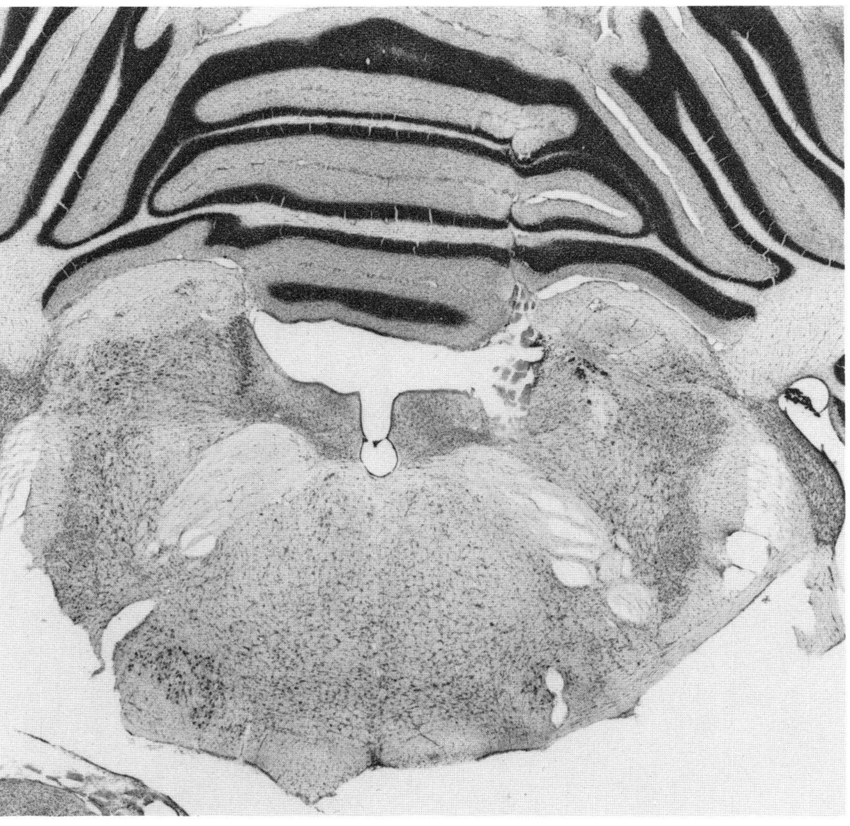

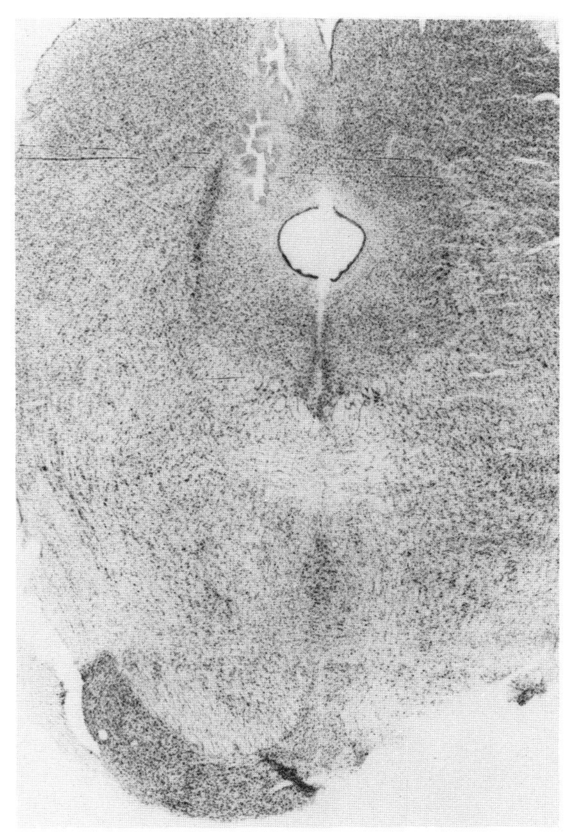

Figure 5. Photograph of rat brain showing site of stimulating electrode in the locus coeruleus (left) and site of microelectrode (recording electrode) in the nucleus raphé dorsalis (right). Klüver stains, approximately 12 X, and 24 X, respectively.

chronic *in vivo* biopsy of these brain areas in the hope of correlating activity in specific chemical systems with behavior of the whole organism.

ACKNOWLEDGEMENTS

The authors wish to thank Dr. Albert A. Manian, Pharmacology Section, Psychopharmacology Research Branch, National Institute of Mental Health, for providing us with 5,6- and 5,7-dihydroxytryptamine. This research was supported by Grants 02211 and 10625, National Institute of Mental Health.

REFERENCES

Baumgarten, H. G., Björklund, A., Lachenmayer, L., Nobin, A., and Stenevi, U. 1971, Long-lasting selective depletion of brain serotonin by 5,6-dihydroxytryptamine, *Acta Physiol. Scand., Suppl.* 373: 1.

Baumgarten, H. G., Lachenmayer, L., and Schlossberger, H. G. 1972, Evidence for a degeneration of indoleamine containing nerve terminals in rat brain, induced by 5,6-dihydroxytryptamine, *Z. Zellforsch.* 125: 553.

Baumgarten, H. G. Victor, S. J., and Lovenberg, W. 1973, Effect of intraventricular injection of 5,7-dihydroxytryptamine on regional tryptophan hydroxylase of rat brain, *J. Neurochem.* 21: 251.

Berman, E. 1973, The effects of locus coeruleus stimulation on raphé unit activity, Master's Thesis, Worcester Polytechnic Institute.

Bloom, F. E., Algeri, S., Groppetti, A., Revuelta, A., and Costa, E. 1969, Lesions of central norepinephrine terminals with 6-OH-dopamine: biochemistry and fine structure, *Science* 166: 1284.

Breese, G. R., and Traylor, T. D. 1970, Effect of 6-hydroxydopamine on brain norepinephrine and dopamine: evidence for selective degeneration of catecholamine neurons, J. Pharmacol. Exp. Ther. 174: 413.

Cordeau, J. P., Moreau, A., Beaulnes, A., and Laurin, C. 1963, EEG and behavioral changes following microinjections of acetylcholine and adrenaline in the brain stem of cats, Arch. Ital. Biol. 101: 30.

Dahlström, A., and Fuxe, K. 1964, Evidence for the existence of monoamine-containing neurons in the central nervous system. I. Demonstration of monoamines in the cell bodies of brain stem neurons, Acta Physiol. Scand., Suppl. 232: 1.

Dahlström, A., and Fuxe, K. 1965, Evidence for the existence of monoamine neurons in the central nervous system. II. Experimentally induced changes in the intraneuronal amine levels of bulbospinal neuron systems, Acta Physiol. Scand., Suppl. 247: 1.

Eränkö, O., and Eränkö, L. 1971, Loss of histochemically demonstrable catecholamines and acetylcholinesterase from sympathetic nerve fibres of the pineal body of the rat after chemical sympathectomy with 6-hydroxydopamine, Histochem. J. 3: 357.

Froment, J., Petitjean, F., Bertrand, N. Cointy, C., and Jouvet, M. 1973, Decrease of sleep and of cerebral indolamines after intraventricular injection of 5-6DHT in the cat, Association for the Psychophysiological Study of Sleep, Annual Meeting.

Fuxe, K., Hökfelt, T., and Ungerstedt, U. 1968, Localization of indolealkylamines in CNS, in "Advances in Pharmacology," (S. Garattini and P. A. Shore, eds.), pp. 235-251, Academic Press, New York.

George, R., Haslett, W. L., and Jenden, D. J. 1964, A cholinergic mechanism in the brainstem reticular formation: induction of paradoxical sleep, Int. J. Neuropharmacol. 3: 541.

Henriksen, S. J., Jacobs, B. L., and Dement, W. C. 1972, Dependence of REM sleep PGO waves on cholinergic mechanisms, Brain Research 48: 412.

Hernández-Peón, R., Chávez-Ibarra, G., Morgane, P. J., and Timo-Iaria, C. 1962, Cholinergic pathways for sleep, alertness and rage in the limbic midbrain circuit, Acta Neurologica Latinoamericana, 8: 93.

Hernández-Peón, R., Chávez-Ibarra, G., Morgane, P. J., and Timo-Iaria, C. 1963, Limbic cholinergic pathways involved in sleep and emotional behavior. Experimental Neurology 8: 93.

Jacobs, B. L., Henriksen, S. J., and Dement, W. C. 1972, Neurochemical bases of the PGO wave, Brain Research 48: 406.

Jouvet, M. 1969, Biogenic amines and the states of sleep, Science 163: 32.

Jouvet, M. 1972a, Veille, sommeil et rêve. Le discours biologique, La Revue de Medicine No. 16-17: 1003.

Jouvet, M. 1972b, The role of monoamines and acetylcholine-containing neurons in the regulation of the sleep-waking cycle. in "Ergebnisse der Physiologie," pp. 166-308, Springer-Verlag, Berlin.

King, C. D., and Jewett, R. E. 1971, The effects of α-methyltyrosine on sleep and brain norepinephrine in cats, J. Pharmacol. Exp. Ther. 177: 188.

Loizou, L. A. 1969, Projections of the nucleus locus coeruleus in the albino rat, Brain Research 15: 563.

Magherini, P. C., Pompeiano, O., and Thoden, U. 1971, The neurochemical basis of REM sleep: a cholinergic mechanism responsible for rhythmic activation of the vestibulo-oculomotor system, Brain Research 35: 565.

McGinty, D. J., Harper, R. M., and Fairbanks, M. K. 1973, 5-HT-containing neurons: unit activity in behaving cats, in "Serotonin and Behavior," (J. Barchas and E. Usdin, eds.), pp. 267-279, Academic Press, New York.

Morgane, P. J., and Stern, W. C. 1973, Chemical anatomy of brain circuits in relation to sleep and wakefulness, in "Advances in Sleep Research," (E. Weitzman, ed.), in press, Spectrum Publications, New York.

Panksepp, J., Jalowiec, J. E., Morgane, P. J., Stern, W. C. and Zolovick, A. J. 1973, Noradrenergic pathways and sleep-waking states in cats, Experimental Neurology, in press.

Petitjean, F., Laguzzi, R., Sordet, F., Jouvet, M. and Pujol, J. F. 1972, Effets de l'injection intraventriculaire de 6-hydroxydopamine. I. Sur les monoamines cérébrales du chat, Brain Research 48: 281.

Shute, C. C. D. and Lewis, P. R. 1963, Cholinesterase-containing systems of the brain of the rat, Nature 199: 1160.

Shute, C. C. D. and Lewis, P. R. 1967, The ascending cholinergic reticular system: neocortical, olfactory and subcortical projections, Brain 90: 497.

Stern, W. C. and Morgane, P. J. 1973, Effects of α-methyltyrosine on REM sleep and brain amine levels in the cat, Biological Psychiatry 6: 301.

Uretsky, N. J. and Iversen, L. L. 1969, Effects of 6-hydroxydopamine on noradrenaline-containing neurones in the rat brain, Nature 221: 557.

Zolovick, A. J., Stern, W. C., Jalowiec, J. E., Panksepp, J., and Morgane, P. J. 1973, Sleep-waking patterns and brain biogenic amine levels in cats after administration of 6-hydroxydopamine into the dorso-lateral pontine tegmentum, Pharmacology, Bio-Chemistry and Behavior, in press.

ABSTRACT:

The indole- and catecholamines have been implicated in various aspects of the regulation of the states of sleep and wakefulness. Approaches to the elucidation of the relative roles of these and other amine systems in maintaining and elaborating the states of vigilance using electrolytic lesioning and psychopharmacological techniques have produced ambivalent and conflicting results. Chemical lesioning in amine-specific pathways may offer a better hope of resolving some of these issues. Also, studying of unit activity in inter-connected areas thought to intimately relate to aspects of waking, slow wave sleep, and REM sleep is a relatively new approach which should shed additional light on these problems. We will summarize several of these types of recent studies in our laboratory. In one group of studies we have employed "chemical lesioning" with 6-hydroxydopamine (6-OHDA), an agent which purportedly produces selective degeneration of catecholamine containing neurons. 6-OHDA was injected bilaterally through chronically implanted cannulae either into the locus coeruleus (n=6, 80 μg as salt per side) or into the ventral noradrenergic pathways (n=4, 12 μg as salt per side) of adult cats, and sleep-waking patterns were recorded for the following 4 weeks after which the animals were sacrificed for measurement of regional brain amines and histology. The neurochemical patterns of serotonin (5-HT) and norepinephrine (NE) depletions in the 9 regions of the brain examined were almost identical in the 2 groups. The average decrease in NE was 50% while that for 5-HT was 30%. Thus, in contrast to the rat, 6-OHDA does not exert catecholamine-specific effects in the cat. In the chronic recording period, starting 1 week after 6-OHDA administration, sleep-waking patterns showed one important difference between the 2 groups. While both injection groups exhibited a tendency toward decreased waking and increased slow wave sleep, the locus coeruleus group showed a 50% decrease in REM time while the ventral noradrenergic pathway group had a 25-50% elevation of REM. One possible explanation for these findings, despite a very similar pattern of neurochemical depletion in different brain regions, is that the locus coeruleus injections of 6-OHDA produced non-specific neural damage in the dorsolateral pontine reticular formation. Such destruction may also have disrupted a cholinergic mechanism of REM sleep. Evidence for such a cholinergic mechanism will also be reviewed. In the ventral pathway group the cholinergic pathways mediating aspects of REM are probably spared after 6-OHDA, and REM time shows the same elevation as we and others have reported to occur after i.p. alpha-methyl-tyrosine, a catecholamine antagonist. Our present studies indicate that, contrary to the commonly accepted view of REM sleep, this state does not appear to be dependent on catecholamine systems. We have also studied the effects of 5,6- and 5,7-dihydroxytryptamine on regional brain chemistry in the rat. These 2 serotonin analogues have been reported to exert prolonged and specific destructive effects on serotonergic neurons. In order to study the specificity of chemical lesioning produced by these agents, they were injected intraventricularly in doses of 50 or 100 μg (25 μl volume). The regional effects on 5-HT and NE levels were determined on post-injection days 2 and 10 (n=6 per group) in the telencephalon, diencephalon, and lower brainstem. 5,6-DHT significantly lowered 5-HT levels to 50-65% of normal on days 2 and 10 in all brain regions while only transiently lowering NE on day 2. The degree of depression in 5-HT was as great or greater on day 10 as day 2. On the other hand, the isomer of 5,6-DHT, 5,7-DHT, produced mixed results: both 5-HT and NE were lowered to 55-80% of normal on day 2 and these effects persisted on day 10. Vehicle injections produced minimal or no effects. These results indicate that, while 5,6-DHT produces relatively specific depleting effects on 5-HT, 5,7-DHT produces less marked and essentially non-specific monoamine depletions in brain tissue in both the short and long-term. The effects of 5,6-DHT on sleep-waking patterns in the rat are currently being examined. In other experiments we have studied the effects of electrical stimulation of the locus coeruleus on anterior raphé unit activity. Many studies have suggested that the raphé nuclei and the locus coeruleus (LC) play significant roles in the sleep-waking cycle. Further, there is histofluorescence evidence of a noradrenergic projection from the LC to the anterior raphé. We examined the possible influence of these LC projections upon raphé unit activity by recording single units from the nucleus raphé dorsalis following electrical stimulation of the LC. Raphé activity was recorded in rats, under chloral hydrate anesthesia, for 2 minutes prior to presentation of unilateral single pulse, monophasic stimulation (0.5 msec, 20 μA). These pulses were given once every 30 seconds via bipolar concentric needle electrodes. To date, a total of 11 raphé units have been recorded (one per rat). Eight units showed inhibition following LC stimulation while 3 showed no effect. The level of inhibition for the first 2 sec. in the 8 cells affected was 30% (baseline spike rate = 1.1 spikes/sec; post-stimulation rate was 0.74 spikes/sec.). A small inhibitory effect was still seen after 15 seconds. Prior to sacrifice 8 raphé units were tested with i.p. LSD (250 μg/kg) and all showed cessation of firing, thereby providing evidence of raphé localization (histology now in progress). Thus, it appears that projections from the LC inhibit the activity of raphé serotonergic units and suggest that one role of the LC is to inhibit serotonin release in terminals of raphé projections. The chemical anatomy of many of these neuronal systems will be discussed in terms of their relative roles in sleep-waking activity. (Supported by NIMH grants 02211 and 10625).

PART IV
NEUROHUMORAL FACTORS AND MENTAL DISORDERS

COMPARATIVE BIOCHEMICAL AND BEHAVIORAL STUDIES OF CHRONIC SCHIZOPHRENIC PATIENTS AND NORMALS

H. E. Himwich and N. Narasimhachari

Thudichum Psychiatric Research Laboratory, Galesburg

State Research Hospital, Galesburg, Illinois 61401

The pathfinding experiments of Kety and his co-workers (Pollin, Cardon and Kety, 1961) revealed that the administration of amino acids, including methionine or tryptophan, with a monoamine oxidase inhibitor, iproniazid (Marsilid), caused behavioral worsening, representing either a "biochemically-induced acute flare-up of a chronic schizophrenic process on the one hand, or a toxic delirium superimposed upon chronic schizophrenia on the other." We used their method but gave the patients L-cysteine as the amino acid and tranylcypromine (Parnate) as the monoamine oxidase inhibitor. We added an additional facet by determining the changes in the urinary contents of these patients that took place simultaneously with the worsening of the symptoms. In all observations, psychopharmacological agents were replaced by similar looking placebos. All diagnoses were made according to the *Diagnostic and Statistical Manual of Mental Disorders* prepared by a committee of the American Psychiatric Association.

BEHAVIORAL CHANGES AND URINARY ALTERATIONS PRODUCED BY TRANYLCYPROMINE (PARNATE) AND L-CYSTEINE IN CHRONIC SCHIZOPHRENIC PATIENTS AND NORMAL CONTROLS

Research Design

One series of observations was made on four chronic male schizophrenic patients, three paranoid, one retarded catatonic, all with active symptoms (Table I, Study 1) (Tanimukai et al., 1970). In this series the substitution of placebo took place from 4 to 6

TABLE I. The Effects of L-cysteine and Tranylcypromine (Parnate) on Urinary Constituents and Behavior in Chronic Schizophrenic Patients and Normals

Patients	Treatment	Biochemical Results (in urine)	Symptoms
Study 1			
4 chronic schizo-phrenics	Rigorous diet L-cysteine Tranylcypromine	N,N-dimethyltrypt-amine (DMT) bufotenin 5-methoxy-N,N-dimethyltryptamine (5-MeODMT)	Exacerbations
Study 2			
6 normals	Rigorous diet L-cysteine Tranylcypromine	No DMT No bufotenin No 5-MeODMT	No psychoto-mimetic symp-toms
2 chronic schizo-phrenics	Rigorous diet L-cysteine Tranylcypromine	DMT bufotenin 5-MeODMT	Exacerbations

weeks prior to the initiation of the study. The patients varied in age from 40 to 58 years. They were placed on a rigorously controlled diet with known amounts of calories and proteins. Certain foods that contained large components of serotonin were withheld (bananas, avocados, eggplant, red plums, tomatoes, walnuts). Collections of 24-hour urinary samples were made daily. After a 7-day control period, the patients were given L-cysteine, starting with 4 grams per day and gradually increasing to 20 grams per day over a period of 24 days. During this time they also received tranylcypromine (Parnate) 10 mg t.i.d. The study was extended for an additional 15 days after the administration of L-cysteine and tranylcypromine had been discontinued.

Methods

Clinical. A team of psychiatrists interviewed the patients weekly and behavioral ratings were made independently by each using the scale of Rockland and Pollin (1965) to evaluate the mental status of the patients. The scale consists of three main categories: general appearance and manner, affect and mood, content of thought and thought processes. Increasing degrees of psychopathology are represented by larger total scores.

Biochemical. A procedure for the concentration and purification of urinary indoleamines and their identification by two-dimensional thin-layer chromatography (TLC) was used (Tanimukai et al., 1970). The trimethylsilyl (TMS) derivatives of the purified indoleamine concentrations from the urinary samples were also analyzed by gas-liquid chromatography (GLC) (Capella and Horning, 1966).

Results

Clinical. The symptoms of the patients became more severe and their urine contained three psychotogenic substances which appeared in the urine with increased frequency as the symptoms worsened. Usually the same symptoms which the patients had presented previously now appeared in an exaggerated form. Sometimes the development of new delusions and new hallucinations was also observed. There is agreement that N,N-dimethyltryptamine (DMT) is four times as potent as mescaline (Snyder and Richelson, 1968; Szara, 1956, 1961, 1967). 5-Methoxy-N,N-dimethyltryptamine (5-MeODMT) is much more active than DMT and much less so than LSD (Shulgin, 1970). Both DMT and 5-MeODMT must be given parenterally. Bufotenin cannot pass the blood-brain barrier but when given intraventricularly, it exerts thought-disturbing effects (Mandell and Morgan, 1971).

Biochemical. The two-dimensional thin-layer analyses revealed the presence of DMT, bufotenin and 5-MeODMT in the urine. In addition bufotenin was identified by a second method, namely, GLC.

We next studied the urine of normal human beings (Table I, Study 2) under the same conditions as the chronic schizophrenic patients. The subjects were six normal young men ranging in age from 19 to 22 years (Narasimhachari *et al.*, 1971a). Simultaneously we observed an additional two chronic schizophrenic patients, 57 and 60 years of age respectively; one was a paranoid and one was a hebephrenic.

Methods

Clinical. The same clinical methods were used as in the previous investigation employing the Rockland and Pollin (1965) scale to evaluate behavior.

Biochemical. Twenty-four hour collections of urine were made throughout the study. These were acidified to pH 2 with 6N HCl and stored frozen at -20°C before use. For the separation of amines from the urine samples, at least 75% of a 24-hour collection (concentrated to 10% of the original volume) was used for either ion exchange chromatography on Dowex 50 (H^+ form), or solvent extraction by ethyl acetate at pH 10-10.4. The basic fractions were then refractionated to separate the primary and secondary amines (Gross and Franzen, 1964) and the final purified concentrates, containing the tertiary amines, were used for thin-layer and gas-liquid chromatographic identification (Narasimhachari *et al.*, 1971b). Two spray reagents, *p*-dimethylaminocinnamaldehyde (DACA) and diazotized *o*-tolidine, were used for identification by the TLC method. For GLC, the free tertiary amines and their TMS derivatives were used. When the TLC and GLC results were positive for the three N,N-dimethyltryptaminic compounds (DMT, bufotenin or 5-MeODMT), the samples were run on preparative TLC or preparative GLC, and the fractions corresponding to these three amines were collected and read on an Aminco Bowman spectrophotofluorometer and fluorescence spectra recorded at activation 295 mu. In this study the results were reported as positive for any of these three dimethylated indoleamines only when they were positive by at least two methods used for their identification.

Results

Despite the fact that the normals received L-cysteine and tranylcypromine and lived under similar environmental conditions as much as possible with two such disparate groups, the normals did not display any psychotomimetic symptoms nor did they excrete any abnormal constituents in their urine. Results of the two chronic schizophrenic patients studied simultaneously were the same as those previously observed in the four chronic schizophrenics (Tanimukai *et al.*, 1970). When they exhibited exacerbations of their symptoms, their urine contained the three psychotogenic substances with increased frequency.

There is supporting enzymatic evidence for such results obtained on the brains of lower animals and humans (Fig. 1). In 1961, Axelrod found in the lung of rabbits an enzyme which could take the normal derivatives of tryptophan, namely tryptamine and serotonin, make a

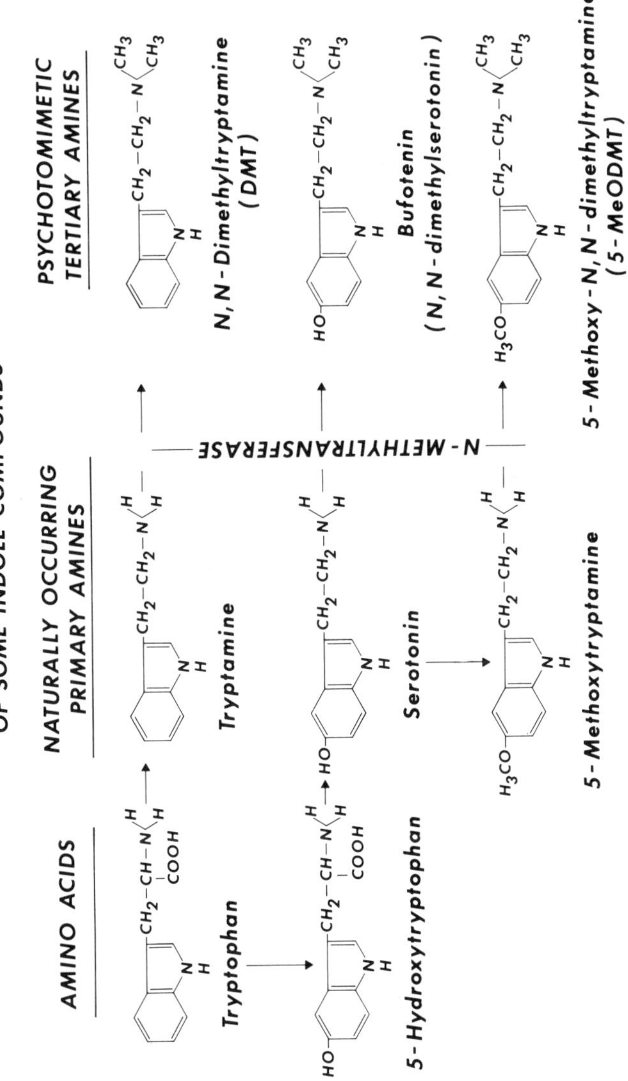

Figure 1. Chemical structures and possible metabolic pathways of some indole compounds

small change in their chemical constitution and render them psychotogenic (substitute the CH_3 for an H atom) and, by a two-step process in each case, form bufotenin from serotonin and DMT from tryptamine. In 1969, Morgan and Mandell found a similar enzyme in the brains of cockerels, sheep and human beings. Saavedra and Axelrod (1972) also found in the brain of rats and human beings an enzyme which could change the normal compound, tryptamine, into the psychotomimetic compound, DMT. Saavedra *et al*. (1973) have shown that enzymatic activity is present in the brains of several species including mouse, guinea pig and frog, as well as in a number of peripheral tissues of the rat. Moreover, they postulate the presence of an endogenous small molecular weight inhibitor of the enzyme N-methyltransferase in the brain.

DRUG-FREE CHRONIC SCHIZOPHRENIC PATIENTS AND NORMALS

In the next investigation we compared the behavior and urinary excretion of the N,N-dimethyltryptamines in chronic schizophrenics free of drugs and on unrestricted diets with those of normal individuals challenged with a diet rich in serotonin-containing foods (Table II, Study 1) (Narasimhachari *et al*., 1972).

Methods

The same clinical and biochemical methods were used as in the previous investigation. While still on drugs, the behavior of the patients was rated according to the scale of Rockland and Pollin (1965). The patients were then placed on the metabolic ward and all psychoactive drugs were eliminated from their regime. After a 4-week drug holiday, we began analyzing 24-hour urinary samples and continued for another 7 weeks while the patients remained without drugs.

Results

The results showed that four of the six patients exhibited two of the psychotogenic derivatives of tryptamine, DMT and bufotenin. The behavior of each of the patients gradually became worse during the drug-free period though to varying degrees in different patients. There were no changes in the normals that could be attributed to their diet nor did they eliminate any of the psychotogenic substances in their urine.

In the next study on drug-free chronic schizophrenic patients (Table II, Study 2) we were joined by Dr. William Carpenter of the National Institute of Mental Health (Narasimhachari *et al*., in press).

TABLE II. Urinary Findings and Behavioral Alterations in Drug-free Schizophrenic Patients and in Normals Under Different Dietary Conditions

Patients	Treatment	Biochemical Results (in urine)	Symptoms
Study 1			
6 chronic schizophrenics	Drugs stopped Uncontrolled diet	DMT bufotenin	Exacerbations
7 normals	Serotonin - rich diet	No DMT No bufotenin	None
Study 2			
6 chronic schizophrenics	Drugs stopped Rigorous diet	DMT bufotenin	Exacerbations

We chose six chronic schizophrenic patients who had never been the subjects of any previous investigation and who exhibited active symptoms despite the administration of tranquilizers.

Methods

Clinical. These patients were placed on the metabolic ward while they were still taking tranquilizing drugs. Weekly behavioral evaluations were made throughout the study using the Rockland and Pollin (1965) rating scale. The patients were taken off all drugs and 24-hour urinary collections were made daily.

After a 4-week drug holiday, the 24-hour urinary collections were analyzed in the laboratory for the three N,N-dimethyltryptamines. Four of the patients remained off drugs for an additional period of 10 weeks and 6 days. The fifth patient joined the group one week after the others and therefore remained drug-free for only 9 weeks and 6 days. The sixth patient's drug-free period ended at 7 weeks and 3 days because he could no longer be managed in a ward environment without drugs.

Biochemical. For the identification and quantitation of bufotenin in urine samples we have used a specific spectrophotofluorometric method after TLC on silica gel G and *o*-phthalaldehyde (OPT) spray (Narasimhachari and Himwich, 1972). OPT is a highly sensitive reagent and is specific for 5-hydroxy- and 5-methoxyindole derivatives. However, for DMT, the spray generally used, DACA, is not as sensitive and levels less than 0.2 μg are not detectable.

Baumann *et al*. (1971) reported high sensitivity for serotonin and bufotenin (5 ng) using cellulose plates for TLC and DACA as the spray reagent. We examined the applicability of the latter method for all three dimethylated tryptamines (Baumann and Narasimhachari, in press) and found that DMT is detectable at 10 ng levels by DACA spray. Thus, a second thin-layer determination on cellulose provides an additional parameter for bufotenin identification and a sensitive method for detection of DMT. The identity of DMT was then confirmed by gas chromatography-mass spectrometry (GC-MS), using the tertiary amine fractions and the TMS derivatives. Details of this method are being reported elsewhere (Narasimhachari and Himwich, in preparation).

Results

Clinical. Without drugs, the behavior of each of the six chronic schizophrenic patients became progressively worse. The behavioral vignettes of patients No. 1 and No. 5 are presented especially because biochemical analyses revealed that each was positive by GC-MS for bufotenin and DMT (Fig. 2).

Patient No. 1: Schizophrenia, chronic undifferentiated. The drugs and dosages administered before the present observations were: thioridazine (Mellaril), 200 mg q.i.d. and chlorpromazine (Thorazine), 75 mg i.m., q. 6 hr PRN.

The patient showed a progressive worsening of his general emotional condition and thought processes during the period of the study. He became more negativistic and anxious, and felt that television programs had reference to his personal life. He disclosed ideas of reference and developed new delusional ideas and believed that his entire family had been killed in an automobile accident. He displayed suspicions concerning the food served him, sometimes throwing it away because he thought that the food was poisoned. He became more and more withdrawn, had difficulty in sleeping, and became more negligent in personal appearance and hygiene. He believed he was involved in a scholarly pursuit of philosophy, mysticism, comparative religions, and his ideas were expressed in a very abstract manner. Loose associations in his thought processes became apparent.

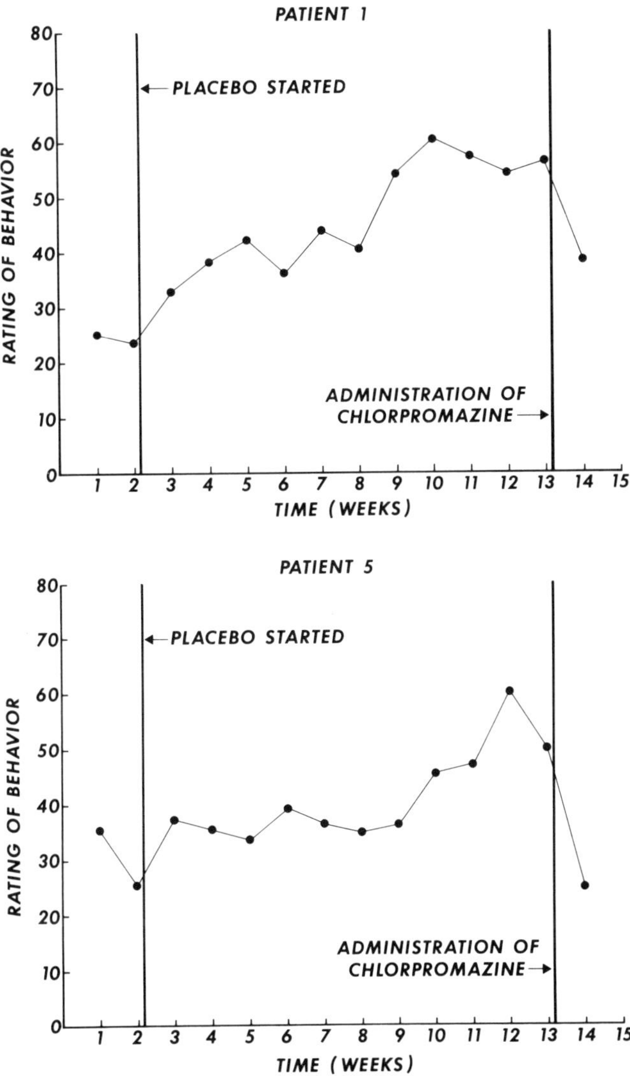

Figure 2. Behavioral ratings according to the Rockland and Pollin (1965) scale

Patient No. 5: Schizophrenia, chronic undifferentiated. The drugs and dosages administered before the present observations were: trifluoperazine (Stelazine), 5 mg b.i.d., trihexyphenidyl (Artane), 2 mg b.i.d., butaperazine (Repoise), 10 mg t.i.d., and chlorpromazine (Thorazine), 100 mg i.m., q. 6 hr PRN.

The patient's behavior became progressively worse during the study. He became more hostile, irritable, negativistic and had no insight into his emotional condition. He had incessant suspiciousness of the interviewer, always questioning in a low, soft tone the reason for the interview and complaining of the repetition of the questions. Though he spent most of the day in bed, he did walk occasionally during which time he had outbursts of laughter and gesticulated. He repeatedly hit himself hard on the jaw, stating that he was hitting his "other mind". His affect was extremely shallow.

Data from the weekly behavioral evaluations of each of the six patients were combined into two scores; one score representing the weeks during which they were still on drugs, and the other representing all placebo weeks after the first 4 weeks on placebo. One set of scores was derived for each rater, yielding a total of six scores per patient, or 36 for the entire population of patients. These scores were subjected to analysis of variance (Winer, 1962). The mean score for the drug period (and standard error of the mean) was 31.3 ± 3.5. The mean score for the placebo period was 49.2 ± 3.9. The increase in scores was significant ($F = 51.6$, $df = 1/25$, $p < .01$). The three raters did not differ from each other.

Biochemical. Bufotenin was identified by two-dimensional TLC in urines of each of the six chronic schizophrenic patients. The samples were positive for bufotenin on both silica gel G TLC and OPT spray and on cellulose TLC and DACA spray. A UV photograph of a thin-layer chromatogram is shown in Fig. 3 (reproduced here in black and white). The two reference spots on the margins are bufotenin and 5-MeODMT. In the sample a fluorescent spot isographic to standard bufotenin is seen. The levels varied from 1 to 3 μg in a 24-hour urinary collection. The tertiary amine fractions from two of these samples of the same patient, patient No. 1, were pooled and then derivatized to TMS derivatives (Narasimhachari *et al.*, 1971c) and run on a GC-MS system. Similar procedures were performed on the pooled tertiary amine fractions from patient No. 5. Both yielded positive results. The column conditions for GC-MS are already reported (Narasimhachari and Himwich, 1973). The GC chart obtained on the total ion monitor is shown in Fig. 4. The urine sample at left showed a peak at the same retention time as the standard bufotenin TMS (5' - 44") and the mass spectrum (Fig. 5) is identical with the mass spectrum of the standard sample. The chart on the right shows an increase in the height of the bufotenin

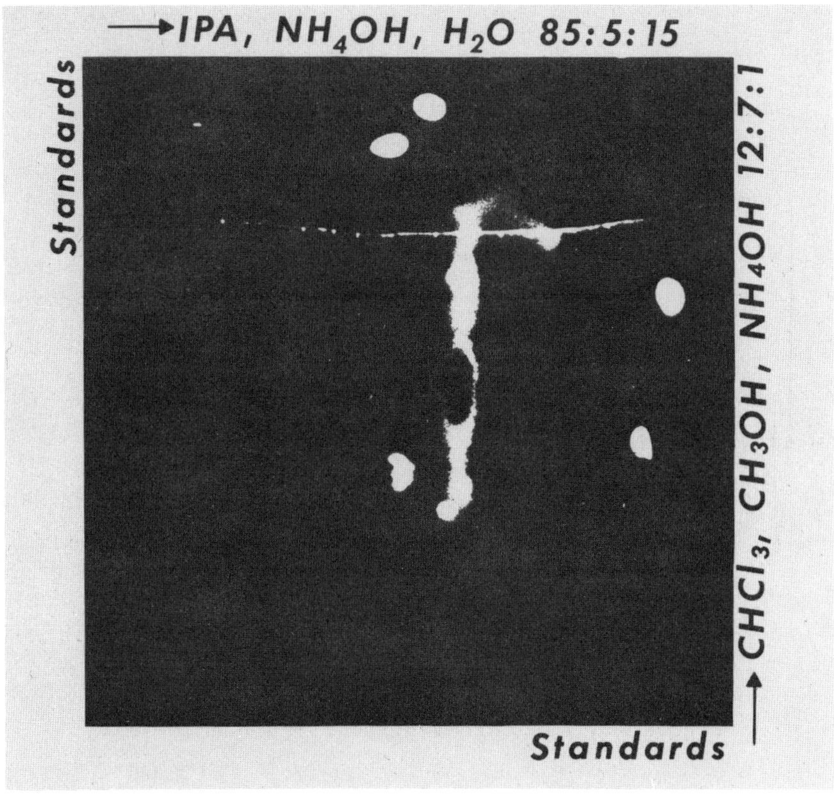

Figure 3. Two-dimensional thin-layer chromatogram of urinary extract positive for bufotenin.

TMS peak when bufotenin TMS was added to the sample TMS. The mass spectrum of the sample (Fig. 5) showed the molecular ion M^{+} at m/e 348, the base peak at m/e 58 (characteristic of tertiary amines) and another peak at m/e 290 $(M-58)^{+}$. The spectra and fragmentations have been reported by Narasimhachari et al. (1971c) (Fig. 6). A preliminary report on the identification of bufotenin in the urine samples of two patients has recently been published (Narasimhachari and Himwich, 1973). When the tertiary amine fractions were screened on cellulose TLC, three of the six patients showed DMT in their urine. The levels were very low (about 1 μg in a 24-hour urine sample). At least three positive samples from the same patient were pooled in the case of two patients and the identity of DMT confirmed by GC-MS. The mass spectra of DMT standard and of the DMT peak in urine sample are shown in Fig. 7.

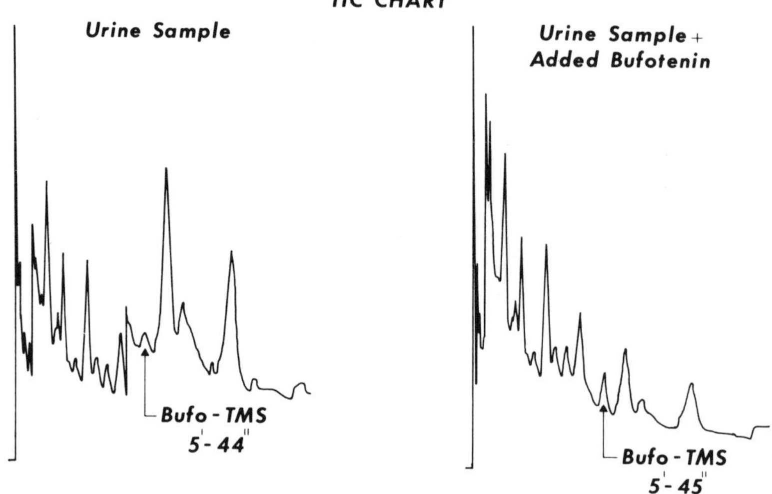

Figure 4. Gas chromatogram of bufotenin-TMS.

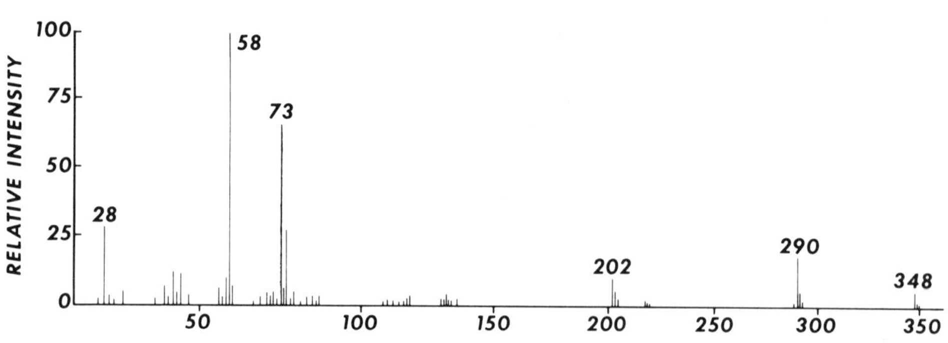

Figure 5. Mass spectrum of bufotenin-TMS.

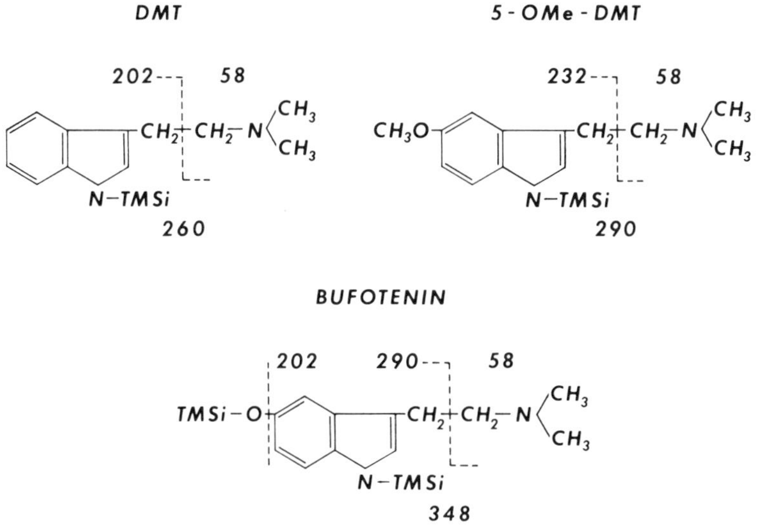

DMT
202
58
CH3
CH2
CH2
N
CH3
N-TMSi
260
5-OMe-DMT
232
58
CH3O
CH2
CH2
N
CH3
CH3
N-TMSi
290
BUFOTENIN
202
290
58
TMSi-O
CH2
CH2
N
CH3
CH3
N-TMSi
348

Figure 6. Fragmentation of bufotenin-TMS, DMT-TMS, and 5-MeODMT-TMS.

DMT STANDARD
28
58
130
188
DMT PEAK
URINE SAMPLE
28
58
130
188

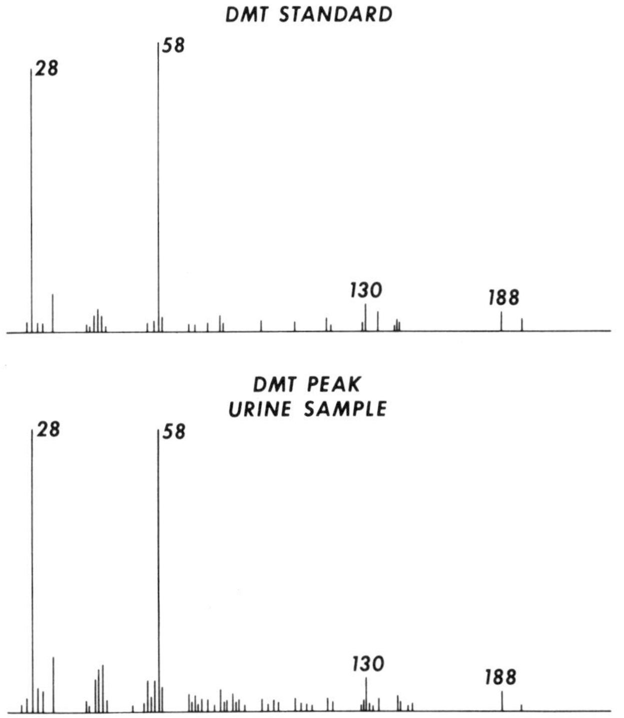

Figure 7. Mass spectra of standard DMT and DMT from urine sample.

STANDARD DMT-TMS + BUFOTENIN-TMS

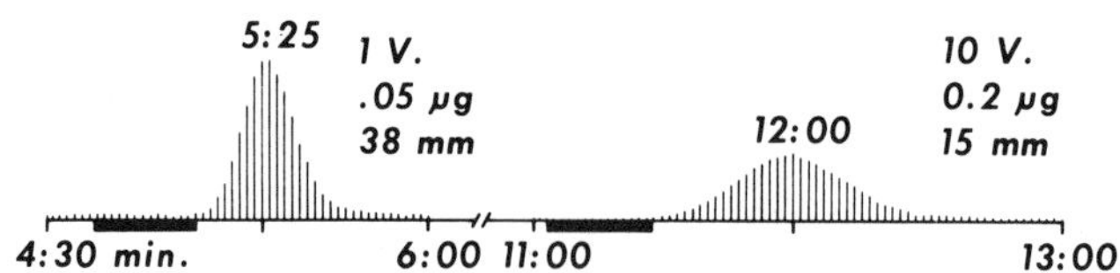

SAMPLE-TMS

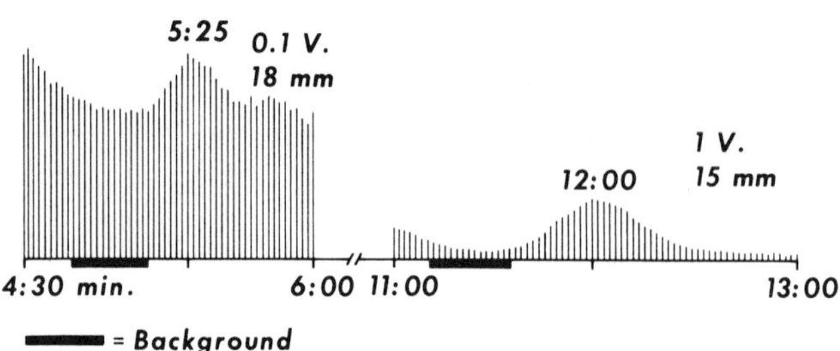

Figure 8. SIM tracings of peak at m/e 58 for standard DMT TMS and bufotenin-TMS and sample TMS.

When we monitored the single ion at m/e 58 with the TMS derivatives, we were able to confirm the positive TLC findings by GC-MS on all the samples tested. Single ion monitor (SIM) tracings of standard and sample TMS derivatives of the peak at m/e 58 are shown in Fig. 8. The levels detected in the urine of drug-free chronic schizophrenics in this study were less than 1 μg for DMT and 3 μg for bufotenin in a 24-hr urinary collection. In rare cases samples negative on TLC showed traces of DMT or bufotenin on GC-MS, possibly because of the higher sensitivity of this method. The peak heights for TMS derivatives in the highest sensitivity scale were 10 mm for 1 ng of DMT TMS and 30 mm for 1 ng of bufotenin TMS. Details of the procedure are reported in a separate communication (Narasimhachari and Himwich, in press).

SUMMARY

Observations were made on four chronic schizophrenic patients maintained under a rigorous diet supplemented by L-cysteine and tranylcypromine. They had exacerbations of their symptoms and excreted N,N-dimethyltryptamine (DMT), bufotenin and 5-methoxy-N,N-dimethyltryptamine (5-MeODMT) in their urine.

This study was repeated with six normals and two chronic schizophrenics. The normals did not excrete DMT, bufotenin and 5-MeODMT, nor did they have any psychotomimetic symptoms, but the two chronic schizophrenic patients exhibited the same behavioral and biochemical changes observed in the first study.

We next used the same methods to compare behavior and the urinary excretions of drug-free chronic schizophrenic patients on an unrestricted diet and normal controls on a diet rich in serotonin-containing foods. After a 4-week drug holiday, we analyzed 24-hour urinary samples throughout the next 7 weeks. Two of the psychotogenic derivatives of tryptamine, DMT and bufotenin, were excreted by four of the six patients as their behavior worsened. There were no changes in the normals either in behavior or urinary contents.

A study was then made on six drug-free chronic schizophrenic patients placed on a rigorous diet. After the fourth week of a drug holiday lasting about 13 weeks, the urinary samples of all of the patients were analyzed. All of the patients showed exacerbations of their symptoms. Bufotenin was identified by two-dimensional thin-layer chromatography (TLC) in the urine samples of all six chronic schizophrenic patients and the identity was confirmed by gas chromatography-mass spectrometry (GC-MS). Three patients showed DMT in the urine. The levels were very low so at least three positive samples from the same patient were pooled for each of two patients and the identification of DMT was confirmed by GC-MS.

We have established by GC-MS the identity of bufotenin in six drug-free chronic schizophrenics and of DMT in three of these six, and have confirmed our earlier findings with TLC and GC by the use of GC-MS. Although we are now using SIM technique for the identification and quantitation of both bufotenin and DMT, the earlier qualitative identification by the two TLC techniques and the fact that we have separated the amine fraction to only tertiary amines (N:N-dimethylated compounds) provides strong support for our findings. Furthermore, the identity was confirmed by obtaining the whole mass spectrum of some pooled samples. As a further confirmation of our results, at least one sample of each subject was run on GC-MS (SIM) on two or more different GC-columns, e.g. 3% SE-30 on GCQ, 3% OV-17 on GCQ and 3% OV-225, where the retention times of different compounds vary distinctly. Under these conditions the chances of our identification being ambiguous are remote.

REFERENCES

Axelrod, J., 1961, Enzymatic formation of psychotomimetic metabolites from normally occurring compounds, Science 134: 343.

Baumann, P., and Narasimhachari, N., 1973, Identification of N:N-dimethyltryptamine, 5-methoxy-N:N-dimethyltryptamine and bufotenin by cellulose TLC, J. Chromatog. 86: 269.

Baumann, P., Scherer, B., Kramer, W., and Matussek, N., 1971, Separation of indole derivatives and catecholamines by thin-layer chromatography, J. Chromatogr. 59: 463.

Capella, P., and Horning, E. C., 1966, Separation and identification of derivatives of biologic amines by gas-liquid chromatography, Anal. Chem. 38: 316.

"Diagnostic and Statistical Manual of Mental Disorders," Second Edition, 1968, The Committee on Nomenclature and Statistics of the American Psychiatric Association, American Psychiatric Association, Washington, D. C.

Gross, H., and Franzen, Fr., 1964, Zur Bestimmung körpereigener Amine in biologischen Substraten, Biochem. Z. 340: 403.

Mandell, A. J., and Morgan, M., 1971, Indole(ethyl)amine N-methyltransferase in human brain, Nature, (New Biology), 230: 85.

Morgan, M., and Mandell, A. J., 1969, Indole(ethyl)amine N-methyltransferase in the brain, Science 165: 492.

Narasimhachari, N., and Himwich, H. E., 1972, The determination of bufotenin in urine of schizophrenic patients and normal controls, J. Psychiat. Research 9: 113.

Narasimhachari, N., and Himwich, H. E., 1973, GC-MS identification of bufotenin in urine samples from patients with schizophrenia or infantile autism, Life Sci. 12 (Part II): 475.

Narasimhachari, N., and Himwich, H. E., GC-MS identification of DMT in urine samples from drug-free chronic schizophrenic patients and its quantitation by the technique of single (selective) ion monitoring, Biochem. Biophys. Res. Commun. in press.

Narasimhachari, N., Heller, B., Spaide, J., Haskovec, L., Fujimori, M., Tabushi, K., and Himwich, H. E., 1971a, Urinary studies of schizophrenics and controls, Biol. Psychiat. 3: 9.

Narasimhachari, N., Plaut, J., and Leiner, K., 1971b, Thin-layer and gas-liquid chromatographic methods for the identification and estimation of indoleamines in urine samples, Biochem. Med. 5: 304.

Narasimhachari, N., Spaide, J., and Heller, B., 1971c, Gas liquid chromatographic and mass spectrometric studies on trimethylsilyl derivatives of N-methyl- and N,N-dimethyltryptamines, J. Chromatog. Sci. 9: 502.

Narasimhachari, N. Avalos, J., Fujimori, M., and Himwich, H. E., 1972, Studies of drug-free schizophrenics and controls, Biol. Psychiat. 5: 311.

Narasimhachari, N., Baumann, P., Pak, H. S., Carpenter, W. T., Zocchi, A. F., Hokanson, L., Fujimori, M., and Himwich, H. E., Gas chromatographic-mass spectrometric identification of urinary bufotenin and dimethyltryptamine in drug-free chronic schizophrenic patients, Biol. Psychiat. in press.

Pollin, W., Cardon, P. V., Jr., and Kety, S. S., 1961, Effects of amino acid feedings in schizophrenic patients treated with iproniazid, Science 133: 104.

Rockland, L. H., and Pollin, W., 1965, Quantification of psychiatric mental status, Arch. Gen. Psychiat. 12: 23.

Saavedra, J. M., and Axelrod, J., 1972, Psychotomimetic N-methylated tryptamines: Formation in brain in vivo and in vitro, Science 175: 1365.

Saavedra, J. M., Coyle, J. T., and Axelrod, J., 1973, The distribution and properties of the nonspecific N-methyltransferase in brain, J. Neurochem. 20: 743.

Shulgin, A., 1970, Discussion, in "Psychotomimetic Drugs", D. H. Efron ed., p. 119, Raven Press, New York.

Snyder, S. H., and Richelson, E., 1968, Psychedelic drugs: Steric factors that predict psychotropic activity, Proc. Nat. Acad. Sci. U.S. 60: 206.

Szara, S., 1956, Dimethyltryptamin: Its metabolism in man; the relation of its psychotic effect to the serotonin metabolism, Experientia 12: 441.

Szara, S., 1961, Hallucinogenic effects and metabolism of tryptamine derivatives in man, Fed. Proc. 20: 885.

Szara, S., 1967, Hallucinogenic amines and schizophrenia (with a brief addendum on N-dimethyltryptamine), in "Amines and Schizophrenia," H. E. Himwich, S. S. Kety, and J. R. Smythies eds., pp. 181-197, Pergamon Press, New York.

Tanimukai, H., Ginther, R., Spaide, J., Bueno, J. R., and Himwich H. E., 1970, Detection of psychotomimetic N,N-dimethylated indoleamines in the urine of four schizophrenic patients, *Brit. J. Psychiat*. 117: 421.

Winer, B. J., 1962, Analysis of Variance, in "Statistical Principals in Experimental Design," McGraw Hill, New York.

PHENYLETHYLAMINE: POSSIBLE ROLE IN DEPRESSION AND ANTIDEPRESSIVE DRUG ACTION

H. C. Sabelli, A. D. Mosnaim, and A. J. Vazquez

University of Health Sciences/The Chicago Medical School

2020 West Ogden Avenue, Chicago, Illinois 60612

The well-chosen title of this conference implies that emotional behavior can be coded in a simple alphabet consisting of a relatively small number of neuromodulators and synaptic transmitters. One of the main difficulties in deciphering this code is that we know only a few of its "letters" [catecholamines (CA), serotonin (5HT), acetylcholine, etc.]. Toman and Sabelli (cf. Sabelli, 1964) proposed that not only is there a multiplicity of transmitters of various presynaptic origins, but there are also many endogenous chemicals that regulate neural function by modulating synaptic transmission mediated by other chemicals and by acting upon extrasynaptic membranes. 2-Phenylethylamine (PEA) may represent one example of this concept of neuromodulators.

On the basis of behavioral experiments, we concluded (Fischer et al., 1967) that the behavioral stimulant action of amphetamine is not due to CA release. We suggested that it mimics the action of endogenous PEA, an amine found (Nakajima et al., 1964) in the brain of animals treated with monoamine oxidase inhibitors (MAOI). Fischer et al. (1968) showed that the urinary excretion of free PEA was reduced in depressed patients. In this paper, we review studies from Fischer's laboratory and from our own which led us to propose a role for PEA and for its metabolite phenylethanolamine (OHPEA) in the modulation of wakefulness, alertness and excitement (a function usually attributed in its entirety to brain CA). This phenylethylamine theory of affective behavior states that a decrease in the brain levels of endogenous PEA may play a major physiopathological role in at least certain forms of endogenous depression, whereas an increase in PEA brain levels, or the activation of specific PEA receptors in brain neurons may be partly responsible for antidepressive and stimulant drug actions (Sabelli and Mosnaim, 1973).

Identification and Distribution of PEA and OHPEA

Because of its liposolubility, PEA readily crosses membranes (e.g. intestinal and blood-brain barriers); the brain accumulates PEA against a concentration gradient (Nakajima et al., 1964). Thus, PEA-sensitive brain neurons may be affected not only by locally released PEA but also by blood-borne PEA. Jackson and Temple (1970) found endogenous PEA in ox blood and spleen and the following year we reported its presence in the brain of mice and rabbits (Sabelli et al., 1971b; Mosnaim and Sabelli, 1971). We conclusively identified PEA and OHPEA in human brain and urine by mass, infrared and ultraviolet spectroscopy (Inwang et al., 1972, 1973; Mosnaim et al., 1973a,b). OHPEA had been previously detected in heart (Creveling et al., 1968). Using a specific and sensitive method for the quantification of PEA in biological samples developed in our laboratory (Mosnaim and Inwang, 1973), we have found endogenous PEA in most peripheral tissues in man, cat, and rabbit. The brain concentrations of PEA in man (Mosnaim et al., 1973a), mouse (Sabelli et al., 1971b), rat (Fischer et al., 1972a) and rabbit (Mosnaim et al., 1973c) are of the same order of magnitude as those of CA and 5HT, about 0.2 to 0.5 μg/g of wet tissue. In rabbits treated with a MAOI (to prevent the destruction of PEA), PEA levels are highest in the caudate (Nakajima et al., 1964). In untreated rabbits, PEA is present in all brain areas including cerebral cortex (0.31 ± 0.11 μg/g) and cerebellum (1.34 ± 0.23 μg/g) (Mosnaim et al. 1973c). It should be noted that the values reported by Nakajima et al. (1964) and Edwards and Blau (1972) were lower.

Synthesis and Metabolism of PEA and OHPEA

PEA is probably formed mainly by the decarboxylation of L-phenylalanine, which is catalyzed by L-aromatic aminoacid decarboxylase (L-AAAD), an enzyme widely distributed in neural as well as in non-neural tissues (Fig. 1). The administration of L-phenylalanine increases the urinary excretion of PEA in man (Oates et al., 1963) and its levels in brain (Sabelli et al., 1971b; Mosnaim et al., 1973c; Edwards and Blau, 1972), and peripheral tissues (Nakajima et al., 1964; Edwards and Blau, 1973). D-phenylalanine also increases the brain levels of PEA (Mosnaim et al., 1973c). Brain synthesis of PEA was demonstrated (Whalley et al., 1973) following the intraventricular injection of labeled L-phenylalanine (Fig. 4). Liver is the main site for accumulation of PEA following the intraperitoneal (ip) injection of labeled L-phenylalanine (Whalley et al., 1973). The rate of PEA synthesis probably depends on: (a) the plasma levels of phenylalanine (Edwards and Blau, 1973). (In a way similar to that in which the rates of brain synthesis of 5HT and of CA appear to depend on the plasma levels of their aminoacid precursors); (b) on tissue aminoacid uptake (an active process in which phenylalanine competes with other aminoacids); (c) the rate of

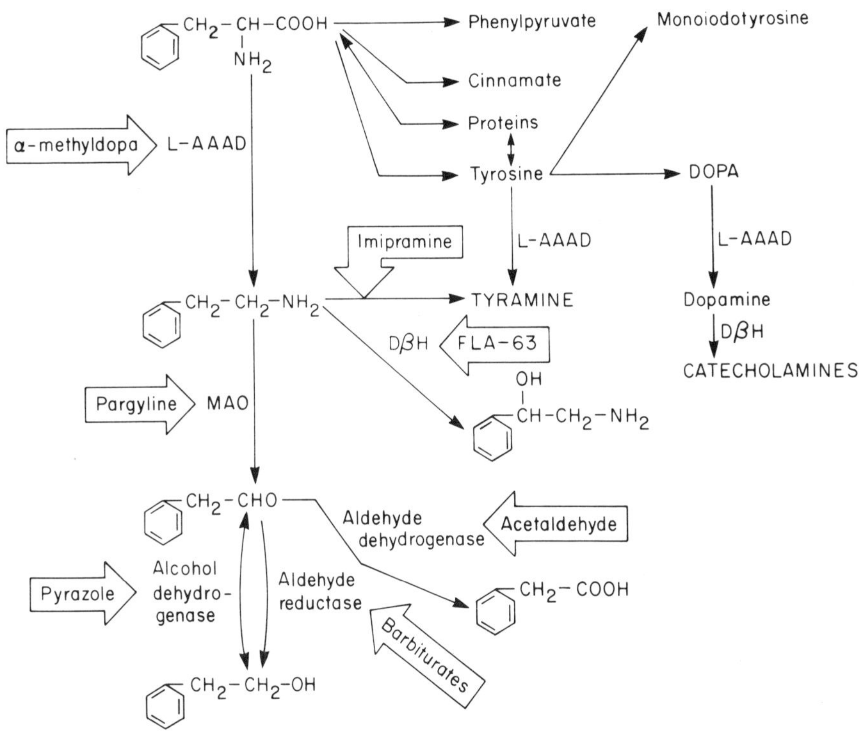

Figure 1 - Metabolism of 2-phenylethylamine (hypothetical)

phenylalanine decarboxylation relative to other pathways for phenylalanine metabolism.

PEA metabolism is currently under study in our laboratory. Presumably, PEA is mainly oxidized [PEA is the specific substrate for MAO type B (Yang and Neff, 1973)] with formation of phenylacetaldehyde, which, in turn, is mainly metabolized to phenylacetate. Phenylacetate has been identified in brain (Mosnaim _et al._, 1973d) and appears to be formed mainly from PEA (Edwards and Blau, 1973). PEA may be p-hydroxylated and it is a substrate for dopamine-β-hydroxylase (DβH) forming OHPEA (Creveling _et al._, 1962; Molinoff _et al._, 1969), a substrate for MAO and for N-methyltransferases.

Pharmacologic Effects of PEA, OHPEA, and their Synthetic Analogs

In most animal species, the systemic administration of PEA induces only a brief period of stimulation followed by behavioral depression. Similarly, depression can be induced by large doses of the PEA deaminated metabolites phenylacetaldehyde and phenylacetic

Table 1. Effects of PEA and OHPEA on Electroencephalogram (EEG), Visual Evoked Responses (VER), Blood Pressure (BP), Maximal Electroshock Seizures (MES), and Behavior of Grouped Mice (Data from Sabelli *et al.*, 1973b)

		RABBIT			MOUSE	
Pretreatment (1)	Treatment (1)	EEG Activation (sec)	VER	BP	MES (2)	Behavior of Grouped Mice
Pargyline	Saline	<10	=	=	100	normal
	PEA	350 - 500	↓	↑	10	fight & jump
	OHPEA	<60	=	↑	85	excitable
Pargyline & Reserpine	Saline	<10	=	=	100	sedated
	PEA	300 - 450	↓	=	25	jumping
	OHPEA	15 - 25	=	=	100	hypoactive

(1) Doses for rabbits: Pargyline 100 mg/Kg, i.p. 48 and 24 hrs. prior; Reserpine 5 mg/Kg, i.p. 24 hrs. prior; PEA 2.5 mg/Kg, i.v.; OHPEA 5 mg/Kg, i.v. Doses for mice: Pargyline 100 mg/Kg, i.p. 24 hrs. prior; Reserpine 10 mg/Kg, i.p. 24 hrs. prior; PEA and OHPEA 50 mg/Kg, i.p.

(2) % of animals with maximal tonic seizures evoked by supramaximal electroshock

acid (Giardina *et al.*, 1970). These compounds have been shown to be biologically active in electrophysiological experiments (Sabelli and Giardina, 1970a; 1973a,b). They may mediate in part the depressant effect of large doses of PEA. In animals pretreated with MAOI's, PEA induces amphetamine-like effects (Table 1) including sympathomimetic effects and central nervous system (CNS) stimulation (Mantegazza and Riva, 1963; Nakajima *et al.*, 1964; Fischer *et al.*, 1967; Sabelli and Giardina, 1973b): increases the exploratory activity of a single mouse and the spontaneous activity of grouped mice, activates the EEG, reduces the amplitude of the slow negative wave component of the visual evoked responses (VER), potentiates the behavioral stimulating effects ("popcorn behavior", fighting) of

DOPA in grouped mice, etc. At large doses, PEA induces fighting and episodes of stereotyped jumping ("popcorn behavior") in grouped mice and antagonizes maximal electroshock seizures. OHPEA is a similar but weaker CNS stimulant (Sabelli _et al._, 1973b).

The CNS effects of PEA, OHPEA, and their α-methylated analogs (amphetamine, ephedrine, etc.) may be partially due to CA release, an action which has been demonstrated not only in the periphery but also in the CNS (Fuxe _et al._, 1967). However, the ability of amphetamines (and of PEA after a MAOI) to restore to normal the posture and activity of animals treated with the CA depleter reserpine, indicates that their CNS actions may not be entirely mediated by CA release. Whereas their sympathomimetic effects are fully prevented by reserpine, this agent only slightly reduces the behavioral stimulant, EEG activating and anticonvulsant effects of PEA (Table 1) and does not prevent the action of PEA on the VER (Fig. 2). Further evidence against the view that the CNS effects of PEA are entirely due to CA release are the many instances in which PEA exerts stimulant actions which are opposite to the depressant effects observed when the CNS levels of CA are increased by the

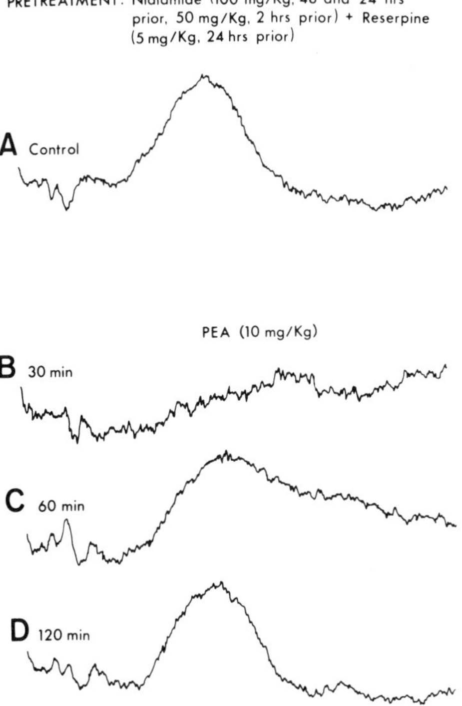

Figure 2 - VER in non-anesthetized rabbits added by Computer of Average Transients. Analysis Time 500 msec.

administration of their aminoacid precursor DOPA, or by the intraventricular injection of CA or by the systemic injection of large doses of CA in mammals or of small doses of CA in newly hatched chicks (lacking a well-developed blood-brain barrier). Whereas PEA mimics alerting stimuli in reducing the amplitude of the slow negative wave of the VER (Sabelli et al., 1971a; Sabelli and Giardina, 1973b), DOPA increases this response (Sabelli et al., 1971a). In chicks, all CA induce sleep whereas PEA induces excitement. The exploratory activity of mice is decreased by CA and DOPA, an effect antagonized by PEA (Fischer et al., 1967).

More direct evidence for a difference between the CNS receptors for PEA and for CA has been obtained using a multiple barrel microelectrode array for microiontophoretic injection and simultaneous extracellular recording of unit activity (Giardina et al., 1973): PEA as well as OHPEA exert effects opposite to those of norepinephrine in approximately 50% of the neurons tested.

Thus, phenylisopropylamines appear to induce behavioral stimulation by releasing endogenous PEA or by mimicking the action of PEA on specific PEA receptors. This is not to exclude the fact that these compounds also release CA and interact with 5HT receptors (Vane, 1960; Vazquez and Krip, 1973). Dewhurst (1968) proposed that non-hydroxylated amines (PEA and tryptamine) are behavioral stimulants whereas hydroxylated amines (5HT and CA) are behavioral depressants. Since 5HT and tryptamine act on the same receptors and induce similar behavioral depressant effects (Sabelli and Giardina, 1970b), the possible action of PEA on tryptamine receptors does not appear to account for its CNS stimulant effect.

PEA and OHPEA as Neuromodulators

Available experimental evidence indicates that PEA and OHPEA meet some of the criteria required for proof that a substance is a chemical transmitter: (a) their presence in neural tissue has been unequivocally demonstrated; (b) the brain contains the enzymatic machinery for their synthesis and metabolism; (c) these compounds have strong biological activity in the CNS; (d) synaptic terminals uptake phenylethylamines (Baldessarini and Vogt, 1971); (e) depletion by denervation has been demonstrated for OHPEA (Saavedra and Axelrod, 1973) etc. Hartman and Udenfriend (1972) demonstrated in brain the presence of two neuronal tracts containing DβH but not norepinephrine. Presumably, the transmitter released by these neurons is a β-hydroxylated phenylethylamine other than norepinephrine, such as octopamine or OHPEA. Thus, OHPEA may be a synaptic transmitter and PEA functions as its precursor. However, it seems that the effects of PEA are for the most part not mediated by its conversion to OHPEA. PEA is much more effective than OHPEA as a behavioral stimulant (an observation that parallels findings with their α-methylated synthetic analogs, the strong stimulant amphetamine and the weaker stimulant norephedrine). Moreover, most of the CNS effects of PEA are not prevented by the DβH inhibitor FLA-63 (Sabelli *et al.*, 1973b). Endogenous PEA and certainly administered PEA are widely distributed in all brain areas, whereas DβH is present only in a limited number of neuronal pathways (Hartman and Udenfriend, 1972), rendering unlikely that most of the effects of PEA be mediated by OHPEA.

There is evidence against the view that PEA is a neurotransmitter: (a) the brain distribution of PEA, if not entirely uniform, is at least suggestive of the presence of PEA in many types of neural (or glial) cells; (b) subcellular fractionation studies (Edwards and Blau, 1973) indicate that PEA is not concentrated in synaptic fractions; (c) PEA lacks hydroxyl groups, a fact which hinders its storage in monoamine granules (Baldessarini and Vogt, 1971); (d) in contrast to all other known transmitters, PEA readily crosses biological membranes. The ability of a true synaptic transmitter to readily cross synaptic diffusional barriers and the blood-brain barrier should certainly conspire against meaningful communication between brain neurons. We thus postulate that PEA is not a synaptic transmitter but rather a modulator with widespread regulatory functions in both brain and peripheral organs, possibly functioning as co-transmitter in some monoaminergic synapses (also OHPEA may be a co-transmitter). In the CNS, PEA may function as one of the modulators responsible for wakefulness, activity and excitement. Jackson and Temple (1970) have proposed that PEA may be an important excitatory modulator of cardiac function; the urinary excretion of PEA is decreased following myocardial infarction (Mosnaim *et al.*, 1973b).

A New View of Dale's Principle

A well accepted postulate in synaptic transmitter theory is that each neuron releases one and only one transmitter (Dale's principle). Thus, neurons are classified as cholinergic, noradrenergic, serotonergic, etc. This concept has received overwhelming experimental support and it appears firmly validated by the recently developed histofluorescence techniques. Mammalian tissues contain many amines other than the classical mediators, such as tryptamine, tyramine, octopamine, etc.; to this list we must now add PEA and OHPEA. Some of these amines replace the CA at the sympathetic nerve terminals in animals treated with MAOI ["false neurotransmitter" concept (Kopin, 1968)]. Moreover, some neuroamines may be synthetized and/or accumulated in neurons considered specific for other amines (for instance, 5HT in CA neurons). Sympathetic neurons release other endogenous phenylethylamines in addition to norepinephrine (Molinoff and Kopin, unpublished). Certain chromaffin neurons of the earthworm ventral cord apparently contain both norepinephrine and dopamine in significant amounts (Ehinger and Myhrberg, 1971). The observation that not only the neuroamines but also their aminoacid precursors DOPA (Vazquez et al., 1973; Sabelli et al., 1973c) and 5-hydroxytryptophan (Pedemonte and Sabelli, 1973), and their deaminated metabolites (Holtz, 1966; Sabelli, 1970; Sabelli et al., 1969, 1970; Sabelli and Giardina, 1970a,b, 1971, 1973a,b; Feldstein et al., 1970) are biologically active, further suggests that a multiplicity of chemicals may act at each synaptic site. A modification of Dale's principle is implicit in the notion of "neuromodulator" proposed by Toman and Sabelli. In our definition, neuromodulators regulate transmission across a synapse that is mediated by some other chemical or electrical mechanism.

Based on the above observations, we propose the following concept: neurons synthetize and release a pool of one main transmitter and several metabolically related co-transmitters (e.g. non-hydroxylated amines, aldehydes, aminoacids, etc.) which together determine the response of the postsynaptic neuron. A limiting case is that in which the transmitter pool includes only one chemical (as it occurs at the neuromuscular junction). The actual content of the transmitter and co-transmitter pool varies according to the physiological state of the neuron. Thus, neurons are only specific in their biopolymer constituents, and hence ought to be classified according to the specific macromolecules they contain, rather than on the basis of the main transmitter they release. Among these functional biopolymers, the presence of specific aminoacids and amine uptake mechanisms, of enzymes for the synthesis and degradation of amines, and of the "storage-receptors" for binding these amines, determine which of the many chemical species becomes the predominant compound mediating transmission at a particular synapse. Hormones, circadian rhythms, and other physiological factors may be capable of altering the relative proportion of co-transmitters in the pool. There is often a functional antagonism between metabolically related compounds [e.g. glutamic acid (excitation) vs GABA (inhibition)]; norepinephrine (mainly α) vs epinephrine (α and β); serotonin vs 5-hydroxyindoleacetic acid [shortening vs prolongation of frog cardiac action potential (Brunner and Sabelli, 1971)].

The basic structural requirement of most neurotransmitters is an ethyl group and a terminal amine. Among β-aromatic ethylamines, efficacy is often increased by hydroxylation. Thus, a typical monoaminergic neuron contains at least 3 types of enzymes: decarboxylases, hydroxylases, and metabolizing enzymes which modify or abolish pharmacological potency by interacting with the amine or the hydroxyl groups. Thus, in addition to its amine transmitter, monoaminergic nerve terminals can also accumulate, synthetize, and presumably release other pharmacologically active agents: non-hydroxylated amine analogs of the amine transmitters (e.g. PEA and OHPEA in CA synapses), their aminoacid precursors, and their corresponding deaminated products (aldehydes, acids, and alcohols). These agents may act as neuromodulators. The fact that MAO is more abundant in presynaptic terminals than in postsynaptic sites suggests that this enzyme may be involved in the synthesis of active products to be released by the nerve terminal (in addition to playing a role in the metabolic disposition of the neuroamines).

This concept of a "biochemical plasticity" of aminergic transmission (which is implicit in Kopin's "false neurotransmitter" concept) is purely speculative but readily testable.

Non-synaptic Receptors for Phenylethylamine and Other Neuroamines

Toman and Sabelli proposed that neuromodulators may act not only at synapses but also at non-synaptic sites. In our view, the basic structure of the neural membrane, synaptic or axonal, is a continuous, essentially uniform, protein sheet. Neuroamine receptors may be scattered throughout the plasma membrane. Secondary folding of the membrane protein, probably triggered by transmitter action, may concentrate receptors in subsynaptic membranes and may account for the difference between synaptic and axonal membranes (Grundfest, 1957). Cholinergic, 5HT and CA receptors (displaying stereospecificity, specific blockade by relevant drugs, etc.) can be demonstrated in peripheral nerves (Sabelli and Gorosito, 1969). In isolated frog nerves, high concentrations of PEA (or its derivatives amphetamine or methamphetamine) block conduction; non-blocking concentrations of PEA facilitate conduction block by chlorpromazine. In contrast, epinephrine does not depress nerve conduction and antagonizes chlorpromazine block. Neither PEA nor epinephrine modifies conduction block by imipramine. Thus, even in synapse-free axonal membranes, one can distinguish the effects of CA from those of PEA, as well as the action of the antipsychotic chlorpromazine from that of the antidepressive imipramine. Neuroamines, as well as neurotropic drugs, appear to affect all types of excitable membranes (not only synapses) and these membrane effects are so qualitatively different that they may well explain the pharmacological differences between phenylethylamines and CA, antipsychotic and antidepressive drugs, etc. Extrasynaptic neuroamine receptors probably play no physiological role in the case of CA and 5HT, since they are separated both by distance and diffusional barriers from the sites of release of these amines. On the other hand, such receptors may play a significant role in the actions of endogenous PEA (since this amine is capable of crossing most biological membranes) and of some neurotropic drugs.

The Role of Endogenous PEA in Drug Action

Biochemical determinations in animal brain as well as in human urine indicate that PEA levels are increased by antidepressive agents and reduced by drugs that induce depression. Brain levels of PEA are increased by pargyline, imipramine (Sabelli *et al.*, 1971b, 1973a; Mosnaim *et al.*, 1973c), and by electroshock treatment (Mosnaim *et al.*, 1972); and reduced by reserpine (Mosnaim and Sabelli, 1971; Fischer *et al.*, 1972a; Mosnaim *et al.*, 1973c). PEA brain levels are not affected by chlorpromazine (Mosnaim *et al.*, 1973c). DOPA, which exerts antidepressive effects in 25% of depressed patients (Goodwin *et al.*, 1970), also increases the brain levels of PEA (Sabelli *et al.*, 1973c). The urinary excretion of PEA is increased by MAOI's (Oates *et al.*, 1963; Fischer *et al.*, 1968) and by tricyclics (Fischer *et al.*, 1972a,b; see also Figure 5), and reduced by α-methyldopa (Oates *et al.*, 1963). These observations, together with the behaviorally stimulant action of PEA, suggest a role for this amine in antidepressive drug action.

Obviously, changes in brain PEA levels or action do not account for all the behavioral effects of any of these agents. An essential feature of the neuronal networks underlying emotional functions is the existence of feedback loops and parallel circuits capable of compensating for isolated failures along any one neuronal pathway. In addition, a multiplicity of neuromodulators acting at each synapse may compensate for failures along any biochemical pathway. It is therefore unlikely for a drug or an endogenous metabolite to selectively alter one behavioral pattern by specifically affecting any one single enzyme or receptor. In our view (Sabelli, 1964; Sabelli and Giardina, 1973a), psychotropic drugs owe their behavioral action to their ability to interact simultaneously with a whole spectrum of neuromodulators. As discussed before, many neurotropic drugs appear to affect all types of excitable membranes; such an action may account for the broad spectrum of their activity upon receptors, amine uptake mechanisms, etc. In support of this view, Seeman (1972) has found that the usual therapeutic blood concentrations of chlorpromazine, haloperidol and other neurotropic drugs appear to be sufficient to block conduction in the small fibers of the CNS.

Role of PEA in the Mode of Action of Tricyclic Antidepressants

Three alternate or additive mechanisms for the antidepressive action of imipramine and related drugs have been proposed:

(i) Enhancement of 5HT synaptic activity by inhibition of 5HT reuptake (Carlsson *et al.*, 1968). This hypothesis does not account for the antidepressive action of desipramine and protriptyline which are not good inhibitors of 5HT reuptake (cf. Sulser and Sanders-Bush, 1971). Moreover, indoleethylamines are CNS depressants cf. Himwich and Alpers, 1970).

(ii) Inhibition of central cholinergic mechanisms (Sabelli *et al.*, 1961; Biel *et al.*, 1962), which can only be an additional mechanism since atropine is not a very effective antidepressant.

(iii) Enhancement of central "adrenergic" synaptic transmission which has been attributed (cf. Sulser and Sanders-Bush, 1971) to inhibition of CA reuptake by the synaptic terminals. This action may contribute to, but it cannot be the sole explanation for, the effectiveness of tricyclics because: (a) cocaine, a well-known inhibitor of CA reuptake, is not an antidepressant; (b) the antidepressant iprindole does not inhibit CA reuptake (cf. Sulser and Sanders-Bush, 1971); (c) tricyclics reduce blood pressure, indicating that they do not facilitate CA transmission in the periphery, although they block CA reuptake both in brain and in peripheral organs; (d) CA appear to depress behavior in animals (Rothballer, 1959; Marley, 1966; Fischer et al., 1967; Himwich and Alpers, 1970); Brittain and others (cf. Himwich and Alpers, 1970) have reported that imipramine antagonizes the behavioral depressant effect of NE in newly hatched chicks. As an alternative explanation, imipramine may enhance central adrenergic transmission by increasing the levels and efficacy of PEA in brain, probably by inhibiting its disposition. (Imipramine affects several metabolic pathways for amine metabolism such as oxidative deamination and p-hydroxylation; cf. Sulser and Sanders-Bush, 1971.) Imipramine blocks the autonomic effects of phenylethylamines (Sabelli, 1960) which are due to CA release, but potentiates their behavioral stimulant effects (cf. Himwich and Alpers, 1970). Imipramine enhances the amount of labeled PEA recovered from rabbit liver following the ip administration of labeled L-phenylalanine (Whalley et al., 1973). Iprindole enhances the CNS effects of phenylethylamines but it produces only small increases in brain PEA levels (Mosnaim et al., 1973c).

From our earlier studies, we were impressed with the multiplicity of actions of imipramine, including its ability to enhance the α and β effects of CA as well as the peripheral effects of 5HT (Sabelli, 1960), to compete with cholinergic drugs (Sabelli et al., 1961) even in peripheral nerves (Sabelli and Gorosito, 1969), etc. Thus, we proposed (Sabelli, 1964; Sabelli and Giardina, 1973a) that imipramine-like antidepressives owe their therapeutic action to their ability to simultaneously enhance the central effects of several monoamines (CA, 5HT, and PEA) and block the actions of acetylcholine and histamine. The ability of imipramine to induce cardiac arrythmias (Sabelli and Toman, 1961) and to affect axonal conduction in a characteristic manner (different from local anesthetics, phenothiazines, etc.)(Toman and Sabelli, 1968; Sabelli and Gorosito, 1969) suggests that the apparent multiplicity of its receptor actions might be due to a single action upon excitable membranes.

Role of PEA in the Mechanism of the Antidepressive Action of MAOI's

As expected, the administration of some MAOI's (but not isocarboxazid) induces a marked increase in the tissue levels of

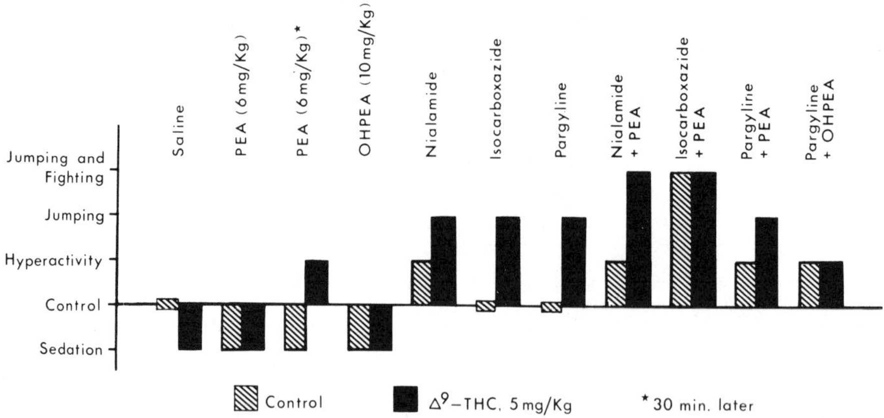

Figure 3 - Behavioral studies in grouped mice (n = 10). PEA, OHPEA, or Δ^9-THC administered by themselves induce sedation in non-pretreated animals, and excitement in mice pretreated 48 and 24 hours prior with MAOI's. PEA also induced excitement when injected 30 min after Δ^9-THC. The simultaneous adminstration of PEA and Δ^9-THC, but not of OHPEA and Δ^9-THC, induced even greater excitement in MAOI pretreated mice.

endogenous PEA (Sabelli _et al._, 1971a, 1973a; Fischer _et al._, 1972a and b; Mosnaim _et al._, 1973c; Nakajima _et al._, 1964; Edwards and Blau, 1972). All of the MAOI's tested (iproniazid, isocarboxazid, nialamide, pargyline, etc.) enhance the amphetamine-like effects of PEA (Fig. 3). Pargyline also affects the action of PEA metabolites (Giardina _et al._, 1970) either because this MAOI exerts other effects unrelated to enzyme inhibition or because the PEA metabolites act by interfering with MAO. The antidepressive action of MAOI probably results from at least three mechanisms:

(i) Enhancement of the brain levels of monoamines, particularly PEA and OHPEA, as suggested by the following observations. Non-hydroxylated phenylethylamines are behavioral stimulants whereas 5HT, tryptamine, and CA exert mainly behavioral depressant effects. Moreover, MAOI's enhance the pharmacological effects of PEA, tyramine, etc., much more than those of CA. Non-hydroxylated amines, such as PEA, may replace CA in the granules (Kopin, 1968). MAOI's increase more the brain levels of PEA than those of CA because of: (a) the greater sensitivity to MAO of non-hydroxylated amines; (b) free CA are mainly disposed of by tissue uptake and by catechol-O-methyltransferase, rather than by MAO; (c) CA synthesis is inhibited by MAOI's (Costa and Neff, 1962).

(ii) Inhibition of the synthesis of aldehydes and other behaviorally depressant deaminated metabolites of the neuroamines (Sabelli _et al._, 1970).

(iii) Inhibition of central CA synapses (Sabelli and Mosnaim, 1973). The anti-hypertensive effect of MAOI's has been attributed to depression of sympathetic synaptic transmission. A similar depression of CA synapses may be expected to occur in the CNS. Since CA are capable of inducing behavioral depressant effects, this inhibition of central CA transmission may contribute to the antidepressive effects of MAOI's.

The Role of PEA in the Stimulant Action of Marihuana

In man and other animals, Δ^9-THC induces sedation together with euphoria and other stimulant effects. In mice and rabbits pretreated with MAOI's, Δ^9-THC induces marked stimulation (Sabelli et al., 1973c) (Fig. 3). This suggests that the stimulant actions of Δ^9-THC may be mediated by brain amines, and that their behaviorally depressant deaminated metabolites contribute to the predominant sedative action of Δ^9-THC. Experiments with selective monoamine depleters indicate that neither CA nor 5HT mediate Δ^9-THC stimulation (Sabelli et al., 1973c). Although endogenous amines may not be essential for Δ^9-THC induced excitement [since the latter can be obtained after inhibition of L-AAAD with α-methyldopa (Sabelli et al., 1973c)], endogenous PEA may contribute to the stimulant action of marihuana: (a) low doses of Δ^9-THC induce a marked and selective increase in PEA brain levels (Sabelli et al., 1973c) (Fig. 4); (b) Δ^9-THC appears to inhibit the

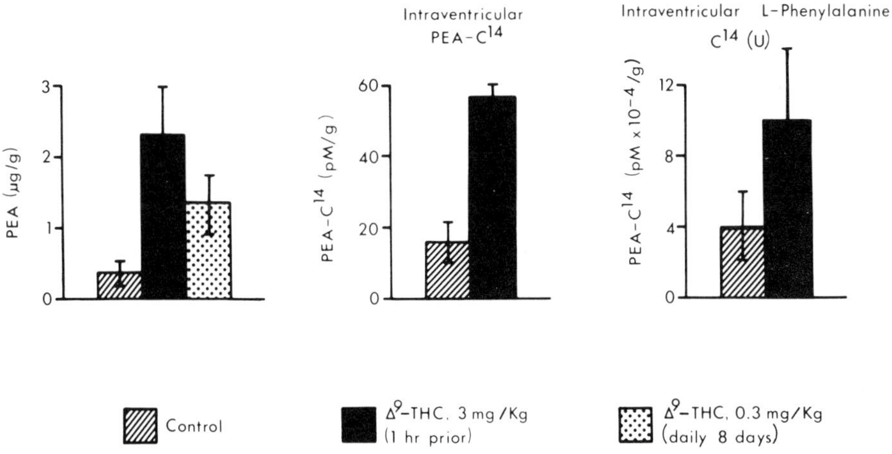

Figure 4 - Δ^9-THC increases rabbit brain levels of endogenous PEA (left) and of endogenously formed PEA following the intraventricular injection of L-phenylalanine (right), probably by inhibiting PEA disposition, as shown for intraventricularly administered PEA (center).

disposition of PEA in brain; it has been shown to decrease DβH levels (Ng et al., 1973) and its administration markedly increases the amount of PEA-C^{14} recovered from brain following the intraventricular injection of PEA-C^{14} or of phenylalanine-C^{14} (Fig. 4) (Mosnaim et al., 1973c; Whalley et al., 1973); (c) high doses of PEA induce similar effects ("popcorn behavior" and aggressiveness) to those induced by Δ^9-THC plus a MAOI; low doses of PEA potentiate the behavioral stimulation induced by Δ^9-THC plus a MAOI (Sabelli et al., 1973c); (d) Δ^9-THC increases the excitatory effect of iontophoretically administered PEA upon cortical unit potentials (Mosnaim et al., 1973d); (e) PEA induces marked behavioral stimulation in mice pretreated with Δ^9-THC (similar to that observed in animals pretreated with a MAOI) but not in non-pretreated mice (Fig. 3).

OHPEA is not as effective as PEA in facilitating the excitement induced by a MAOI plus Δ^9-THC; at certain doses OHPEA actually antagonizes it (Sabelli et al., 1973b,c). Notwithstanding, experiments with DβH inhibitors indicate that also a β-hydroxylated amine contributes to the stimulant effect of Δ^9-THC in MAOI-treated mice.

PEA Deficit in the Physiopathology of Endogenous Depression

Using semi-quantitative chromatographic procedures, Jepson et al. (1960) found PEA in human urine and reported that its urinary excretion was not modified by diet. Using a similar method, Fischer et al. (1968) found PEA to be absent in the urine of depressed subjects. We (Mosnaim et al., 1973a) found that control subjects (n = 27) excrete 453 ± 50 μg of free PEA/24 hours, and 433 ± 47 μg of conjugated PEA/24 hours. Most of the conjugate is a β-glucuronide (Inwang et al., 1973b). PEA excretion was reduced in depressed patients (n = 24) (Fig. 5): 352 ± 85 μg of free PEA/24 hours and 270 ± 83 μg of conjugated PEA/24 hours. The daily urinary excretion of PEA was below the lowest value found in normal subjects in a large percent of depressed subjects: 29% for free PEA, 62% for conjugated PEA, 71% for free or conjugated PEA, and 46% for total PEA (free and conjugated). The modes of the control group (350 μg free PEA/24 hours and 350 μg conjugated PEA/24 hours) markedly differed from the modes of the group of depressed patients (50 μg free PEA/24 hours and 50 μg conjugated PEA/24 hours). The differences between the two groups were statistically significant.

Antidepressive drug treatment does not account for the reduction in urinary PEA; tricyclics increase PEA urinary excretion (Fig. 5; see also Fischer et al., 1972b) as expected from their ability to increase brain PEA levels. Our results are in agreement with those obtained by other investigations in patients with endogenous depression (Fischer et al., 1968, 1972a; Boulton and Milward, 1971). Fischer et al. (1972a,b), and Rodriquez Casanova and Fernandez Labriola (1972) found that the urinary excretion of PEA was normal in secondary depressions and increased in schizo-

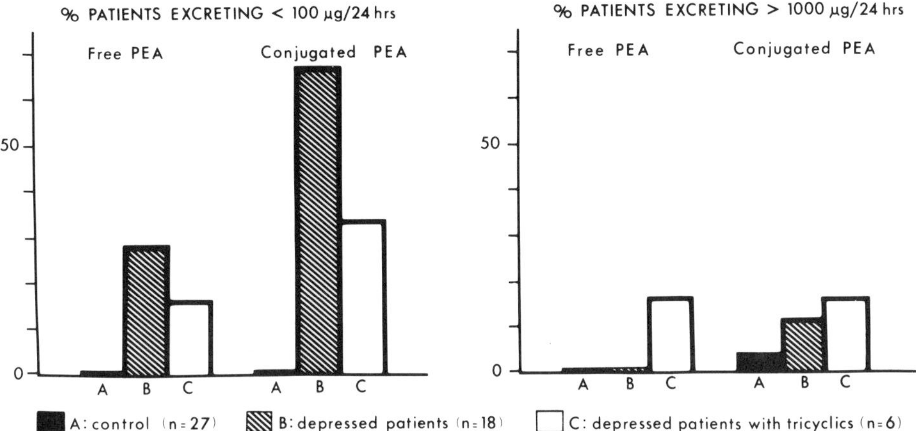

Figure 5 - Urinary excretion of PEA in man (from Mosnaim et al., 1973a)

affective depression. Thus, the reduction in PEA excretion does not appear to be the consequence of the depressive symptomatology; rather, the deficit in urinary PEA would be the reflection of a deficit in brain PEA which, in turn, would contribute to the pathophysiology of some forms of endogenous depression. Such speculation is based on the fact that the urinary levels of PEA reflect plasma levels of this amine, which in turn, probably are in equilibrium with brain PEA (since this amine readily crosses the blood-brain barrier). Thus, in the case of PEA, it seems meaningful to measure its urinary levels as an estimate of its tissue concentration. A reduction in the brain levels of PEA may result from either increased metabolism (leading to excessive synthesis of behaviorally depressant metabolites) or from faulty synthesis. Either may occur in the CNS or in peripheral organs (brain levels of PEA might be determined by its plasma concentration). Total body decarboxylation (Coppen, 1967) and brain levels of MAO (Sandler and Youdim, 1972) do not seem to be altered in depression. Plasma MAO activity was found increased in premenopausal depressed women (Klaiber et al., 1972). If this form of depression is the consequence of the excessive oxidation of a biogenic amine by plasma MAO, then the amine in question must be capable of crossing the blood-brain barrier.

In contrast to depression, there is an increase in PEA excretion in mania (Fischer et al., 1972a). That some opposite biochemical changes must occur in mania and depression is obvious, notwithstanding, these two disorders must also have a common biochemical characteristic because: (a) the existence of bipolar disease; (b) the coexistence of manic and depressive symptoms in patients treated with DOPA (Goodwin, 1971); (c) the ability of lithium to prevent both mania and depression. [Here again we have an example of the dialectic law of the interpenetration of the polar opposites.] Kety (1970) and Prange (1973) have suggested that an indoleethylamine deficit may underlie both types of affective disorders.

Other Medical Studies

Further support for the view that PEA is a neuromodulator is provided by the observation that its urinary excretion is altered in several medical entities other than affective disorders. In the urine of phenylketonurics there is an increased concentration of PEA (Jepson _et al._, 1960; Oates _et al._, 1963) and a neurotoxic Schiff base formed by PEA and pyridoxal phosphate (Loo, 1967). Excessive tissue levels of PEA might deplete vitamin B_6 and play a role in the pathophysiology of phenylketonuria (Edwards and Blau, 1973). Schizophrenics excrete in the urine high amounts of PEA and of several indoleethylamines (Fischer _et al._, 1972b). An excess in brain PEA may contribute to the partial overlaps in symptomatology of manic disorders, schizophrenias and chronic amphetamine intoxications.

The urinary output of PEA is reduced in myocardial infarction, in coma due to cerebrovascular accidents (Mosnaim _et al._, 1973b), and in Parkinsonism (Fischer _et al._, 1968; Sabelli _et al._, 1973a). A deficiency in brain PEA may explain the depressive symptoms characteristic of myocardial infarction and commonly found in Parkinsonian patients. The low PEA excretion in Parkinsonism, the high concentration of PEA in striatum, the ability of PEA and of amphetamine to induce choreic-like movements in animals, to potentiate the CNS stimulant effects of DOPA, and to antagonize the reserpine-induced Parkinson-like syndrome in mice, suggest that PEA may modulate extrapyramidal functions (Sabelli _et al._, 1973a).

The Phenylethylamine and the Catecholamine Theories of Affect

Many investigators in biological psychiatry are pursuing the hypothesis that affective disorders are diseases involving primarily brain phenylethylamines. In its original formulation, this hypothesis referred specifically to dihydroxylated phenylethylamines (CA). This theory had a remarkable heuristic value; as all other viable hypotheses, the experimental results stemming from it indicate the need for its revision. Alterations in PEA brain levels and function appear to account for affective changes which are not readily explained by the CA theory: (1) the administration of PEA induces clearly amphetamine-like stimulant effects, whereas the effects of the CA precursor DOPA are a mixture of stimulation and depression. In fact, a large percentage of Parkinson patients treated with DOPA develop affective depression (Jenkins and Groh, 1970). (2) A large percentage of depressed patients were shown, in all the published studies, to excrete in the urine low amounts of free and/or conjugated PEA, whereas changes in the excretion of CA metabolites are less consistent (Schildkraut, 1973; Maas *et al.*, 1972). (3) PEA, but not CA, appears to play a role in the stimulant actions of Δ^9-THC which in animals are enhanced by CA depletion. (4) Imipramine increases the brain levels of PEA whereas it does not significantly alter CA levels. (5) MAO, the target for the antidepressant action of many drugs, preferentially metabolizes PEA; CA are mainly disposed of by other mechanisms.

In summary, the experimental evidence indicates the need to further explore the function of PEA and OHPEA as ergotropic neuromodulators. A deficiency in brain PEA may be one essential component of the spectrum of biochemical alterations underlying endogenous depression. Changes in brain PEA may also be necessary, but are not the only factors responsible for the stimulant action of many drugs.

Acknowledgement: This research was supported by NIMH Grant MH-14110 and by Grant 310-11-RD from the State of Illinois, Department of Mental Health. The authors wish to thank Mr. S. Myles and Ms. Ann Maslanka for their invaluable assistance. We also wish to acknowledge Dana Flavin, W.J. Giardina, E.E. Inwang, U.P. Madubuike, W.A. Pedemonte, C. Whalley, and other colleagues who contributed to this research project.

References

Baldessarini, R.J., and Vogt, M., 1971, The uptake and subcellular distribution of aromatic amines in the brain of the rat, J. Neurochem. 18:2519.

Biel, J.H., Nuhfer, P.A., Hoya, W.K., Leiser, H.A., and Abood, L.G., 1962, Cholinergic blockage as an approach to the development of new psychotropic agents, Ann. N.Y. Acad. Sci. 96:251.

Boulton, A.A., and Milward, L., 1971, Separation, detection and quantitative analysis of urinary β-phenylethylamine, J. Chrom. 57:287.

Brunner, E.L., and Sabelli, H.C., 1971, The effect of 5-hydroxyindoleacetic acid on the action potential of the frog ventricle, Pharmacologist 13:267.

Carlsson, A., Fuxe, K., and Ungerstedt, U., 1968, The effect of imipramine on central 5-hydroxytryptamine neurons, J. Pharm. Pharmacol. 20:150.

Coppen, A., 1967, The biochemistry of affective disorders, Brit. J. Psychiat. 113:1237.

Costa, E., and Neff, N.H., 1966, Isotopic and non-isotopic measurements of the rate of catecholamine biosynthesis, in "Biochemistry and Pharmacology of the Basal Ganglia," (E. Costa, L.J. Côté, and M.D. Tahr, eds.), pp. 141-156, Raven Press, New York.

Creveling, C.R., Daly, J.W., Witkop, B., and Udenfriend, S., 1962, Substrates and inhibitors of dopamine-β-oxidase, Biochem. Biophys. Acta 64:125.

Creveling, C.R., Daly, J.W., Tokuyama, T., and Witkop, B., 1968, The combined use of α-methyltyrosine and threo-dihydroxyphenylserine-selective reduction of dopamine levels in the central nervous system, Biochem. Pharmacology 17:65.

Dewhurst, W.G., 1968, New theory of cerebral amine function and its clinical application, Nature 218:1130.

Edwards, D.J., and Blau, K., 1972, Analysis of phenylethylamines in biological tissues by gas-liquid chromatography with electron-capture detection, Analyt. Biochem. 45:387.

Edwards, D.J., and Blau, K., 1973, Phenylethylamines in brain and liver of rats with experimentally induced phenylketonuria-like characteristics, Biochem. J. 132:95.

Ehinger, B., and Myhrberg, H.E., 1971, Neuronal localization of dopamine, noradrenaline, and 5-hydroxytryptamine in the central and peripheral nervous system of Lumbricus terrestris (L.), Histochemie 28:265.

Feldstein, A., Chang, G.H., and Kuchorki, J.M., 1970, Tryptophol, 5-hydroxytryptophol and 5-methoxytryptophol induced sleep in mice, Life Sciences 9:323.

Fischer, E., Ludmer, R.I., and Sabelli, H.C., 1967, The antagonism of phenylethylamine to catecholamines on mouse motor activity, Acta Physiol. Lat. Amer. 17:15.

Fischer, E., Heller, B., and Miro, A.N., 1968, β-Phenylethylamine in human urine, Arzneim.-Forsch. 18:1486.

Fischer, E., Spatz, H., Heller, B., and Reggiani, H., 1972a, Phenylethylamine content of human urine and rat brain, its alteration in pathological conditions and after drug administration, Experientia 28:307.

Fischer, E., Spatz, H., Saavedra, J.M., Reggiani, H., Miró, A.H., and Heller, B., 1972b, Urinary elimination of phenylethylamine, Biol. Psychiat. 5:139.

Fuxe, K., Grobecker, H., and Jonsson, J., 1967, The effect of β-phenylethylamine on central and peripheral monoamine-containing neurons, Europ. J. Pharmacol. 2:202.

Giardina, W.J., Sabelli, H.C., and Aderman, M., 1970, Non-hydroxylated and deaminated derivatives of catecholamines on mouse exploratory behavior: A hypothesis of antidepressive action,

Proc. 78th Ann. Conv. Amer. Psychol. Assoc., Miami Beach, Sept. 3-8, pp. 811-812.

Giardina, W.J., Pedemonte, W.A., and Sabelli, H.C., 1973, Iontophoretic study of the effect of norepinephrine and 2-phenylethylamine on single cortical neurons, Life Science 12:153.

Goodwin, F.K., 1971, Psychiatric side effects of Levodopa in man, J.A.M.A. 218:1915.

Goodwin, F.K., Murphy, D.L., Brodie, H.K.H., and Bunney, Jr., W.E., 1970, L-Dopa, catecholamines and behavior: a clinical and biochemical study in depressed patients, Biol. Psychiat. 2:341.

Grundfest, H., 1957, General problems of drug actions on bioelectrical phenomena, Ann. N.Y. Acad. Sci. 66:537.

Hartman, B.K., and Udenfriend, S., 1972, The application of immunological techniques to the study of enzymes regulating catecholamine synthesis and degradation, Pharmacol. Rev. 24:311.

Himwich, H.E., and Alpers, H.S., 1970, Psychopharmacology, in "Annual Reviews of Pharmacology," Vol. 10, (H.W. Elliot, W.C. Cutting, and R.H. Dreisback, eds.), pp. 313-334, Annual Reviews, Inc., Palo Alto, California.

Holtz, P., 1966, Introductory remarks, Pharmacol. Review 18:85.

Inwang, E.E., Sugerman, J.H., DeMartini, W.J., Mosnaim, A.D., and Sabelli, H.C., 1972, Ultraviolet spectrophotometric determination of β-phenylethylamine-like substances in biological samples and its possible correlation with depression, presented at the Annual Convention of the Society of Biological Psychiatry, Dallas, April 28-30, 1972.

Inwang, E.E., Madabuike, U.P., and Mosnaim, A.D., 1973a, Evidence for the excretion of 2-phenylethylamine glucuronide in human urine, Experientia (in press).

Inwang, E.E., Mosnaim, A.D., and Sabelli, H.C., 1973b, Isolation and characterization of phenylethylamine and phenylethanolamine from human brain, J. Neurochem. 20:1469.

Jackson, D.M., and Temple, D.M., 1970, β-Phenylethylamine as a cardiotonic constituent of tissue extracts, Comp. Gen. Pharmac. 1:155.

Jenkins, R.B., and Groh, R.H., 1970, Mental symptoms in Parkinsonian patients treated with L-DOPA, The Lancet 2:177.

Jepson, J.B., Lovenberg, W., and Zaltzman, P., 1960, Amine metabolism studied in normal and phenylketonuric humans by monoamine oxidase inhibition, J. Biochem. 74:5.

Kety, S., 1970, Brain amines and affective disorders, in "Brain Chemistry and Mental Disease," (B.T. Ho, and W.M. McIsaac, eds.), pp. 237-244, Plenum Press, New York.

Klaiber, E.L., Broverman, D.M., Vogel, W., Kobayashi, Y., and Moriarty, D., 1972, Effects of estrogen therapy on plasma MAO activity and EEG driving responses of depressed women, Amer. J. Psychiat. 128:1492.

Kopin, I.J., 1968, Identification of "true" and "false" transmitters in psychopharmacology, Public Health Service Publication 1836, (D.H. Effron, ed.), Washington, pp. 57-60.

Loo, Y.H., 1967, Characterization of a new phenylalanine metabolite in phenylketonuria, J. Neurochem. 14:813.

Maas, J.W., Fawcett, J., and Dekirmenjian, H., 1972, Catecholamine metabolism, depressive illness and drug response, Arch. Gen. Psychiat. 26:252.

Mantegazza, P., and Riva, M., 1963, Amphetamine-like activity of β-phenylethylamine, J. Pharm. 15:472.

Marley, E., 1966, Behavioral and electrophysiological effects of catecholamines, Pharm. Rev. 18:753.

Molinoff, B.P., Landsberg, L., and Axelrod, J., 1969, An enzymatic assay for octopamine and other β-hydroxylated phenylethylamines, J. Pharmac. Exp. Ther. 170:253.

Mosnaim, A.D., and Inwang, E.E., 1973, A spectrophotometric method for the quantification of 2-phenylethylamine in biological specimens, Anal. Biochem. (in press).

Mosnaim, A.D., and Sabelli, H.C., 1971, Quantitative determination of the brain levels of a β-phenylethylamine-like substance in control and drug-treated mice, Pharmacologist 13:283.

Mosnaim, A.D., Inwang, E.E., and Sabelli, H.C., 1972, Phenylethylamine-like substance(s) in mammalian brain and their increase by repeated electroshock seizures, presented at the Society for Neuroscience, Houston, October 8-11, 1972.

Mosnaim, A.D., Inwang, E.E., Sugerman, J.H., DeMartini, W.J., and Sabelli, H.C., 1973a, Ultraviolet spectrophotometric determination of 2-phenylethylamine in biological samples and its possible correlation with depression, Biological Psychiatry 6(3):235.

Mosnaim, A.D., Inwang, E.E., Sugerman, J.H., and Sabelli, H.C., 1973b, Identification of 2-phenylethylamine in human urine by infrared and mass spectroscopy and its quantification in normal subjects and cardiovascular patients, Clinica Chimica Acta (in press).

Mosnaim, A.D., Sabelli, H.C., and Inwang, E.E., 1973c, The influence of psychotropic drugs on the levels of endogenous 2-phenylethylamine in brain, presented at the Society of Biological Psychiatry, Montreal, June 8-10, 1973.

Mosnaim, A.D., Whalley, C., Pedemonte, W.A., Vazquez, A.J., and Sabelli, H.C., 1973d, Further evidence for a role of 2-phenylethylamine as a mediator for the stimulant action of Δ^9-tetrahydrocannabinol, to be presented at the Society for Neuroscience, San Diego, Nov. 1973.

Nakajima, T., Kakimoto, Y., and Sano, I., 1964, Formation of phenylethylamine in mammalian tissues and its effect on motor activity in the mouse, J. Pharm. 143:319.

Ng, L.K.Y., Lamprecht, F., Williams, R.B., and Kopin, I.J., 1973, Δ^9-Tetrahydrocannabinol and ethanol: differential effects on sympathetic activity in differing environmental setting, Science 180:1368.

Oates, J.A., Nirenberg, P.Z., Jepson, J.B., Sjverdrma, A., and Undenfriend, S., 1963, Conversion of phenylalanine to phenylethylamine in patients with phenylketonuria, Proc. Soc. Exptl. Biol. and Med. 112:1078.

Pedemonte, W.A., and Sabelli, H.C., 1973, Possible direct modulator role of serotonin and its metabolites on sleep mechanisms in newly-hatched chicks, to be presented at the Society for Neuroscience, San Diego, Nov. 1973.

Prange, Jr., H.J., 1973, The use of drugs in depression: its theoretical and practical basis, Psychiatric Annals 3:56.

Rodriguez Casanova, E., and Fernandez Labriola, R., 1972, Estudios sobre la Fenetilaminuria en depresiones, Ann. Biol. Psychiat. 1:5.

Rothballer, A.B., 1959, The effects of catecholamines on the central nervous system, Pharmacol. Rev. 11:494.

Saavedra, J.M., and Axelrod, J., 1973, Demonstration and distribution of phenylethanolamine in brain and other tissues, Proc. Nat. Acad. Sci. (USA) 70:769.

Sabelli, H.C., 1960, Pressor effects of adrenergic agents and serotonin after the administration of imipramine, Arzneim.-Forsch. 10:935.

Sabelli, H.C., 1964, A pharmacological strategy for the study of central modulator linkages, in "Recent Advances in Biological Psychiatry," Vol. 6, (J. Wortis, ed.), pp. 145-182, Plenum Press, New York.

Sabelli, H.C., 1970, Optic evoked potential changes induced by deaminated metabolites of serotonin: 5-Hydroxytryptophol and 5-hydroxyindoleacetic acid, Experientia 26:58.

Sabelli, H.C., and Giardina, W.J., 1970a, CNS effects of the aldehyde products of brain monoamines, Biol. Psychiat. 2:119.

Sabelli, H.C., and Giardina, W.J., 1970b, Tryptaldehydes (Indoleacetaldehydes) in serotonergic sleep of newly hatched chicks, Arzneim.-Forsch. 20:74.

Sabelli, H.C., and Giardina, W.J., 1971, The influence of inhibitors of amine metabolism on the effects of serotonin and its metabolites on the photic evoked potential in rabbits, Experientia 27-64.

Sabelli, H.C., and Giardina, W.J., 1973a, Evidence for biological activity of deaminated metabolites of catecholamines, presented at the Society of Biological Psychiatry, Montreal, June 8-10.

Sabelli, H.C., and Giardina, W.J., 1973b, Amine modulation of affective behavior, in "Chemical Modulation of Brain Function," (H.C. Sabelli, ed.), pp. 225-259, Raven Press, New York.

Sabelli, H.C., and Gorosito,M., 1969, Evidence for biogenic amine receptors in toad sciatic nerves, Int. J. Neuropharmacol. 8:495.

Sabelli, H.C., and Mosnaim, A.D., 1973, Phenylethylamine hypothesis of affective behavior, presented at the American Psychiatric Association, Honolulu, May 7-13, 1973.

Sabelli, H.C., and Toman, J.E.P., 1961, Efectos electrocardiograficos de agentes antidepresivos, Rev. Soc. Argent. Biol. 37:205.

Sabelli, H.C., Levin, J.J., and Toman, J.E.P., 1961, Mecanismos colinérgicos y agentes antidepresivos, Rev. Soc. Argent. Biol. 37:87.

Sabelli, H.C., Giardina, W.J., Alivisatos, S.G.A., Seth, P.K., and Ungar, F., 1969, Indoleacetaldehydes: Serotonin-like effects on the central nervous system, Nature (Lond.) 223:73.

Sabelli, H.C., Giardina, W.J., and Alivisatos, S.G.A., 1970, Influence of serotonin and related substances upon photic evoked potentials of rabbit: Evidence for biological activity of the aldehyde derivative, Arzneim.-Forsch. 20:68.

Sabelli, H.C., Giardina, W.J., and Bartizal, F., 1971a, Photic evoked cortical potentials: Interaction of reserpine, monoamine oxidase inhibition and DOPA, Biol. Psychiat. 3:273.

Sabelli, H.C., Giardina, W.J., Mosnaim, A.D., and Sabelli, N.H., 1971b, A comparison of the functional roles of norepinephrine, dopamine and phenylethylamine in the central nervous system, Proceedings of the Satellite Symposium "Central and Peripheral Adrenergic Systems" XXV International Congress of Physiological Sciences, Warsaw, August 3-6, 1971.

Sabelli, H.C., Mosnaim, A.D., and Vazquez, A.J., 1973a, Possible role of 2-phenylethylamine and L-DOPA in extrapyramidal disorders and their drug therapy, to be presented at the American Society for Pharmacology and Experimental Therapeutics, Michigan, August 19-23, 1973.

Sabelli, H.C., Vazquez, A.J., and Flavin, D.F., 1973b, Differential effects of two putative neuromodulators: 2-phenylethylamine and 2-phenylethanolamine, to be presented at the Society for Neuroscience, San Diego, Nov. 1973.

Sabelli, H.C., Mosnaim, A.D., Vazquez, A.J., and Madrid-Pedemonte, L., 1973c, (-)-Trans-Δ^9-tetrahydrocannabinol-induced increase in brain 2-phenylethylamine: its possible role in the psychological effect of marihuana, in "Drug Addiction: Behavioral Aspects," (J. Singh, ed.), Futura Publishing Co., New York, in press.

Sandler, M., and Youdim, M.B., 1972, Multiple forms of monoamine oxidase: functional significance, Pharmacol. Rev. 24:331.

Schildkraut, J.J., 1973, Norepinephrine metabolism in the pathophysiology and classification of depressive and manic disorders, in "Psychopathology and Psychopharmacology," (J.O. Cole, A.Freedman, and A. Friedhoff, eds.), John Hopkins Press, Baltimore, in press.

Seeman, P., 1972, The membrane actions of anesthetics and tranquilizers, Pharmacol. Rev. 24:583.

Sulser, F., and Sanders-Bush, E., 1971, Effect of drugs on amines in the CNS, Ann Rev. Pharmacol. 11:209.

Toman, J.E.P., and Sabelli, H.C., 1968, Neuropharmacology of earthworm giant fibers, Int. J. Neuropharmacol. 7:543.

Vane, J.B., 1960, The actions of sympathomimetic amines on tryptamine receptors, in "Adrenergic Mechanisms (A CIBA Foundation Symposium)", (J.R. Vane, G.E. Wolstenholme, and M. O'Connor, eds.), pp. 356-372, J. and A. Churchill, Ltd., London.

Vazquez, A.J., and Krip, G., 1973, Interactions between amphetamine and serotonergic agents on cat's isolated cerebral cortex, Biol. Psychiatry 7:11.

Vazquez, A.J., Giardina, W.J., Madrid-Pedemonte, L., Mosnaim, A.D., and Sabelli, H.C., 1973, Direct role of DOPA in the central nervous system, to be presented at the Society for Neuroscience, San Diego, Nov. 1973.

Whalley, C., Mosnaim, A.D., and Sabelli, H.C., 1973, 2-Phenylethylamine in rabbit brain and liver: its synthesis, metabolism and role in the action of imipramine, pargyline, and marihuana, to be presented at the American Society for Pharmacology and Experimental Therapeutics, Michigan, August 19-23, 1973.

Yang, H-YT., and Neff, N.H., 1973, Monoamine oxidase: I. a natural substrate for type B enzyme, Fed. Proc. 32:797.

ABSTRACT:

2-Phenylethylamine (PEA) is formed by L-phenylalanine decarboxylation, a reaction that may occur throughout the body and has been demonstrated in the brain. Using mass, infrared and ultraviolet spectroscopy, gas and thin-layer chromatography, etc., we have found PEA (and its metabolite phenylethanolamine) in the central nervous system (CNS) -- including cerebellum and cerebral cortex -- and most viscerae (cat, rabbit, and man). PEA brain levels are similar to those of catecholamines. PEA readily crosses the blood brain barrier. PEA is mainly metabolized by MAO, forming phenylacetaldehyde, which is a CNS depressant that may mediate, in part, the effects of PEA given alone. In animals treated with an inhibitor of MAO, PEA effects are similar to those of amphetamine (sympathomimetic effects, EEG alerting, excitement, extrapyramidal stereotipias, antagonism against reserpine-induced Parkinson-like syndrome). The sympathomimetic effects of PEA are mainly due to catecholamine release. The CNS effects of PEA are for the most part not mediated by phenylethanolamine: this amine is not as effective as PEA as a CNS stimulant, it can be formed only in those areas containing dopamine-β-hydroxylase, and moreover, inhibitors of this enzyme enhance PEA stimulant effects. The CNS effects of PEA may be due in part to catecholamine release,but they are mostly direct because they are not prevented by catecholamine depleters, and they are not mimicked by the catecholamine precursor DOPA. PEA increases the behavioral and neurological stimulant effects of DOPA and antagonizes DOPA-induced akinesia. Moreover, PEA and norepinephrine (iontophoretic) exert different effects in 50% of the optic cortex neurons (rabbits) suggesting that catecholamine and PEA receptors are different. Amphetamine may act in part by releasing catecholamines and PEA, and in part as an agonist on specific PEA receptors; phenothiazines and haloperidol may act in part by blocking PEA. PEA brain levels (mice, rats, rabbits) are increased by anti-depressive treatments: imipramine (which does not alter catechol-amine levels), MAO inhibitors and electroshock. Reserpine, a drug that causes depression, depletes PEA. Very low doses of Δ^9-tetra-hydrocannabinol (Δ^9-THC) markedly increases brain PEA levels and slow down PEA disposition by brain tissue without affecting catecholamine or serotonin levels. Δ^9-THC potentiates the excitatory effects of iontophoretic PEA on cortical units. The aggressive excitement induced in mice by Δ^9-THC plus MAO inhibitors may be mediated by PEA. The urinary excretion of PEA in human subjects is markedly decreased in patients with endogenous depression or with Parkinsonism, normal in patients with secondary depression, and is increased in mania, schizophrenia and phenylpyruvic oligophrenia. With E. Fischer, we have proposed that PEA is one of the modulators of affective and extrapyramidal functions. A deficit in brain PEA may be one of the biochemical lesions in affective depression and Parkinsonism,whereas an excess in the brain levels or turnover of PEA may be associated with mania and with chorea. The mood-lifting effects of antidepressive drugs and of marihuana may result at least in part from their ability to increase brain PEA.

NEUROCHEMICAL, NEUROENDOCRINE AND BIORHYTHMIC ASPECTS OF SLEEP IN MAN: RELATIONSHIP TO CLINICAL PATHOLOGICAL DISORDERS

Roger Broughton*

Faculties of Medicine and Psychology

University of Ottawa, Ottawa, Canada

Like all living organisms, each of us exhibits marked variations in activity throughout the 24-hour period. In humans these are manifestly mental as well as physical. At the levels of vigilance, perception, motoricity, physiology, biochemistry and probably even in structure, we therefore are different individuals at different times (Luce, 1970). It follows that excessive consistency cannot be demanded.

These variations are not entirely random. Profound alterations, of course, accompany the daily recurrence of the sleep-wakefulness cycle. Less evident are the cyclic swings of consciousness, perception and motor activity in wakefulness which express what Kleitman (1963) termed the basic rest-activity cycle (BRAC). These repeat every 80-120 minutes in normal adult man and more frequently in childhood. The more obvious alternation in mammals or birds of so-called non-rapid-eye-movement (NREM) sleep, or quiet sleep, with the rapid-eye-movement (REM) sleep, or active sleep, is the continuation of this phylogenetically older BRAC into the hypnic state.

There is experimental evidence relating these two sleep states to fluctuations in neurochemical substances, including presumed neurotransmitters or neuromodulators, and relating the sleep-wakefulness cycle to neuroendocrine changes. Much of the recent emphasis on the neurochemical aspects has centered upon cerebral monoamines and has been extensively reviewed amongst others by Koella (1967), Hartmann (1967), Jouvet (1969), Cordeau (1970), Morgane

* Associate, Medical Research Council of Canada

and Stern (1972). Both the neurochemical and neuroendocrine aspects of sleep are treated in the recent volume The Sleeping Brain, edited by Chase (1972). My task is to attempt to integrate this already vast neurochemical and neuroendocrine literature with the similarly voluminous literature concerning the medical aspects of sleep knowing that, at any rate, much of what can be said today will be quite out of date in five years.

Neurochemical data on sleep are largely derived from research on cat and rat. Numerous studies suggest that 5-hydroxytryptamine (5-HT, serotonin), which is particularly heavily stored in brain stem raphé nuclei cells (Dahlstrom and Fuxe, 1964), is involved in NREM sleep. This tentative conclusion is based upon the effects of enzymatic blockage, the administration of precursors, histofluorescence studies, and the effects of ablation of these serotonergic cells (cf. reviews above). There is also evidence that serotonergic posterior raphé cells are important in the initiation of REM sleep, and perhaps in restricting the phasic ponto-geniculo-occipital (PGO) spikes to REM sleep.

Catecholamines, in particular norepinephrine, are highly concentrated in the locus coeruleus complex. Based on studies using similar techniques, this complex appears to be involved in producing the tonic descending, mainly motor inhibitory, and the ascending activating manifestations of REM sleep (Jouvet, 1969). Circulating catecholamines have long been known to excite the ascending reticular system and so sustain arousal (Rothballer, 1956). Recent evidence by Harner and Dorman (1970) suggests that dopamine is related to awakening from sleep. Also γ-aminobutyric acid (GABA) is an active inhibitory agent and is probably active in the regulation of vigilance. Other amino acids such as glutamic acid are also probably involved (Jasper et al., 1965; Jasper and Koyama, 1969). Indeed, even the short-chain polypeptides may be important in relationship to control of sleep (Monnier, this symposium) and perhaps to repair functions, as has recently been shown for the liver (Pickart and Thaler, 1973) and has been postulated as occurring in REM sleep (Oswald, 1969a).

There is also, of course, much evidence involving cholinergic mechanisms in sleep. Hernández-Péon (1965, 1967) mapped out a cholinergic sleep system throughout the brain stem, diencephalon and telencephalon. Jouvet (1962) has given a role to ACh in initiating REM sleep. Jasper and co-workers have shown that increased ACh is liberated from neocortex during the electrocortical activation that accompanies arousal from sleep (Celesia and Jasper, 1966) and during REM sleep (Jasper and Tessier, 1971).

It seems probable, in fact, that all chemical constituents of the brain (and perhaps of the body?), whether protein carbohydrate, fat, fluid or electrolyte, will change in some way with such marked general biological alterations as occur between sleep and wakefulness. Indeed, circadian and even ultradian general and regional changes of cerebral monoamines have already been documented by Reis and collaborators (Reis and Wurtman, 1968; Reis et al. 1969) and by others (Dixit and Buckley, 1967; Friedman and Walker, 1968).

To complete this very brief review of the neurochemical aspects of vigilance it must be stressed that a number of problems limit its direct application to all species, and in particular to man. First, it is the neurocircuitry rather than the presence of the implicated chemical substances which relates to the phylogenetic appearance of the two basic sleep states (Broughton, 1972a). Secondly, the postulated neurotransmitters may not be necessary. For example, when serotonin synthesis is blocked by parachlorophenylalanine (pCPA) to levels where there is no histofluorescently identifiable 5-HT, normal amounts of non-REM and NREM sleep nevertheless return (Dement et al., 1969). But most important of all, the correlations made in "lower" animals simply do not hold true for man, even when essentially identical manipulations are performed.

Thus, although pCPA in cats produces marked insomnia with greater reduction of NREM than REM sleep, in man it mainly reduces REM sleep (Wyatt et al., 1969), an effect that can be reversed by giving the immediate precursor of serotonin, 5-hydroxytryptophan or 5-HTP (Wyatt et al., 1970a). Moreover, d,l 5HTP (600 mg orally) to untreated subjects increases REM sleep significantly from baseline levels, but has no effect on NREM sleep (Wyatt et al., 1971). In man, therefore, serotonergic mechanisms appear mainly related to REM sleep, not NREM sleep.

Catecholamine mechanisms have also been studied in man. Using the tyrosine hydroxylase inhibitors α-methyl-d-l-paratyrosine (AMPT) and α-methylphenylalanine (AMPA) and the catecholamine precursor DOPA, Wyatt's group (cf. Wyatt, 1972) has shown that low CA levels appear to increase REM sleep, and high CA levels reduce it. Catecholamine and serotonin therefore seemingly produce opposing effects upon REM sleep. Moreover, Wyatt found that reduction of CA synthesis by AMPT produced a dissociative REM state with persistant muscle tone in the presence of the other features of REM sleep. Similar dissociations, as we shall see, have been reported during sleep in certain psychiatric disorders.

The general conclusion from the monoamine studies is that animal models generally remain inappropriate for interpretation of human data. Moreover, as Wyatt (1972) has pointed out, there are a number of presently necessary simplifying assumptions which may not be true. To quote: "Researchers studying the relationship of biogenic amines to sleep generally make the implicit assumption that amines are neural transmitters, that the neurones containing the amines have unitary functions, and that the neuronal activity in question is present during some stage of sleep or wakefulness and not at other times". Further complexities of the unsettled amine issue are indicated by Oswald (1972) in the same symposium: "Consider the drugs imipramine and desipramine; imipramine is an excellent blocker of 5-HT uptake mechanisms and desipramine is an excellent blocker of NE uptake. There is no difference between these drugs in their effects on human sleep, either acutely or over a period of weeks. If the simple idea of effects on receptors or uptake mechanisms were correct, we would expect the effects to be very different". Finally, to paraphrase Kripke (1973), even if we presently knew the neurochemical mechanisms of the different sleep states, which we do not, we still have no understanding of the chemical mechanisms of the cyclicity itself.

The endocrine aspects of sleep, unlike the neurochemical, have largely been researched in man. An exciting story it is indeed. Since the original reports of Takahashi *et al.* (1968), Honda *et al.* (1969), Sassin *et al.* (1969) and Parker *et al.* (1969), at least fifteen studies have confirmed that in adult man a major peak of growth hormone (GH) secretion occurs some 1-2 hours after sleep onset, usually in stage 3-4 sleep. This peak is not seen in infants below three months of age, who exhibit continuous high GH secretion levels lacking circadian variation (Vigneri and D'Agata, 1971). But it is present in older children (Illig *et al.*, 1971; Mace *et al.*, 1972) and disappears in the elderly (Karacan *et al.*, 1971; Carlson *et al.*, 1972). The sleep-related secretion is considered a better index of growth hormone status than are the GH responses to insulin or arginine (Illig *et al.*, 1971).

Most important is that this hypnic GH secretion is sleep induced and not simply a circadian rhythm entrained by sleep. Thus, with a 180° shift of the sleep-wakefulness cycle, GH secretion again occurs shortly after sleep onset, now in the day-time. If stages 3 and 4 sleep are physiologically reduced by lightening sleep whenever a subject shows slow waves, the GH secretion is decreased and retarded (Sassin *et al.*, 1969; Karacan *et al.*, 1971); moreover, it shows no compensatory rebound the following nights (Sassin *et al.*, 1969).

Luteinizing hormone (LH) during puberty (Boyar *et al.*, 1972), but not in pre- or post-pubertal subjects (Rubin *et al.*, 1972), also shows an hypnic increase, but in the second rather than the first half of sleep: and it is also apparently sleep induced rather than simply nocturnal circadian in distribution. Marked sleep related increases of LH have also been reported in preovulatory female subjects (Kapen *et al.*, 1973). Prolactin (Sassin *et al.*, 1972) and testosterone (Evans *et al.*, 1971) secretion exhibit increases at night; and both are sleep induced (Sassin *et al.*, 1973; Evans *et al.*, 1971).

Plasma cortisol levels increase towards the second half of night, usually being maximum from 0400-0800 hours (Weitzman et al., 1966). Interestingly, cortisol (and also LH and prolactin) has been shown to be secreted episodically from a zero or very low baseline (Weitzman et al., 1968). This implies the total inadequacy of obtaining circadian data from only 2-4 samples per day, as well as the inappropriateness of simple negative feedback models of secretion. Infants 1-15 weeks of age do not show a circadian rhythm of cortisol secretion, although their cortisol response to stress during crying is intact (Anders et al., 1970). It is important to note that cortisol, unlike GH, LH and probably testosterone, is a circadian rhythm entrained by sleep, and is not truly sleep induced. Thus, 180° reversal of the sleep-wakefulness cycle does not lead to similar reversal of cortisol secretion until the second week or later (Weitzman et al., 1968).

Even from this brief review, it is evident that endocrine and sleep functions are highly interrelated and that endocrinology text books will have to be rewritten in the next few years to include these fundamental biorhythm and episodic secretion aspects.

The foregoing leads naturally to a discussion of biological rhythms in general, a field now often called by the term chronobiology introduced by Franz Halberg (1969). A biological rhythm can be described by its frequency or period, its amplitude, and its phase relation to other rhythms or to clock time. Rhythms with a period of 20 to 28 hours are described as circadian (literally, about a day), those with a period below 20 hours as ultradian, and those longer than 28 hours as infradian. Circahoral and circannual rhythms are ultradian and infradian, respectively. A time-cue, such as sleep-wakefulness or light-darkness, may act as a synchronizer or so-called Zeitgeber, thereby entraining a rhythm to a slightly different frequency than when it is "free-running". One biological rhythm may modulate the frequency of another, as in the case of the sleep-wakefulness circadian cycle's apparent modulation of the 80-120 min so-called BRAC, which appears generally nearer 90 min during sleep and 100-110 min during wakefulness (Kripke, 1973).

The sleep-wakefulness cycle is of course only one of many circadian rhythms. Extensive reviews of the many biological variables showing a 24-hour periodicity are available by Richter (1965), Mills (1966), Aschoff (1967), de Ajuriaguerra (1968) and Luce (1970). As mentioned earlier, these include vigilance, performance, perception, gross motricity, autonomic status, body temperature, neuroendocrine, serum and electrolyte, and hematological status and numerous others. Many interesting studies on circadian variations in performance are included in the volume *Aspects of Human Efficiency* recently edited by Colquhoun (1972).

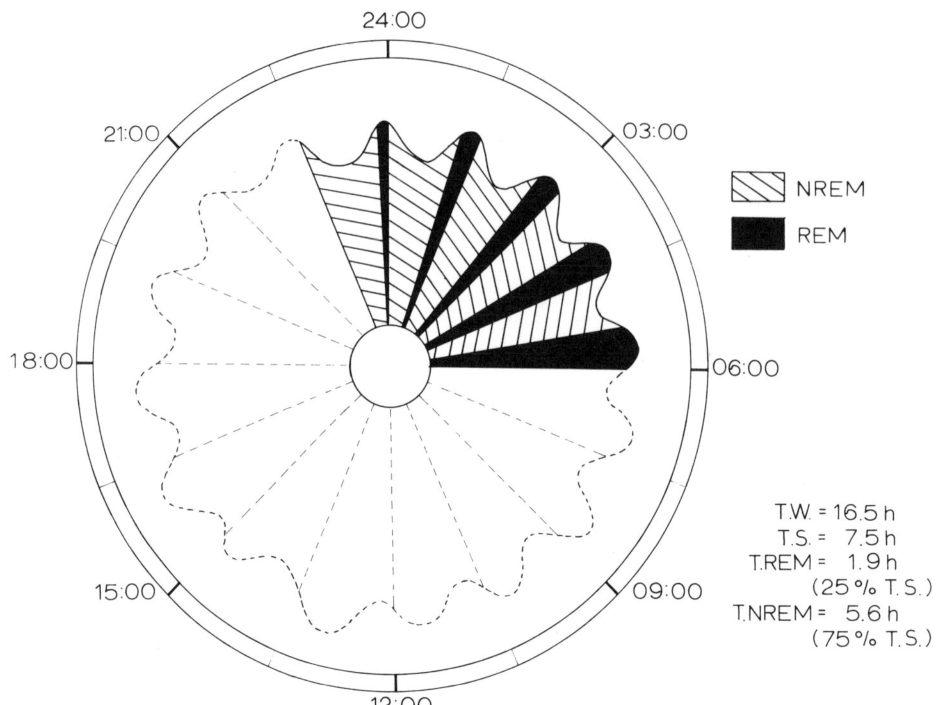

Figure 1. Nocturnal sleep cycles and the basic rest-activity cycle (BRAC) around the 24-hour clock. The idealized subject illustrated goes to bed at 22:30 hours and has a REM period every 90 minutes thereafter, awakening at 06:00. He then shows waking activity peaks in phase with the nocturnal REM periods which recur at approximately 90 minute intervals during the day. There is a relatively prolonged post prandial rest period from about 13:30 - 14:45. The diagram postulates lowest rest/activity ratios in the BRAC during the mid-afternoon, when body temperature is usually highest. (NREM sleep is hashed, REM sleep is solid black.)

Ultradian rhythms in wakefulness related to the basic rest-activity cycle are presently under intensive investigation. This rhythm lengthens from 40-45 min in the neonate to 80-120 min in the adult. As Kleitman (1969) has pointed out, we tend to live a social 90-100 minute day. A typical day might be: awaken at 7:30 A.M., drive to work at 9:00, coffee break at 10:30, lunch at 12:00, back to work at 1:30 P.M., more coffee at 3:00, leave at 4:30, and so on. In fact, very early studies by Wada (1922) reported gastric contractions recurring at about 90-100 min intervals, a finding recently replicated by Hiatt and Kripke (personal communication).

Using a wide variety of techniques, waking ultradian rhythms with this periodicity have already been claimed in man for: oral behavior (Friedman and Fisher, 1967; Oswald et al., 1970), day-dreaming (Kripke and Sonnenschein, 1973), errors in performance tests (Globus, 1972), perception (Orr, 1973), cerebral responsiveness measured as evoked potentials (Tanguay et al., 1973), and cardiac rate and variability (Orr, 1973). Moreover, sudden incoherent changes in speech content temporally related to preceding REM sleep stages have been described (Destrooper and Broughton, 1969). There is also evidence that neuroendocrine functions during sleep may show an ultradian modulation during the NREM-REM alterations for at least: GH during stages 3-4 in the first part of the night (Parker et al., 1969); ADH, as reflected in decreased urinary volume in REM sleep (Mandell et al., 1966a); LH in REM sleep during puberty (Boyar et al., 1972); circulating catacholamines, as reflected in urinary vanillomandelic acid levels (Mandell et al., 1966b); and plasma testosterone levels (Evans et al., 1971).

An increasingly well-defined picture is emerging therefore of superimposed rapid ultradian and circadian (together illustrated in Fig. 1), and indeed monthly, seasonal, annual, and age-related tides of human behavior and their associated neurochemical and neuroendocrine mechanisms or correlates. How this complex process deviates from normal in disease is, I believe, a major challenge of medical investigation for the next decade or two. To emphasize the point, I would mention that the sick body is obviously ill in sleep as well as in wakefulness; that many illnesses are exaggerated by, or only appear in, sleep (in this instance sleep is a threat); that sleep disorders per se are very frequent in medicine; and that severe prolonged sleep deprivation usually leads to deterioration of a medical disorder. We will now consider the clinical pathological states.

Disorders of, or involving, sleep may be classified according to their main symptomatology into five major groups as those of "excessive" sleep, those of "inadequate" sleep, and those with predominantly psychosensory, motor or autonomic symptoms (Table 1, from Broughton 1972b). Conversely, they may be classed pathophysiologically according to the type of sleep mainly involved, as disorders

TABLE I

SLEEP DISORDERS ACCORDING TO SYMPTOMATOLOGY*

I "Excessive" Sleep
1. Sleep attacks
2. Persistent sleepiness (simple hypersomnia)
3. Hypersomnia with periodic respiration
a) pure
b) Ondine's
c) mixed
4. Hypersomnia with megaphagia

II "Inadequate" Sleep
1. Insomnia
a) induction
b) maintenance
c) mixed
2. Pseudo-insomnia

III Psychosensory
1. Hypnagogic hallucinationa (excessive)
2. Terrifying dreams
3. Sleep terrors
4. Sleep drunkeness
5. Peduncular hallucinosis

IV Motor
1. Cataplexy
2. Sleep paralysis
3. Restless legs
4. Jactatio capitis nocturna
5. Somnambulism
6. Somniloquy
7. Snoring, Bruxism, etc.

V Autonomic
1. Enuresis
2. Periodic respiration
3. Painful erections

---

* from Broughton, R., 1972, A proposed classification of sleep disorders, in "Sleep Research", (M.H. Chase, W.C. Stern and P.L. Walter, eds.), pp. 146, Brain Information Service/ Brain Research Institute, UCLA, Los Angeles.

of NREM sleep, REM sleep, sleep "in toto", the hypnagogic state, and so forth (Table 2). This paper follows a more usual order of presentation and considers in turn; narcolepsy, hypersomnia, insomnia, psychobiological attacks in sleep (terrifying dreams, sleep terrors, somnambulism, enuresis, and confusional episodes), and certain psychiatric conditions (anxiety, depression, manic-depressive illness, schizophrenia, drug dependency and drug psychoses).

Narcolepsy only infrequently presents as sleep attacks alone. The attacks then usually consist of NREM sleep (Hishikawa and Kaneko 1965; Roth _et al._, 1968; Berti Ceroni _et al._, 1968; and others) and respond to amphetamines (Roth _et al._, 1968), perhaps particularly the levo-rotary form (Passouant _et al._, 1964). Most narcoleptic attacks are associated with the auxiliary symptoms of cataplexy, sleep paralysis or vivid hypnagogic hallucinations (together comprising the narcolepsy "tetrad") and are based, like these auxiliary symptoms, upon REM sleep mechanisms (cf. Broughton 1971, 1972c). They do not respond favorably to stimulants: rather, stimulants may cause deterioration (Broughton, 1972c; Zarcone 1972). All four REM sleep based symptoms, however, respond to REM suppressant "antidepressant" medication, either tricyclics, such as imipramine or desipramine, or MAO inhibitors, such as phenelzine. The opposing effects of stimulant versus antidepressant drugs upon NREM and REM sleep phenomena surely indicates that these sleep processes have distinct neurochemical substrates in man, even if these are not yet fully known.

The 90-120 minute BRAC relates to the timing of daytime REM sleep narcoleptic attacks, as these tend to take place in phase with nocturnal REM periods, when such patients are placed in a quiet dim room (Passouant _et al._, 1969). The overall stability of circadian rhythms is important to the health of such patients, as randomized shift work makes pharmacological control of the attacks much more difficult (Broughton, 1971 and 1972c). Takahashi _et al._ (1971) have shown that the sleep onset REM periods which characterize the nocturnal sleep of patients with compound narcolepsy (Rechtschaffen _et al._, 1963) are not associated with the sleep induced growth hormone secretion normally observed at that time during stage 3-4 sleep. Finally, Evans and Oswald (1966) noted that l-tryptophan, the dietary precursor of serotonin, doubles the duration of the sleep onset REM period of narcoleptics.

Hypersomnia, by definition, consists of excessively prolonged or deep sleep, rather than relatively brief attacks of overwhelming sleep. The etiologies are multiple and include idiopathic, post-traumatic, infectious, tumoral, degenerative and endocrine varieties. NREM sleep or sleep "in toto" are almost always involved (cf. reviews of Broughton 1971 and 1972c). Etiological treatment aside, these NREM conditions respond best to stimulants, as do the isolated (NREM) narcoleptic attacks. In both instances the improvement is

TABLE II

INCOMPLETE LIST OF SLEEP DISORDERS ACCORDING TO SLEEP STATES MAINLY INVOLVED

		References
1.	NREM Sleep	
	Hypersomnia	
	- simple	Roth *et al.*, (1968, 1969)
	- Kleine-Levin	Barotini & Zappoli, (1968); Popoviciu & Cortariu, (1972)
	- periodic respiration	Jung & Kuhlo, (1965); Gastaut *et al.*, (1966)
	Narcolepsy (isolated; rare form)	Hishikawa & Kaneko, (1965); Roth *et al.*, (1968)
	Enuresis nocturna	Pierce *et al.*, (1961); Broughton & Gastaut, (1964)
	Somnambulism	Gastaut & Broughton, (1965); Jacobson *et al.*, (1965)
	Sleep terrors	Gastaut & Broughton, (1965); Fisher *et al.*, (1970)
	Sleep drunkeness	Gastaut & Broughton, (1965)
	Snoring	Albert & Bollas, (1973)
	Bruxism	Gastaut & Broughton, (1965); Reding *et al.*, (1968)
2.	REM Sleep	
	Hypersomnia (rare form)	Broughton & Guzman, (1973)
	Narcolepsy (compound)	Passouant *et al.*, (1964); Hishikawa & Kaneko, (1965); Dement *et al.*, (1966)
	Cataplexy	Dement *et al.*, (1966)
	Sleep paralysis	Dement *et al.*, (1966); Schwartz & Escande, (1968)
	Hypnagogic dreams (hallucinations)	Rechtschaffen *et al.*, (1963)
	Terrifying dreams	Oswald, (1962)
	Painful erections	Karacan, (1971); Fisher *et al.*, (1972)
	Alcoholic hallucinations	Gross *et al.*, (1966); Greenberg & Pearlman, (1967)
	Drug dependency	Oswald, (1969); Kales *et al.*, (1969)
	Insomnia (REM interruption type)	Greenberg, (1967)

TABLE II (cont'd)

3. Sleep "in toto"	
Hypersomnia	Rechtschaffen & Roth, (1969)
Jacitatio capitis nocturna	Oswald, (1964); Gastaut & Broughton, (1965)
Somniloquy	Rechtschaffen *et al.*, (1962)
Insomnia	Fujiya *et al.*, (1963); Schwartz *et al.*, (1963)
4. Hypnagogic State and Light Sleep	
Terrifying hypnagogic hallucinations	Gastaut & Broughton, (1965)
Jacitatio capitus praenocturna	Gastaut & Broughton, (1965)
Excessive nocturnal myoclonus	Oswald, (1962); Loeb *et al.*, (1964); Lugaresi *et al.*, (1965)
Restless legs	Lugaresi *et al.*, (1965); Coccagna *et al.*, (1966)

ostensibly due to the unspecific arousing effect of stimulant medication. Recently, a case report of Broughton and Guzman (1973) documented the possibility of durable hypersomnia with rapid eye movements. The condition was refractory to stimulants, but disappeared after a few days on tricyclic "antidepressants", thereby opposing as in narcoleptic attacks, drug responsiveness of pathological REM, versus NREM, sleep phenomena.

Insomnia, that is the subjective sensation of insufficient sleep, is most frequently due to stress, environmental disturbances or, increasingly, disruption of biorhythms. Rarely, an organic cause may be present.

The biology of stress has a very large literature. Suffice it to say that circulating catecholamines and cortisol, amongst other substances, are released during stress and that these can have an anxiogenic, insomnia producing effect. There is evidence that prolonged stressful demands on performance following sleep deprivation exaggerate the 90-120 min ultradian fluctuations of perception, cardiac function and other variables (Orr, 1973), and that stress alters numerous circadian rhythms (cf. Mills, 1966), for example of heart rate following coronary infarcts (Broughton and Baron, 1973). Fröberg *et al.* (1972) have shown that after 75 hours of experimental sleep deprivation marked circadian swings in fatigue correlated negatively with performance and with serum epinephrine, and positively with norepinephrine levels. A very particular mechanism of stress induced insomnia is the lack of sleep following excessive exercise to anaerobic levels (J. Desjardins, personal communication) with consequent increase in plasma lactic acid, a known CNS stimulant. Environmental disturbances such as excessive noise (Otto, 1970; Scott, 1972; LeVere *et al.*, 1972) or high temperatures (Parmeggiani, 1968) are, of course, well known to disturb sleep.

A perhaps more interesting mechanism of insomnia is the increasingly common primary disruption of circadian rhythms produced by rapid air travel across time-zones, by shift work in factories, hospitals or elsewhere, and, more exotically, by space travel or locations such as caves or other enclosures which lack external Zeitgebers and so lead to "free running" biorhythms. Such situations differ in important ways. These include the presence after jet lag of light-dark and social activity cues which are helpful for adapting to the newly demanded sleep-wakefulness cycle versus their presence as conflicting synchronizers in shift work, the vestibular and other effects of weightlessness in space travel (Hanley and Adey, 1971), and so forth. But they all lead to at least transitory disturbances of sleep, the neurochemical or neuroendocrine bases of which are not yet determined.

Animal models of sleep disturbances which are directly applicable to man are exceedingly rare. But Mitler *et al.*, (1973) have shown that similar insomnia can be produced and studied in cats by manipulating light-dark cues. Weitzman (personal communication, 1973) has studied a patient who tended to "free-run" in his sleep patterns and go to bed an hour or so later each night. It was as if his body disregarded the normal synchronizers for a 24-hour sleep-wakefulness cycle. The sleep patterns made normal social living difficult and were associated with serious personality changes.

Organically induced insomnias are rare. Nevertheless, lesions in the basal forebrain area, caudal brain stem inhibitory area, or raphé nuclei may produce insomnia (cf. Broughton, 1972c). An interesting case report illustrating the importance of careful studies of such patients is that of Guillemineault et al., (1973). Their patient had persistant marked insomnia of 2-3 hours sleep per night, mild right leg spasticity and slight third cranial nerve impairment for six years after a car accident. Sleep was increased to a more normal 5½-7 hour pattern by 1500 mg d,1 5-HTP orally before retiring. The effect was on both NREM and REM sleep. The drug also increased the CSF levels of the serotonin metabolite 5-hydroxy indole acetic acid (5-HIAA) from abnormally low values of 10 ng/ml to 50 ng/ml. Interestingly, initial normal HVA levels of 30 ng/ml were also increased, to 80 ng/ml. It was believed that the physiopathogenesis involved partial destruction of serotonergic anterior raphé nuclei cells in the brain stem, as a similar insomnia with this pharmacological response has been produced experimentally in cats by such lesions (Jouvet et al., 1967; Pujol et al., 1971).

Terrifying dreams, sleep terrors, somnambulism, enuresis nocturna, nocturnal confusional attacks with or without violence, and other disorders which Gastaut and Broughton (1965) and Gastaut et al. (1965) have described as "episodic phenomena" in sleep must have important neurochemical and neuroendocrine substrates. These, however, are almost entirely unexplored. Apart from the terrifying dreams of REM sleep, the attacks all tend to arise in stage 3-4 sleep, usually in the first cycle. They therefore have powerful relationships to circadian rhythms.

Terrifying or anxious REM dreams have been shown in a number of studies to be related to intensification of the phasic activity of REM sleep such as the REM bursts themselves, myoclonus, and cardiorespiratory irregularity. They also are associated with elevated plasma fatty acid levels (Gottschalk et al., 1966). Although cortisol secretion is maximum in the latter half of the night (Weitzman et al., 1966) when most REM sleep occurs, the possible relationship between the two has not been studied, apart from the effects of withdrawal of REM suppressant drugs to be considered later.

Sleep terrors, or the classical incubus attack, occur cataclysmically in stage 3-4 sleep (Broughton and Gastaut, 1965; Gastaut et al., 1965; Broughton, 1968; Broughton, 1970; Fisher et al., 1970) and seemingly must be related to undocumented outpouring of catecholamines and cortisol into the blood stream. During sleep-walking, by comparison, the individual simply cannot wake up from the stage 3-4 sleep in which they occur (Gastaut and Broughton, 1965; Gastaut et al., 1965; Jacobson et al., 1965; Broughton, 1968): but the responsible constitutional neurochemical, or possibly psychoendocrine, defects are unknown.

Childhood enuresis also occurs preferentially in stage 4 sleep (Pierce et al., 1961; Broughton and Gastaut, 1964a and 1964b). In slow wave sleep glomerular filtration and thereby bladder filling is much greater than in REM sleep, during which concentration apparently from ADH secretion takes place (Mandell et al., 1966a). Moreover, the normal slowing of heart rate in slow-wave sleep does not occur in enuretics (Broughton, 1968); and their general nocturnal cardiac (Broughton, 1968) and bladder (Broughton and Gastaut, 1964b) functions are altered. Enuresis, therefore, apparently involves both neuroendocrine and biorhythm mechanisms.

The nocturnal confusional arousals with disorientation, poor reactivity, and retrograde amnesia also occur mainly in stage 3-4 awakenings (Broughton and Gastaut, 1965; Kales and Jacobson, 1967; Broughton, 1968). The memory aspects of sleep and arousal have been discussed elsewhere (Broughton and Gastaut, 1973). As in sleep-walking mentioned above, one wonders about the neurochemical substrate of durable confusion. Are such subjects relatively lacking in brain-stem catecholamines or in cortical acetylcholine? Conversely, do they perhaps have excessive amounts of inhibitory substances in cerebral systems necessary for full vigilance? As cerebral monoamines and ACh are supposedly involved in arousal and, as discussed in this symposium and elsewhere (1971) by Reis, also in aggression, this gives a speculative possible common neurochemical basis for those rare, but impressive, slow wave confusional arousals associated with violence. It is of note, however, that confusional awakenings in the elderly are more common in REM sleep arousals (Feinberg, 1968).

Psychiatric disorders recently have been increasingly considered in terms of their neurochemical (Schildkraut and Kety, 1957; Kety, 1959a and 1959b), neuroendocrine (Curtis, 1972), sleep (St-Laurent, 1971; Lowy, 1970; Snyder, 1972) and general biorhythm (Richter, 1965; Halberg, 1968) aspects. This impetus is gradually placing mental health and disease on a sounder biological basis. Snyder's careful and recent (1972) review of the sleep changes in psychiatric disease is exemplary and comes to the tentative conclusion that there are no unique patterns of disturbed sleep specific for a given mental disorder. Nevertheless, sleep disturbances were found to be important, essentially ubiquitous, and sometimes even generative of some of the clinical symptomatology. His review includes little neurochemical, neuroendocrine or biorhythm data.

Anxiety, interestingly, has not been studied extensively in relationship to sleep. Generally, the findings have included more frequent awakenings, reduced stages 3 and 4, and increased autonomic levels (Lester, 1967). In the case of laboratory adaptation or so-called "first night" effect (Rechtschaffen and Verdone, 1964; Agnew _et al_., 1966; Mendels and Hawkins, 1967), REM sleep is also reduced.

Anxiety and intra-psychic conflict are part of many psychiatric disorders and are undoubtedly the basis of the shallow, fragmented sleep so frequently seen. In simple depression, for example, numerous studies have documented such patterns (cf. Snyder, 1972). Moldofsky _et al_. (1973) have reported in normals that physiological suppression of stage 3 and 4 sleep by sound stimulation without awakening reduces the waking pain threshold and produces musculoskeletal aches and pains. This is probably the mechanism of these symptoms, which characterize so many psychiatric and general medical problems associated with poor sleep. Neuroendocrine correlates of stress during sleep and wakefulness are contained in a recent study by Sachar _et al_. (1973) which compared 24-hour cortisol secretion in psychotic depressives and normals. They showed that the depressives secreted almost twice as much cortisol per day, had 11 or 12 major secretory episodes compared to the 7 to 9 of the normals, secreted during evening and early sleep periods, when it was essentially absent in normals, and secreted actively for about 8½ hours per day versus 6 1/3 for normals. All inter-group differences except total secretory time renormalized quickly with treatment.

A combined pharmacological-neurochemical approach of Kupfer and Bowers (1972) has given quite interesting insights into sleep and depression. They found an inverse pretreatment relationship between CSF-HVA levels and REM latency, and a positive relationship between CSF-5HIAA levels and amount of stage 3-4 sleep both before and after treatment with the MAO inhibitor phenelzine. Moreover, suppression of REM sleep by phenelzine led to reduced CSF-HVA levels, but not 5-HIAA levels; and REM rebound during phenelzine withdrawal, present in 4 of 5 subjects, was associated with increased CSF-HVA levels towards normal. These findings give some support to the role of serotonin and catecholamines in human slow wave sleep and REM sleep, respectfully.

The biorhythmic aspect of mental illness is perhaps best exemplified by manic-depressive disease, as the patient often shows an infradian periodicity of the behavioral poles, usually of about 4-10 weeks. Hartmann (1968) found normal or elevated REM percentages in the depressive phase, and severe insomnia during the manic phase. Snyder (1968) has also documented marked insomnia preceding psychotic manic episodes. And Kupfer _et al_. (1972) have confirmed classical reports of hypersomnia during the depressive phase. Neuroendocrine and neurochemical studies are lacking.

R. Gjessing's early studies of periodic catatonia indicated that the beginning and middle of the psychotic phases were regularly associated with severe sleep disturbances (Gjessing, 1932). The detailed longitudinal investigation by Takahashi and L. Gjessing (1972) of three cases of periodic catatonia, including two with episodes of extreme excitement, showed high urinary levels of epinephrine and norepinephrine immediately after the onset of the active phase in association with severe reduction of REM sleep, but little change in stages 3 and 4. Norepinephrine and dopamine urinary levels remained high throughout the psychotic phase and waned with clinical improvement, at which time REM percent and other sleep indices renormalized. REM sleep was recovered slower than in REM deprivation studies in normals, and actually tended to peak about one week before the next active phase.

In schizophrenia, sleep research has had the added impetus of examining the hypothesis of J. Hughlings Jackson and William James that psychotic hallucinations can be considered as intrusions of dreaming into wakefulness. Numerous studies reviewed by Snyder (1972), some including efforts to precipitate psychosis by REM deprivation, have not substantiated this concept, although it is applicable to certain alcohol and drug withdrawal states discussed below. There nevertheless remain the interesting observations that waking hallucinatory episodes may be associated with rapid eye movements and recur at about 90 min intervals (West et al., 1962) and that total sleep deprivation may precipitate psychotic episodes in schizophrenics (Koranyi and Lehmann, 1960). It has also been shown that stages 3 and 4 are generally reduced in schizophrenia (Caldwell and Domino, 1967; Feinberg, 1967; Caldwell, 1969; and others) and that, during acute schizophrenic episodes and in actively symptomatic chronic schizophrenia, REM deprivation does not lead to the normal rebound phenomenon (Azumi et al., 1967; Zarcone et al., 1968; Gillin et al., 1973). Although these features are not specific to these conditions, they may be interpreted as suggestive support for the hypothesis of a defect of serotonin metabolism in schizophrenia (Wyatt, 1972).

An interesting preliminary report concerning acute schizophrenia of Watson et al. (1973) described changes in the periorbital phasic integrated potentials (PIPs) of Rechtschaffen et al. (1971). These are one of the phasic components usually present mainly in REM sleep and may reflect PGO spikes in man. They were markedly increased in NREM sleep being present in 60-85% of scored 30 sec epochs, at which time REM sleep was reduced. As the acute initial phase waned, PIPs reduced to 10-15% of NREM epochs, still above normal, and REM sleep percentage increased. These changes recall the experimental effects of reserpine (Delorme et al., 1965) and to a lesser extent pCPA (Dement et al., 1969), both of which cause "spill over" of PGO spikes from REM into NREM sleep and even into wakefulness, and may be considered a breakdown of the normal boundaries between the three basic biological states.

Further clinical evidence of such breakdown includes: reports in acute schizophrenics of marked fluctuations in REM sleep of skin potentials (Wyatt *et al.*, 1970c), which normally are maximum in stages 3 and 4 and inhibited in REM sleep (Asahina, 1962; Broughton *et al.*, 1965; Johnson and Lubin, 1966); increased amplitude in autistic children of evoked potentials in REM sleep (Ornitz *et al.*, 1968), when they are normally reduced; and the frequent appearance in schizophrenics of an "intermediate stage" of sleep having characteristics of both REM sleep and stage 2 (Koresko *et al.*, 1963; Lairy *et al.*, 1965; de Barros-Ferreira *et al.*, 1973).

Drug dependency and drug induced psychosis are increasingly recognized as closely involving altered sleep mechanisms. Many centrally active drugs such as ethanol, barbiturates, tricyclic and MAO inhibiting antidepressants, and narcotics reduce REM sleep (cf. reviews of Kales *et al.*, 1969; Oswald, 1969b; Hartmann, 1969) and, when withdrawn, may lead to REM rebound with percentages of 30-55% of total sleep or more. This rebound is often associated with disrupted sleep and frequent awakenings, intensification of the already mentioned phasic components of REM sleep, and anxious or frankly terrifying dream content (Oswald and Priest, 1965; Kales *et al.*, 1968; Evans *et al.*, 1968; Lewis and Oswald, 1969). The subject may be unable to tolerate these withdrawal symptoms and so go back to taking alcohol, barbiturates, morphine, or whichever drug is involved.

The REM rebound phenomenon may be encountered over a very short time scale, and can even occur in a single night with short acting barbiturates (Kales *et al.*, 1970). Or it may persist over several weeks (Oswald and Priest, 1965; Oswald, 1969a; Haider and Oswald, 1970; Dunleavy *et al.*, 1972). Oswald (1969b) has pointed out that CNS active drugs which do not produce dependence generally also do not show this REM rebound phenomenon, examples being diphenylhydantoin, chlorpromazine and fenfluramine.

Ogunremi *et al.* (1973), in a sophisticated pharmacological-neuroendocrine study combined with tests of mood and performance, have shown that 5 weeks of clinical doses of sodium amytal or benzoctamine (a synthetic hypnotic) in normals was associated with fragmented sleep in the second half of the night, normal nocturnal GH secretion, and reduced plasma cortisol levels. Following withdrawal, the period of REM rebound was associated with increased nocturnal GH secretion and plasma cortisol levels which peaked at the time of maximum daytime anxiety and impairment of mental concentration. And, as mentioned above, withdrawal from the MAO inhibitor phenelzine has been shown by Kupfer and Bowers (1972) to lead to increased CSF levels of HVA, suggesting heightened catecholamine turnover related to REM rebound.

These recent pioneering studies of the neurochemical and neuroendocrine aspects of drug withdrawal REM rebound are particularly important, because such rebound can be sufficiently intense to dissolve the barriers between REM sleep and wakefulness to produce the agitation and hallucinoses of alcohol withdrawal (Gross *et al.*, 1966; Greenberg and Pearlman, 1967; Gross and Goodenough, 1968; Johnson *et al.*, 1970; Maxion and Schneider, 1971) and of phenelzine withdrawal (Fisher, 1972). This intrusion of "waking dreams" also no doubt holds true for other less well documented withdrawal states and confirms the mechanism postulated by Jackson and James for at least these psychoses.

Amphetamines also reduce REM sleep initially (Oswald and Thacore, 1963; Rechtschaffen and Maron, 1964). But repeated intake leads to tolerance with progressive increase of REM sleep to the normal 20-25% of total sleep (Oswald and Thacore, 1965), as also can be seen with alcohol. During withdrawal, REM rebound may occur and, in particular, early REM periods appear shortly after falling asleep (Oswald and Thacore, 1963). These withdrawal changes may last up to two months, and have suggested to Oswald (1969a) that they are correlated with human brain protein synthesis. There must be further differences from withdrawal of either barbiturates or antidepressants, however, as amphetamine withdrawal subjects are somnolent and depressed rather than over-aroused. The depression is associated with decreased urinary excretion of 3-methoxy-4-hydroxy phenylglycol, a metabolite of norepinephrine, and with increased REM densities (Watson *et al.*, 1972). Amphetamines in rats produce marked disturbances of circadian rhythms (Davis *et al.*, 1972), an aspect seemingly not yet studied in man.

Reserpine and psychotomimetics such as LSD have very different effects. In man, reserpine in doses ranging from 0.01 to 0.14 mg/kg has been shown to increase REM sleep (Tissot, 1965; Hartmann, 1966; Hoffman and Domino, 1969). LSD also increases REM sleep (Toyoda, 1964; Green, 1965; Muzio *et al.*, 1966), lengthens the initial REM period (Muzio *et al.*, 1966) and is said to shorten the interval between REM periods (Torda, 1968). These studies are of interest because of the drugs' known psychotropic effects and because both drugs are related to serotonin metabolism, reserpine apparently increasing serotonin turnover rates (Brodie *et al.*, 1966), and LSD blocking serotonin receptors (Gaddum, 1953). Moreover, reserpine given experimentally to the cat has been shown to suppress REM sleep and markedly increase PGO spikes in NREM sleep and wakefulness (Delorme *et al.*, 1965), and LSD also increases PGO spikes outside of REM sleep (Stern *et al.*, 1972). Unfortunately, the neurochemical and neuroendocrine aspects of these drugs have not been studied in man, nor to my knowledge have withdrawal effects been documented.

Finally, there are a number of extremely important further relationships relevant to this paper, but space requirements preclude their consideration. These include cardiac and vascular accidents, organic memory defects, mental retardation, epilepsy, abnormal movement syndromes, endocrine diseases, immune system disorders, and so forth. They have in part been reviewed by Kales and Kales (1970), Lowy (1970), Broughton (1971 and 1972c) and elsewhere. Interesting and pragmatic recent findings in relationship to epilepsy alone include decreases in waking amygdalar spike discharges recorded by sphenoidal electrodes recurring approximately every 90-100 min in relationship to the BRAC (Stevens et al., 1972) and the association of the seizures of waking epilepsy with troughing of increasingly lower anticonvulsant blood levels during sleep (Rowan et al., 1973).

The relationship of these various medical conditions and neurochemical and endocrine changes to sleep is relevant to considerations of the functions of sleep. Most would accept the conclusions of Oswald (1969a) that NREM sleep has to do with anabolic somatic functions, as human GH is mainly secreted at that time (see above), strenuous exercise in athletes can increase stages 3 and 4 sleep (Baekland and Lasky, 1966) and also sleep induced GH secretion (Ogunremi et al., 1973), hyperthyroidism increases slow sleep (Dunleavy and Oswald, 1972), and anorexia with weight loss from chronic fenfluramine treatment also increases these stages (Lewis et al., 1971), as does starvation (Lewis et al., 1973). Similarly, Oswald's conclusion that REM sleep is involved in cerebral tissue repair is also largely substantiated.

Equally interesting is the possibility that amongst their numerous functions, which should be considered as complementary (Broughton, 1972d), release of the stress accumulated during the daytime may be taking place, perhaps selectively in the two sleep states and in relation to phasic events. Thus, in NREM sleep, increased and phasic spike discharges are observed in the amygdala and related allocortex in the cat (Jacobs and McGinty, 1971), in the phylogenetically related cortex of amphibia (Flanigan et al., 1973), and probably also in man. There are well known connections between the so-called basolateral amygdala and the ventromedial hypothalamus, stimulation of the former producing evoked activity in the latter (Dreifuss et al., 1968) and, in rats, a release of growth hormone which can be blocked by preceding doses of pCPA or aMPT (Martin, 1973). GH, like prolactin and cortisol, is considered a stress hormone (Mason, 1968; Brown et al., 1971). It seems reasonable to consider that this allocortical amygdalar spiking in NREM sleep could be related to release of stress involving limbic system functions and drive the hypothalamus to be expressed as GH secretion, marked electrodermal activity, and various other NREM sleep phenomena. Similarly, the phasic PGO spiking of REM sleep involving brain stem, dorsal thalamic and neocortical functions might similarly be related in a general way to stress release via these neural systems and be expressed as increased secretion of cortisol, LH, and prolactin (all are stress hormones), plus testosterone.

Support for the hypothesis of stress release in sleep can be found in the findings of Brewer and Hartmann (1973) that sleep need is least when "everything is going well". Moreover, Chase (personal communication) has found that cats who were rewarded for producing a sensorimotor rhythm (enhanced by muscle relaxation) by feedback to self-stimulating electrodes in cerebral "pleasure areas" would continue self-stimulation at high rates for long periods exceeding several days without sleep and without any apparent behavioral or neurological deficit.

Another interesting theoretical approach relates ultradian functions to survival of the individual and the species. Thus recurrent oral activity, gastric contractions and, in sleep, gastric secretion are manifestations of the 90 min BRAC: and they ensure that the individual repeatedly obtains food. In REM sleep, recurrent penile erections (Fisher et al., 1965; Karacan et al., 1966; Jovanovic, 1967) in males, and vulvar temperature increases (Bokert et al., 1966) in females, may produce recurrent cyclic activation of sex organs and help ensure species survival. It has in fact been shown that male impotence (Jovanovic and Nippert, 1971) and normal aging (Kahn and Fisher, 1969) are associated with reduced or absent REM sleep erections: and impotence is a frequent side-effect of REM suppressing antidepressant drugs.

In conclusion, considerable progress has been made in our knowledge of the neurochemical and neuroendocrine aspects of sleep and the basic rest-activity cycle in man. It has deepened our understanding of the neurohumoral coding of human mental functions and has extended it temporally around the twenty-four hour clock. Moreover, such research has clarified the biology of a number of serious medical conditions including narcolepsy, the hypersomnias, disorders of arousal from slow-wave sleep, various psychiatric disorders, drug dependency and the withdrawal psychoses. Finally, important studies of therapeutic or prognostic importance have been made. The physician therefore can confidently expect this rapidly developing area to help increasingly in the care of his patients.

REFERENCES

Agnew, H.W., Webb, W.B., and Williams, R.L., 1966, The first night effect: an EEG study of sleep, Psychophysiol. 2:263.

Ajuriaguerra, J. de (ed.), 1968, "Cycles Biologiques et Psychiatrie", Masson, Paris.

Albert, J.B., and Ballas, N.C., 1973, Electroencephalographic and temporal correlates of snoring, Bull. Psychonom. Soc. 1:169.

Anders, T.F., Sachar, E.J., Kream, J., Roffwarg, H.P., and Hellman, L., 1970, Behavioral state and plasma cortisol response in the human newborn, Pediatrics 46:532.

Asahina, K., 1962, Paradoxical phase and reverse paradoxical phase in human sleep, J. Physiol. Soc. (Japan) 24:443.

Aschoff, J., 1967, Human circadian rhythms in activity, body temperature and other functions, in "Life Sciences and Space Research," (A.H. Brown and F.G. Favorite, eds.), vol. 5, pp. 159-173, North Holland, Amsterdam.

Azumi, K., Takahashi, S., Takahashi, K., Maruyama, N., and Kikuti, S., 1967, The effects of dream deprivation on chronic schizophrenics and normal adults: a comparative study, Folia Psychiat. Neurol. Jap. 21:205.

Baekeland, F., and Lasky, R., 1966, Exercise and sleep patterns, Percept. Mot. Skills 23:1203.

Barotini, F., and Zappoli, R., 1968, A case of Kleine-Levin syndrome. Clinical and polygraphic study, in "The Abnormalities of Sleep in Man," (H. Gastaut, E. Lugaresi, G. Berti Ceroni, and G. Coccagna, eds.), pp. 239-245, Aulo Gaggi, Bologna.

de Barros-Ferreira, M., Goldsteinas, L., and Lairy, G.C., 1973, REM sleep deprivation in chronic schizophrenics: effects on the dynamics of fast sleep, Electroenceph. Clin. Neurophysiol. 34:561.

Berti Ceroni, G., Coccagna, G., and Lugaresi, E., 1968, Twenty-four hour polygraphic recordings in narcoleptics, in "The Abnormalities of Sleep in Man," (H. Gastaut, E. Lugaresi, G. Berti Ceroni, and G. Coccagna, eds.), pp. 235-238, Aulo Gaggi, Bologna.

Bokert, E., Ellman, S., Fiss, H., and Klein, G.S., 1966, Temperature changes in the female genital area during sleep. Paper presented at the Association for the Psychophysiological Study of Sleep, Gainesville, Florida, March, 1966.

Boyar, R., Finkelstein, J., Roffwarg, H., Kapen, S., Weitzman, E., and Hellman, L., 1972, Synchronization of augmented luteinizing hormone secretion with sleep during puberty, New Engl. J. Med. 287:582.

Brewer, V., and Hartmann, E., 1973, Variable sleepers: when is more or less sleep required? Presented at the Association for the Psychophysiological Study of Sleep, San Diego.

Brodie, B.B., Comer, M.S., Costa, E., and Dlabac, A., 1966, The role of brain serotonin in the mechanism of the central action of reserpine, J. Pharm. Exp. Therap. 152:340.

Broughton, R., 1968, Sleep disorders: disorders of arousal? Science 159:1070.

Broughton, R., 1970, The incubus attack, Intern. Psychiat. Clin. 7:188.

Broughton, R., 1971, Neurology and sleep research, Can. Psychiat. Assoc. J. 16:283.

Broughton, R., 1972a, Phylogenetic evolution of sleep systems, in "The Sleeping Brain," (M. Chase, ed.), pp. 2-7, Brain Information Service/Brain Research Institute, UCLA, Los Angeles.

Broughton, R., 1972b, A proposed classification of sleep disorders, in "Sleep Research," (M. Chase, W. Stern, and P. Walter, eds.), vol. 1, pp. 146, Brain Information Service/Brain Research Institute, UCLA, Los Angeles.

Broughton, R., 1972c, Sleep and neurological states, in "The Sleeping Brain," (M. Chase, ed.), pp. 363-376, Brain Information Service/Brain Research Institute, UCLA, Los Angeles.

Broughton, R., 1972d, Active sleep and the phylogenetic evolution of brain, Psychophysiol. 9:95.

Broughton, R., and Baron, R., 1973, Sleep of acute coronary patients in an intensive care unit. Presented at the Association for the Psychophysiological Study of Sleep.

Broughton, R., and Gastaut, H., 1964a, Polygraphic sleep studies of enuresis nocturna, Electroenceph. Clin. Neurophysiol. 16:625.

Broughton, R., and Gastaut, H., 1964b, Further polygraphic sleep studies of enuresis nocturna (intra-vesicular pressure), Electroenceph. Clin. Neurophysiol. 16:626.

Broughton, R., and Gastaut, H., 1973, Memory and sleep, a clinical review, *in* "Proceedings of the First Annual Meeting of the European Society for Sleep Research," S. Karger, Basel, (in press).

Broughton, R., and Guzman, A., 1973, Hypersomnia with rapid eye movements, *in* "Proceedings of the First Annual Meeting of the European Society for Sleep Research, S. Karger, Basel, (in press).

Broughton, R.J., Poire, R., and Tassinari, C.A., 1965, The electrodermogram (Tarchanoff effect) during sleep, *Electroenceph. Clin. Neurophysiol.* 18:691.

Brown, G.M., Schalch, D.S., and Reichlin, S., 1971, Patterns of growth hormone and cortisol responses to psychological stress in the squirrel monkey, *Endocrinology* 88:956.

Caldwell, D.F., 1969, Differential levels of stage IV sleep in a group of clinically similar chronic schizophrenics, *Biol. Psychiat.* 1:131.

Caldwell, D.F., and Domino, E.F., 1967, Electroencephalographic and eye movement patterns during sleep in chronic schizophrenic patients, *Electroenceph. Clin. Neurophysiol.* 22:414.

Carlson, H.E., Gillin, J.C., Gorden, P., and Snyder, F., 1972, Absence of sleep-related growth hormone peaks in aged normal subjects and in acromegaly, *J. Clin. Endocrin. Metab.* 34:1102.

Celesia, G.G., and Jasper, H.H., 1966, Acetylcholine released from cerebral cortex in relation to state of activation, *Neurology* 16:1053.

Chase, M.H. (ed.), 1972, "The Sleeping Brain", Perspectives in the Brain Sciences, vol. 1, Brain Information Service/Brain Research Institute, UCLA, Los Angeles.

Coccagna, G., Lugaresi, E., Tassinari, C.A., and Ambrosetto, C., 1966, La sindrome della gambe senza riposo (restless legs), *Omnia Med. Ther.* 44:619.

Colquhoun, W.P. (ed.), 1966, "Aspects of Human Efficiency," The English Universities Press Limited, London.

Cordeau, J.P., 1970, Monoamines and the physiology of sleep and waking, *in* "L-Dopa and Parkinsonism," (A. Barbeau, and M.H. McDowell, eds.), pp. 369-383, Davis and Co., Philadelphia.

Curtis, G.C., 1972, Psychosomatics and chronobiology: possible implications of neuroendocrine rhythms, *Psychosom. Med.* 34:235.

Dahlstrom, A., and Fuxe, K., 1964, Evidence for the existence of monoamine containing neurons in the central nervous system. I. Demonstration of monoamines in the cell bodies of brain stem neurons, Acta Physiol. Scand. Suppl 232, 62:1.

Davis, J.A., Ancill, R.J., and Redfern, P.H., 1972, Hallucinogenic drugs and circadian rhythms, Prog. Brain Res. 36:79.

Delorme, F., Froment, J.L., and Jouvet, M., 1965, Effets remarquables de la réserpine sur l'activité EEG phasique ponto-géniculo-occipitale, C.R. Soc. Biol. (Paris) 159:900.

Dement, W., Rechtschaffen, A., and Gulevich, G., 1966, The nature of the narcoleptic sleep attack, Neurology 16:18.

Dement, W., Zarcone, V., Ferguson, J., Cohen, H., Pivik, T., and Barchas, J., 1969, Some parallel findings in schizophrenic patients and serotonin-depleted cats, in "Schizophrenia: Current Concepts and Research," (D.V. Siva Sankar, ed.), pp. 775-881, PJD Publ. Ltd, Hicksville, New York.

Destrooper, J., and Broughton, R., 1969, REM awakening latencies and a possible REM breakthrough phenomenon, Psychophysiol. 6:231.

Dixit, B.M., and Buckley, J.P., 1967, Circadian changes in brain 5HT and plasma corticosterone in the rat, Life Sci. 6:755.

Dreifuss, J.J., Murphy, J.T., and Gloor, P., 1968, Contrasting effects of two identified amygdaloid efferent pathways on single hypothalamic neurons, J. Neurophysiol. 31:237.

Dunleavy, D.L.F., and Oswald, I., 1972, Sleep and thyrotoxicosis in "Sleep Research," (M. Chase, W. Stern, and P. Walter, eds.), vol. 1, p. 182, Brain Information Service/Brain Research Institute, UCLA, Los Angeles.

Dunleavy, D.L.F., Brezinova, V., Oswald, I., Maclean, A.W., and Tinker, M., 1972, Changes during weeks in effects of tricyclic drugs on the human sleeping brain, Brit. J. Psychiat. 120:663.

Evans, J.I., and Oswald, I., 1966, Some experiments in the chemistry of narcoleptic sleep, Brit. J. Psychiat. 112:401.

Evans, J., Lewis, S.A., Gibb, I.A.M., and Cheetham, M., 1968, Sleep and barbiturates: some experiments and observations, Brit. Med. J. 4:291.

Evans, J.I., Maclean, A.W., Ismail, A.A.A., and Love, D., 1971, Concentrations of plasma testosterone in normal men during sleep, Nature 229:261.

Feinberg, I., 1967, Sleep electroencephalographic and eye-movement patterns in patients with schizophrenia and with chronic brain syndrome, Res. Publ. A.R.N.M.D. 45:211.

Feinberg, I., 1968, Sleep in organic brain conditions, in "Sleep; Physiology and Pathology," (A. Kales, ed.), pp. 131-147, Lippincott, Philadelphia and Toronto.

Fisher, C., Gross, J., and Zuch, J., 1965, Cycle of penile erection synchronous with dreaming (REM) sleep, Arch. Gen. Psychiat. 12:29.

Fisher, C., Byrne, J.V., Edwards, A., and Kahn, E., 1970, REM and NREM nightmares, Intern. Psychiat. Clin. 7:183.

Fisher, C., Kahn, E., Edwards, A., and Davis, D., 1972, Total suppression of REM sleep with the MAO inhibitor Nardil in a subject with painful nocturnal REMP erection, in "Sleep Research," (M. Chase, W. Stern, and P. Walter, eds.), vol. 1, p. 159, Brain Information Service/Brain Research Institute, UCLA, Los Angeles.

Flanigan Jr., W.F., Wilcox, R.H., and Rechtschaffen, A., 1973, The EEG and behavioral continuum of the crocodilian, Caiman sclerops, Electroenceph. Clin. Neurophysiol. 34:521.

Friedman, S., and Fischer, C., 1967, On the presence of a rhythmic diurnal, oral instinctive drive cycle in man, J. Amer. Psychoanal. Assoc. 15:317.

Friedman, A.H., and Walker, C.A., 1968, Circadian rhythms in rat mid-brain and correlate nucleus biogenic amine levels, J. Physiol. (London) 197:77.

Fröberg, J., Karlsson, C.G., Leir, L., and Lidberg, L., 1972, Circadian variations in performance, psychological ratings, catecholamine excretion and diuresis during prolonged sleep deprivation, Intern. J. Psychobiol. 2:23.

Fujiya, Y., Toyoda, J., Sasaki, K., and Narita, S., 1963, A study on the polygraphic findings in all-night sleep recordings of patients with insomnia, Psychiat. Neurol. Jap. 65:292. (in Jap.)

Gaddum, J.H., 1953, Tryptamine receptors, J. Physiol. (London) 119:363.

Gastaut, H., and Broughton, R., 1965, A clinical and polygraphic study of episodic phenomena during sleep. Rec. Adv. Biol. Psychiat. 7:197.

Gastaut, H., Batini, C., Broughton, R., Fressy, J., and Tassinari, C.A., 1965, Etude électroencéphalographique des phénomènes épisodiques non-épileptiques au cours du sommeil, *in* "Le Sommeil de Nuit Normal et Pathologique," (H. Fischgold, ed.), pp. 215-236, Masson, Paris.

Gastaut, H., Tassinari, C.A., and Duron, B., 1966, Polygraphic study of the episodic diurnal and nocturnal (hypnic and respiratory) manifestations of the Pickwick syndrome, *Brain Research* 1:167.

Gillin, J.C., Buchsbaum, M., Jacobs, L.S., Fram, D., Williams, R., Mellon, E., Snyder, F., and Wyatt, R.J., 1973, Partial REM deprivation, schizophrenia and field articulation, Presented to the Association for the Psychophysiological Study of Sleep.

Gjessing, R., 1932, Beiträge zur Kenntnis der Pathophysiologie des katatonen Stupors: I Mitteilung. Über periodisch Rezwierenden Katatonen Stupor mit kritischem Beginn und Abschluss. *Arch. Psychiat. Nervenkr.* 96:319.

Globus, G., 1972, Discussion, *in* "The Sleeping Brain," (M. Chase, ed.), pp. 211-213, Brain Information Service/Brain Research Institute, UCLA, Los Angeles.

Gottschalk, L.A., Stone, W.N., Gleser, G.C., and Iacono, J.M., 1966, Anxiety levels in dreams: relation to changes in free fatty acids, *Science* 153:654.

Green, W.J., 1965, The effect of LSD on the sleep-dream cycle. An exploratory study, *J. Nerv. Ment. Dis.* 140:417.

Greenberg, R., 1967, Dream interruption insomnia, *J. Nerv. Ment. Dis.* 144:18.

Greenberg, R., and Pearlman, C., 1967, Delirium tremens and dreaming, *Amer. J. Psychiat.* 124:133.

Gross, M.M., and Goodenough, D.R., 1968, Sleep disturbances in the acute alcoholic psychoses, *Psychiat. Res. Reports* 24:132.

Gross, M.M., Goodenough, D., Tobin, M., Halpert, E., Lepore, D., Perlstein, A., Sirota, M., DiBianco, J., Fuller, R., and Kishner, I., 1966, Sleep disturbances and hallucinations in the acute alcoholic psychoses, *J. Nerv. Ment. Dis.* 142:493.

Guillemineault, C., Cathala, J.P., and Castaigne, P., 1973, Effects of 5-hydroxytryptophan on sleep of a patient with a brain-stem lesion, *Electroenceph. Clin. Neurophysiol.* 34:177.

Haider, I., and Oswald, I., 1970, Late recovery processes after drug overdose, *Brit. Med. J.* 2:318.

Halberg, F., 1968, Physiological correlations underlying rhythmometry, with special reference to emotional illness, *in* Symposium Bell-Air III, pp. 73-126, Masson, Paris.

Halberg, F., 1969, Chronobiology, *Ann. Rev. Physiol.* 31:675.

Hanley, J., and Adey, W.R., 1971, Sleep and wake states in the biosatellite III monkey. Visual and computer analysis of telemetered electroencephalographic data from earth orbital flight, *Aerosp. Med.* 42:304.

Harner, R.N., and Dorman, R.M., 1970, A role for dopamine in control of sleep and wakefulness, *Trans. Amer. Neurol. Assoc.* 95: 252.

Hartmann, E.L., 1966, Reserpine: its effect on the sleep-dream cycle in man, *Psychopharmacologia* 9:242.

Hartmann, E., 1967, "The Biology of Dreaming", Chas. C. Thomas, Springfield.

Hartmann, E., 1968, Longitudinal studies of sleep and dream patterns in manic-depressive patients, *Arch. Gen. Psychiat.* 19:312.

Hartmann, E., 1969, The biochemistry and pharmacology of the D-state (dreaming sleep), *Exp. Med. Surg.* 27:105.

Hartmann, E., Chung, R., and Chien, C.P., 1971, L-tryptophan and sleep, *Psychopharmac.* 19:114.

Hernández-Péon, R., 1965, Central neuro-humoral transmission in sleep and wakefulness, *Prog. Brain Res.* 18:96.

Hernández-Péon, R., 1967, Neurophysiology, phylogeny, and functional significance of dreaming, *Exper. Neurol.* Suppl. 4:106.

Hishikawa, Y., and Kaneko, Z., 1965, Electroencephalographic study on narcolepsy, *Electroenceph. Clin. Neurophysiol.* 18:249.

Hoffman, J.S., and Domino, E.F., 1969, Comparative effects of reserpine on the sleep cycle of man and cat, *J. Pharmacol. Exptl. Therap.* 170:190.

Honda, Y., Takahashi, K., Takahashi, S., Azumi, K., Irie, M., Sakuma, M., Tsushima, T., and Shizuma, K., 1969, Growth hormone secretion during nocturnal sleep in normal subjects, *J. Clin. Endocrinol. Metab.* 29:20.

Illig, R., Stahl, M., Henrichs, I., and Hecker, A., 1971, Growth hormone release during slow-wave sleep, *Helv. Paed. Acta* 26:665.

Jacobs, B.L., and McGinty, D.J., 1971, Amygdala unit activity during sleep and waking, *Exper. Neurol.* 33:1.

Jacobson, A., and Kales, A., 1967, Somnambulism: all-night EEG and related studies, *Res. Publ. A.R.N.M.D.* 45:424.

Jacobson, A., Kales, A., Lehmann, D., and Zweizig, J.R., 1965, Somnambulism: all-night electroencephalographic studies, *Science* 148:975.

Jasper, H.H., and Koyama, I., 1969, Rate of release of amino acids from the cerebral cortex in the cat as affected by brain stem and thalamic stimulation, *Can. J. Physiol. Pharmacol.* 47:889.

Jasper, H.H., and Tessier, J., 1971, Acetylcholine liberation from cerebral cortex during paradoxical (REM) sleep, *Science* 172:601.

Jasper, H.H., Khan, R.T., and Elliott, K.A.C., 1965, Amino acids released from the cerebral cortex in relation to its state of activation, *Science* 147:1448.

Johnson, L.C., and Lubin, A., 1966, Spontaneous electrodermal activity during waking and sleeping, *Psychophysiol.* 3:8.

Johnson, L.C., Burdick, J.A., and Smith, J., 1970, Sleep during alcohol intake and withdrawal in the chronic alcoholic, *Arch. Gen. Psychiat.* 22:406.

Jouvet, M., 1962, Recherches sur les structures nerveuses et les méchanismes responsables des différentes phases du sommeil physiologique, *Arch. Ital. Biol.* 100:125.

Jouvet, M., 1969, Biogenic amines and the states of sleep, *Science* 163:32.

Jouvet, M., Bobillier, P., Pujol, J., and Renault, J., 1967, Suppression du sommeil et diminution de la serotonine cérébrale par lésion du système du raphé chez le chat, *C.R. Acad. Sc.* (Paris) 264:360.

Jovanovic, U.J., 1967, Erecktionen im Schlaf, *Arch. Psychiat. Nervenkr.* 210:220.

Jovanovic, U.J., and Nippert, M., 1971, Erecktionen im Schlaf und Sexualität, *J. Neurovisc. Relat.* Suppl. 10:580.

Jung, R., and Kuhlo, W., 1965, Neurophysiological studies of abnormal night sleep and the Pickwickian syndrome, *Brain Research* 18:140.

Kahn, E., and Fisher, C., 1969, REM sleep and sexuality in the aged, J. Geriat. Psychiat. 2:181.

Kales, A., and Kales, J., 1970, Evaluation, diagnosis and treatment of clinical conditions related to sleep, J.A.M.A. 213:2229.

Kales, A., Malmstrom, E.J., Rickles, W.H., Hanley, J., Ling Tan, T., Stadel, B., and Hoedemaker, F.S., 1968, Sleep patterns of a pentobarbital addict; before and after withdrawal, Psychophysiol. 5:208.

Kales, A., Malmstrom, E.J., Scharf, M.B., and Rubin, R.T., 1969, Psychophysiological and biochemical changes following use and withdrawal of hypnotics, in "Sleep: Physiology and Pathology," (A. Kales, ed.), pp. 331-343, Lippincott, Philadelphia and Toronto.

Kales, A., Preston, T.A., Tan, T.-L., and Allen, C., 1970, Hypnotics and altered sleep-dream patterns. I. All-night EEG studies of glutethimide, methylprylon and pentobarbital, Arch. Gen. Psychiat. 23:219.

Kapen, S., Boyar, R., Hellman, L., and Weitzman, E., 1973, Episodic release of luteinizing hormone at mid-menstrual cycle in normal adult women, J. Clin. Endocrin. Met. 36:724.

Karacan, I., 1971, Painful nocturnal penile erections, J.A.M.A. 215: 1831.

Karacan, I., Goodenough, D.R., Shapiro, A., and Starker, S., 1966, Erection cycle during sleep in relation to dream anxiety, Arch. Gen. Psychiat. 15:183.

Karacan, I., Rosenbloom, A.L., Williams, R.L., Finley, W.W., and Hursch, C.J., 1971, Slow wave sleep deprivation in relation to plasma growth hormone concentration, Behav. Neuropsychiat. 2:11.

Kety, S.S., 1959a, Biochemical theories of schizophrenia. Part I. Science 129:1528.

Kety, S.S., 1959b, Biochemical theories of schizophrenia. Part II. Science 129:1590.

Kleitman, N., 1963, "Sleep and Wakefulness", The University of Chicago Press, Chicago and London.

Kleitman, N., 1969, Basic rest-activity cycle in relation to sleep and wakefulness, in "Sleep: Physiology and Pathology," (A. Kales, ed.), pp. 33-38, J.P. Lippincott, Philadelphia and Toronto.

Koella, W.P., 1967, "Sleep, its Nature and Physiological Organization", Chas. C. Thomas, Springfield.

Koranyi, E.K., and Lehmann, H.E., 1960, Experimental sleep deprivation in schizophrenic patients, Arch. Gen. Psychiat. 2:76.

Koresko, R.L., Snyder, F., and Feinberg, I., 1963, "Dream time" in hallucinating and nonhallucinating schizophrenic patients, Nature (London) 199:1118.

Kripke, D., 1973, Ultradian rhythms in sleep and wakefulness, in "Advances in Sleep Research," (E.D. Weitzman, ed.), Spectrum Publications, New York (in press).

Kripke, D., and Sonnenshein, D., 1973, A 90-minute daydream cycle. Presented to the Association for the Psychophysiological Study of Sleep, San Diego.

Kupfer, D.J., and Bowers, M.B., 1972, REM sleep and central monoamine oxidase inhibition, Psychopharmacologia (Berl.) 27:183.

Kupfer, D.J., Wyatt, R.J., Scott, J., and Snyder, F., 1970, Sleep disturbance in acute schizophrenic patients, Amer. J. Psychiat. 126:1213.

Kupfer, D.J., Himmelhoch, J.M., Swartzburg, M., Anderson, C., Byck, R., and Detre, T.P., 1972, Hypersomnia in manic-depressive disease, Diseases of the Nervous System 33:720.

Lairy, G.C., Cor-Mordret, M., Faure, R., and Ridjanovic, S., 1965, Sommeil de nuit des malades mentaux: étude des bouffées délirantes, in "Le Sommeil de Nuit Normal et Pathologique," (H. Fischgold, ed.), pp. 353-381, Masson, Paris.

Lester, B.K., Burch, N.R., and Dossett, R.C., 1967, Nocturnal EEG-GSR profiles: the influence of presleep states, Psychophysiol. 3:238.

Lewis, S., and Oswald, I., 1969, Overdose of tricyclic anti-depressants and deductions concerning their cerebral action, Brit. J. Psychiat. 115:1403.

Lewis, S.A., Oswald, I., and Dunleavy, D.L.F., 1971, Chronic fenfluramine administration: some cerebral effects, Brit. Med. J. 3:67.

Lewis, S.A., MacFadyen, U.M., and Oswald, I., 1973, Starvation and human slow wave sleep, Presented to the Psychophysiological Study of Sleep, San Diego.

LeVere, T.E., Bartus, R.T., and Hart, F.D., 1972, Electroencephalographic and behavioral effects of nocturnally occurring jet aircraft noise, Aerosp. Med. 43:384.

Loeb, C., Massazza, G., Sacco, G., and Arone, A., 1964, Etude polygraphique des myoclonies hypniques chez l'homme, Rev. Neurol. 110:258.

Lowy, F.L., 1970, Recent sleep and dream research: clinical applications, C.M.A.J. 102:1069.

Luce, G., 1970, "Biological Rhythms in Psychiatry and Medicine", U.S. Department Health, Education and Welfare, Chevy Chase, Maryland.

Lugaresi, E., Tassinari, C.A., Coccagna, G., and Ambrosetto, C., 1965, Particularités cliniques et polygraphiques du syndrome d'impatience des membres inférieurs, Rev. Neurol. 113:545.

Lugaresi, E., Coccagna, G., Gambi, D., Berti Ceroni, G., and Poppi, M., 1966, A propos de quelques manifestations nocturnes myocloniques, (Nocturnal myoclonus of Symonds), Rev. Neurol. 115:547.

Mace, J.W., Gotlin, R.W., and Beck, P., 1972, Sleep related human growth hormone (GH) release: a test of physiologic growth hormone secretion in children, J. Clin. Endocrinol. Metab. 34:339.

Mandell, A.J., Chaffey, B., Brill, P., Mandell, M.P., Rodnick, J., Rubin, R.T., and Sheff, R., 1966a, Dreaming sleep in man: changes in urine volume and osmolarity, Science 151:1558.

Mandell, A.J., Brill, P.L., Mandell, L.P., Rodnick, J., Rubin, R.T., Sheff, R., and Chaffey, B., 1966b, Urinary excretion of 3-methoxy-4-hydromandelic acid during dreaming sleep in man, Life Science 5:169.

Martin, J.B., 1973, Functions of the hypothalamus and amygdala in regulation of growth hormone (GH) secretion. Presented at joint meeting, American Neurological Association and Canadian Congress of Neurological Sciences, Montreal.

Mason, J.W., 1968, "Over-all" hormonal balance as a key to endocrine organization, Psychosom. Med. 30:791.

Maxion, H., and Schneider, E., 1971, Alcoholdelir und Traumschlaf, Arch. Psychiat. Nervenkr. 214:116.

Mendels, J., and Hawkins, D.R., 1967, Sleep laboratory adaptation in normal subjects and depressed patients ("first night effect"), Electroenceph. Clin. Neurophysiol. 22:556.

Mills, J.N., 1966, Human circadian rhythms, Physiol. Rev. 46:128.

Mitler, M.M., Sokolove, P.G., Pittendrigh, C.S., and Dement, W.C., 1973, Activity-inactivity and wakefulness-sleep rhythms in the hamster under three lighting conditions, Presented to the Association for the Psychophysiological Study of Sleep, San Diego.

Moldofsky, H., Scarisbrick, P.N., England, R.S., and Smythe, H., 1973, Musculo-skeletal symptoms and EEG sleep disturbance. Presented to the Association for the Psychophysiological Study of Sleep, San Diego.

Morgane, P.J., and Stern, W.C., 1972, Relationship of sleep to neuroanatomical circuits, biochemistry and behavior, Ann. N.Y. Acad. Sci. 193:95.

Muzio, J.W., Roffwarg, H.P., and Kaufman, E., 1966, Alterations in the nocturnal sleep cycle resulting from LSD, Electroenceph. Clin. Neurophysiol. 21:313.

Ogunremi, O.O., Adamson, L., Brezinova, V., Hunter, W.M., Maclean, A.W., Oswald, I., and Percy-Robb, I.W., 1973, Two anti-anxiety drugs: a psychoneuroendocrine study, Brit. Med. J. 301:202.

Ornitz, E.M., Ritvo, E.R., Panman, L.M., Lee, Y.H., Carr, E.M., and Walter, R.D., 1968, The auditory evoked response in normal and autistic children during sleep, Electroenceph. Clin. Neurophysiol. 25:221.

Orr, W.C., 1973, Sleep loss, sustained performance, and the basic rest-activity cycle (BRAC), Presented to the Association for the Psychophysiological Study of Sleep, San Diego.

Oswald, I., 1962, "Sleeping and Waking", Elsevier, Amsterdam.

Oswald, I., 1964, Rocking at night, Electroenceph. Clin. Neurophysiol. 16:577.

Oswald, I., 1969a, Human brain protein, drugs and dreams, Nature (London) 223:893.

Oswald, I., 1969b, Sleep and dependence on amphetamine and other drugs, in "Sleep; Physiology and Pathology," (A. Kales, ed.), pp. 317-330, Lippincott, Philadelphia and Toronto.

Oswald, I., 1972, Amine mechanisms in sleep, "The Sleeping Brain," (M. Chase, ed.), pp. 170-172, Brain Information Service/Brain Research Institute, UCLA, Los Angeles.

Oswald, I., and Priest, R.G., 1965, Five weeks to escape the sleeping pill habit, Brit. Med. J. 2:1093.

Oswald, I., and Thacore, V.R., 1963, Amphetamine and phenmetrazine addiction: physiological abnormalities in the abstinence syndrome, Brit. Med. J. 2:427.

Oswald, I., Merrington, J., and Lewis, H., 1970, Cyclical "on demand" oral intake by adults, Nature 225:959.

Otto, E., 1970, Einfluss von Schalbreizen auf EEG-Aktivität, Herz-Periodendauer und Atemmechanische Messiverte in Schlaf, Deutsch. Gesundl. 25:1661.

Parker, D.C., Sassin, J.F., Mace, J.W., Gotlin, R.W., and Rossman, L.G., 1969, Human growth hormone release during sleep: electroencephalographic correlation, J. Clin. Endocrinol. Metab. 29:871.

Parmeggiani, R.P., 1968, Modification of sleep phases as a result of fast adaptation to environmental temperature, in "The Abnormalities of Sleep in Man," (H. Gastaut, E. Lugaresi, G. Berti Ceroni, and G. Coccagna, eds.), pp. 67-70, Aulo Gaggi, Bologna.

Passouant, P., Schwab, R.S., Cadhillac, J., and Baldy-Moulinier, M., 1964, Narcolepsie-cataplexie. Etude de nuit et du sommeil de jour. Traitement par une amphétamine lévogyre, Rev. Neurol. 111:415.

Passouant, P., Halberg, F., Genicot, R., Popoviciu, L., and Baldy-Moulinier, M., 1969, La périodicité des accès narcoleptiques et le rhythme ultradian du sommeil rapide, Rev. Neurol. 121:155.

Pickart, L., and Thaler, M.M., 1973, Tripeptide in human serum which prolongs survival of normal liver cells and stimulates growth in neoplastic liver, Nature New Biology 243:85.

Pierce, C.M., Whitman, R.M., Mass, J.W., and Gay, M.L., 1961, Enuresis and dreaming, Arch. Gen. Psychiat. 4:166.

Popoviciu, L., and Corfariu, O., 1972, Etude clinique et polygraphique au cours du nycthémère d'un cas de syndrome de Kleine-Levin-Critchley, Rev. Roum. Neurol. 9:221.

Pujol, J.F., Buguet, A., Froment, J.L., Jones, B., and Jouvet, M., 1971, The central metabolism of serotonin in the cat during insomnia. A neurophysiological and biochemical study after administration of p-chlorophenylalanine or destruction of the raphé system, Brain Research 29:195.

Reding, G.R., Zepelin, H., Robinson, J.E., Zimmerman, S.O., and Smith, V.H., 1968, Nocturnal teeth-grinding: all night psychophysiologic studies, J. Dent. Res. 47:786.

Rechtschaffen, A., and Maron, L., 1964, The effect of amphetamine on the sleep cycle, Electroenceph. Clin. Neurophysiol. 14:438.

Rechtschaffen, A., and Roth, B., 1969, Nocturnal sleep of hypersomnias, Activ. Nerv. Sup. (Praha) 11:229.

Rechtschaffen, A., and Verdone, P., 1964, Amount of dreaming: effect of incentive adaptation to laboratory, and individual differences. Percept. Mot. Skills 19:947.

Rechtschaffen, A., Goodenough, D., and Shapiro, A., 1962, Patterns of sleep talking, Arch. Gen. Psychiat. 7:418.

Rechtschaffen, A., Wolpert, E.A., Dement, W.C., Mitchell, S.A., and Fisher, C., 1963, Nocturnal sleep of narcoleptics, Electroenceph. Clin. Neurophysiol. 1963, 22:465.

Rechtschaffen, A., Molinari, S., Watson, R., and Wincor, M.Z., 1971, Extra-ocular potentials: a possible indicator of PGO activity in man, Psychophysiol. 7:336.

Reis, D.J., 1971, Brain monoamines in aggression and sleep, Clin. Neurosurg. 18:471.

Reis, D.J., and Wurtman, R.J., 1968, Diurnal changes in brain noradrenalin, Life Sciences 7:91.

Reis, D.J., Corvelli, A., and Conners, J., 1969, Circadian and ultradian rhythms of serotonin regionally in cat brain, J. Pharmacol. Exp. Therap. 167:328.

Richter, C.R., 1965, "Biological Clocks in Medicine and Psychiatry", Chas. C. Thomas, Springfield.

Roth, B., Bruhova, S., and Lehovsky, M., 1968, On the problem of pathophysiological mechanisms of narcolepsy, hypersomnia and dissociated sleep disturbances, in "The Abnormalities of Sleep in Man," (H. Gastaut, E. Lugaresi, G. Berti Ceroni, and G. Coccagna, eds.), pp. 191-204, Aulo Gaggi, Bologna.

Roth, B., Bruhova, S., and Lehovsky, M., 1969, REM sleep and NREM sleep in narcolepsy and hypersomnia, Electroenceph. Clin. Neurophysiol. 26:176.

Rothballer, A.B., 1956, Studies on the adrenalin-sensitive component of the reticular activating system, Electroenceph. Clin. Neurophysiol. 8:603.

Rowan, A.J., Pippinger, C.E., McGregor, P.A., and French, J.H., 1973, Studies of anticonvulsant blood levels and their temporal relationship to seizure activity, Presented to the Eastern Society of Electroencephalographers, Mont Gabriel, Quebec.

Rubin, R.T., Kales, A., Adler, R., Fagan, T., and Odell, W., 1972, Gonadotrophin secretion during sleep in normal adult men, Science 175:196.

Sachar, E.J., Hellman, L., Roffwarg, H.P., Halpern, F.S., Fukushima, D.K., and Gallagher, T.F., 1973, Disrupted 24-hour patterns of cortisol secretion in psychotic depression, Arch. Gen. Psychiat. 28:19.

Sassin, J.F., Parker, D.C., Johnson, L.C., Rossman, L.G., Mace, J.W., and Gotlin, R.W., 1969, Effects of slow wave sleep deprivation on human growth hormone release in sleep: preliminary study, Life Sciences 8:1299.

Sassin, J.F., Parker, D.H., Mace, J.W., Gotlin, R.W., Johnson, L.C., and Rossman, L.G., 1969, Human growth hormone release: relation to slow wave sleep and sleep-waking cycles, Science 165:513.

Sassin, J.F., Frantz, A.G., Weitzman, E.D., and Kapen, S., 1972, Human prolactin: 24-hour pattern with increased release during sleep, Science 177:1205.

Sassin, J.F., Frantz, A.G., Kapen, S., and Weitzman, E.D., 1973, The nocturnal release of prolactin is dependent upon sleep, J. Clin. Endocrinol. Metab. (in press).

Schildkraut, J.J., and Kety, S.S., 1967, Biogenic amines and emotions, Science 156:21.

Schwartz, B.A., and Escande, J.P., 1968, Respiration hypnique pickwickienne, in "The Abnormalities of Sleep in Man," (H. Gastaut, E. Lugaresi, G. Berti Ceroni, and G. Coccagna, eds.), pp. 204-214, Aulo Gaggi, Bologna.

Schwartz, B.A., Gilbaud, G., and Fischgold, H., 1963, Etudes electroencephalographiques sur le sommeil de nuit. I. "L'insomnie chronique", Presse Med. 71:1474.

Scott, T.D., 1972, The effects of continuous, high density, white noise on the human sleep cycle, Psychophysiol. 9:227.

Snyder, F., 1968, Sleep disturbance in relation to acute psychosis, in "Sleep; Physiology and Pathology," (A. Kales, ed.), pp. 170-182, Lippincott, Philadelphia and Toronto.

Snyder, F., 1972, Electroencephalographic studies of sleep in psychiatric disorders, in "The Sleeping Brain," (M. Chase, ed.), pp. 376-

393, Brain Information Service/Brain Research Institute, UCLA, Los Angeles.

St-Laurent, J., 1971, Contributions to psychiatry of recent studies on sleep, Can. Psychiat. Assoc. J. 16:327.

Stern, W., Morgane, P.J., and Bronzino, J.D., 1972, LSD: effects on sleep patterns and spiking activity in the lateral geniculate nucleus, Brain Research 41:199.

Stevens, J.R., Kodama, H., Lonsbury, B., and Mills, L., 1972, Ultradian characteristics of spontaneous seizure discharges recorded by radio telemetry in man, Electroenceph. Clin. Neurophysiol. 31:313.

Takahashi, S., and Gjessing, L.R., 1972, Studies of periodic catatonia III; longitudinal sleep study with urinary secretion of catecholamines, J. Psychiat. Res. 9:123.

Takahashi, Y., Kipnis, D.M., and Daughaday, W.H., 1968, Growth hormone secretion during sleep, J. Clin. Invest. 47:2079.

Takahashi, K., Takahashi, S., Azumi, K., Honda, Y., and Utena, H., 1971, Changes of plasma growth hormone during nocturnal sleep in normals and in hypersomnic patients Adv. Neurol. Sci. (Tokyo) 14:743. (in Jap.)

Tanguay, P.E., Ornitz, E.M., Forsythe, A.B., Lee, J.C.M., and Hartman, D., 1973, Basic rest-activity cycle rhythms in the human auditory evoked response, Electroenceph. Clin. Neurophysiol. 34:593.

Tissot, R., 1965, The effects of certain drugs on the sleep cycle in man, Prog. Brain Res. 18:175.

Torda, C., 1968, Contribution to the serotonin theory of dreaming (LSD infusion), N.Y. State J. Med. 68:1135.

Toyoda, J., 1964, The effects of chlorpromazine and imipramine on the human nocturnal sleep electroencephalogram, Folia Psychiat. Neurol. Japan. 18:199.

Vigneri, R., and D'Agata, R., 1971, Growth hormone release during the first year of life in relation to sleep-waking periods, J. Clin. Endocrinol. Metab. 33:561.

Wada, T., 1922, An experimental study of hunger and its relation to activity, Arch. Psychol. 8:1.

Watson, R., Hartmann, E., and Schildkraut, J.J., 1972, Amphetamine withdrawal: affective state, sleep patterns, and MHPG excretion, Amer. J. Psychiat. 129:39.

Watson, R., Liebmann, K., and Watson, S., 1973, Periorbital phasic integrated potentials in acute schizophrenia. Presented to the Association for the Psychophysiological Study of Sleep, San Diego.

Weitzman, E.D., Schaumburg, H., and Fishbein, W., 1966, Plasma 17-hydroxycorticosteroid levels during sleep in man, J. Clin. Endocrinol. Metab. 26:121.

Weitzman, E.D., Goldmaker, D., Kripke, D., MacGregor, P., Kream, J., and Hellman, L., 1968, Reversal of sleep-waking cycles: effect on sleep stage patterns and certain neuroendocrine rhythms, Trans. Amer. Neurol. Assoc. 93:153.

West, L.J., Janszen, H.H., Lester, B.K., and Cornelisoon Jr., F.S., 1962, The psychosis of sleep deprivation, Ann. N.Y. Acad. Sci. 96:66.

Wyatt, R.J., 1972, Serotonergic and adrenergic systems in human sleep, in "The Sleeping Brain," (M. Chase, ed.), pp. 168-170, Brain Information Service/Brain Research Institute, UCLA, Los Angeles.

Wyatt, R.J., Engelman, K., Kupfer, D.J., Scott, J., Sjoerdsma, A., and Snyder, F., 1969, Effects of para-chlorophenylalanine on sleep in man, Electroenceph. Clin. Neurophysiol. 27:529.

Wyatt, R.J., Chase, T.N., Engelman, K., Kupfer, D.L., Scott, J., Sjoerdsma, A., and Snyder, F., 1970a, Reversal of parachlorophenylalanine (pCPA) REM suppression in man by 5-hydroxytryptophan (5-HTP), Psychophysiol. 7:318.

Wyatt, R.J., Engelman, K., Kupfer, D.J., Fram, D., Sjoerdsma, A., and Snyder, F., 1970b, Effects of 1-tryptophan (a natural sedative) on human sleep, Lancet 2:842.

Wyatt, R.J., Stern, M., Fram, D.H., Tursky, B., and Grinspoon, L., 1970c, Abnormalities in skin potential fluctuations during sleep in acute schizophrenics, Psychosom. Med. 32:301.

Wyatt, R.J., Zarcone, V., Engelman, K., Dement, W.C., Snyder, F., and Sjoerdsma, A., 1971, Effects of 5-hydroxytryptophan on the sleep of normal human subjects, Electroenceph. Clin. Neurophysiol. 30:505.

Zarcone, V., 1972, Discussion in "The Sleeping Brain," (M. Chase, ed.), pp. 394-395, Brain Information Service/Brain Research Institute, UCLA, Los Angeles.

Zarcone, V., Gulevich, G., Pivik, T., and Dement, W., 1968, Partial REM phase deprivation and schizophrenia, Arch. Gen. Psychiat. 18:194.

ABSTRACT:

In birds and mammals two types of sleep, quiet sleep and active sleep, often referred to in man as non-rapid eye movement (NREM) and rapid eye movement (REM) sleep, have been defined. There is increasing evidence that they represent the hypnic equivalent of the quiescent and activity poles of Kleitman's basic rest-activity cycle (BRAC), which in adult man has a periodicity of some 90 - 100 min.

These two sleep states have separate neurophysiological and neurochemical substrates. It is generally believed that the maintenance of wakefulness reflects ascending reticulo-cortical activation, largely catecholaminergic; that quiet sleep reflects active inhibition of such arousal by basal forebrain, mid-pontine and other areas requiring an intact raphé nuclei system, this state probably involving mainly serotonergic mechanisms; and that active sleep reflects mainly an activation of pontine structures, probably via a combination of cholinergic, catecholaminergic and perhaps GABA transmitter mechanisms. It has recently been found that endocrine status is to a great extent temporally related to the circadian sleep-waking tides and also to the 90 min. ultradian BRAC. Thus growth hormone, cortisol, prolactin, luteinizing hormone during parathyroid, and possibly, testosterone all exhibit circadian rhythmicity which is sleep related. Many clinical disorders involving sleep can be increasingly correlated with these neuro-chemical and neuroendocrine mechanisms.

Narcolepsy, a disorder characterized by involuntary sleep attacks, when associated with attacks of weakness from emotional stimuli (cataplexy), sleep paralysis or vivid hypnagogic hallucinations (i.e. the narcolepsy "tetrad"), has been shown to involve mainly inappropriate REM sleep mechanisms, has daytime sleep attacks phase related to nocturnal REM sleep, and responds to drugs which inhibit the catecholaminergic mechanisms of REM sleep, that is to tricyclic "antidepressants" and to monamine oxidase inhibitors. Sleep attacks of NREM mechanism, however, respond best to CNS stimulants, i.e. amphetamines or methylphenidate. Similarly, pathologically prolonged or deep sleep (hypersomnia) may relate mainly to either NREM or, very rarely, REM sleep mechanisms and responds best to the respective suppressors of each of these two states.

Insomnia may reflect primarily stress, a disturbance of biological rhythms or even organic brain disease. Thus, brain damage (eg., post-infectious, post-traumatic) in the basal forebrain or mid-pontine (and lower) areas initiating sleep, or in the raphé system necessary for quiet sleep, lead to pathological loss of sleep and may be reflected in altered detectable metabolites of the brain amine systems involved. A disturbance of the normal temporal organization of circadian and ultradian biorhythms can be induced by forcing the body into new time sequence demands, as in jet-lag from east-west travel across time zones, in shift-work or in space travel, and lead to severe insomnia. And severe stress, by altering circulating catecholamine and endocrine levels and by disrupting normal biorhythmicity, may also lead to sleep disturbances and be reflected in brain amine and neuroendocrine variables.

Also of relevance are sleep apnea, enuresis nocturna, sleep walking, sleep terrors, terrifying dreams, and a number of other conditions including, in particular, drug dependency.

THE CHEMICAL CODING VIA THE CHOLINERGIC SYSTEM: ITS ORGANIZATION AND BEHAVIORAL IMPLICATIONS [1)]

A.G. Karczmar

Department of Pharmacology and Institute for Mind, Drugs and Behavior, Loyola University Medical Center, Maywood, Illinois 60153

The central cholinergic synapses which constitute the subject of this paper are those at which the transmission between the presynaptic and the postsynaptic neuron - or, more precisely, between the presynaptic nerve terminal and the postsynaptic membrane - is mediated by acetylcholine (ACh); a complication may arise from the fact that at certain, non-cholinergic synapses ACh may play a modulatory, facilitatory rather than transmissive role. In view of the wide occurrence (cf. below) of the cholinergic synapses or modulations, the cholinergic system must participate significantly in brain functions and behavioral processes. In fact, several participants of this Symposium discussed the cholinergic participation in the appetitive and thermal control (R.D. Myers), in conditioning and learning (H. Brust-Carmona), in aggression (D.J. Reis), and in certain phases of sleep (P. Morgane); related aspects of the cholinergic system were described by others (Domino, 1968; Domino et al., 1968; Bovet-Nitti, 1965; Aprison, 1965; Stein, 1968; Karczmar et al., 1972; Karczmar, 1971).

My present goal, however, will be to conceptualize as to the general, organizational characteristics of the central cholinergic system rather than to stress its specific functions; indeed, I like to speculate that this general organizational aspect of the central cholinergic system which is concerned with its participation in the inhibitory circuitry underlies the specific phenomena

1) The published and unpublished results from our laboratories described in this paper were supported in part by the USA-NIH Research Grants NS06672 and NS06455, USA-NSF Research Grant GB8718, and USA-NIH Training Grant GM77.

referred to above. In presenting this argument I will stress the complexities of the cholinergic synapses and of the cholinergic modulations, describe their involvement with synapses activated by other than ACh transmitters illustrating this involvement in terms of neurochemical, neuropharmacological and behavioral data, and speculate how these involvements may concern inhibitory circuitry, brain excitability and brain rhythms.

CENTRAL DISTRIBUTION OF THE COMPONENTS OF THE CHOLINERGIC SYSTEM AND ITS RELATIONSHIP TO THE EXISTENCE OF CENTRAL CHOLINERGIC SYNAPSES

The function of cholinergic synapses depends on the presence of ACh, of its synthesizing system which includes cholineacetylase (ChAc), and of its catabolic enzymes, particularly acetylcholinesterase (AChE). Actually, any mapping of the brain for these three substances shows a superimposition, although this does not mean that at all brain sites there is a constant relationship between AChE, ChAc and ACh. On the whole, however, concentrations of all three components of the cholinergic system go hand in hand. The differential distributions of these components is generally emphasized; indeed, while the central afferent pathways, such as the classical lemniscal, reticulo-cerebellar, diffuse (reticulo-thalamico-cortical) and direct projection (reticulo-cortical) pathways, as well as the rhinencephalic circuitry and the relay stations such as the caudate, the thalamus and the hypothalamus contain high levels of ACh, AChE and ChAc, certain portions of the motor (pyramidal and extrapyramidal system) and of the sensory pathways are poor in these substances (Karczmar, 1967; Koelle, 1963). What however should be stressed is that very few systems contain uniformly little AChE, ACh or ChAc: while within a system certain neurons may be very poor in these substances, other neurons may exhibit high concentrations of the latter (Koelle, 1963; Lewis and Shute, 1966), although the pertinent studies were frequently confined to the measurement of AChE alone. For reasons that will become apparent, I prefer to stress the _wide-spread_ rather than the _differential_ distribution of the components of the cholinergic system.

The presence of even all the three components of the cholinergic system is not a criterion of that of the cholinergic synapses (Karczmar, 1969) or even of cholinoceptivity: certain neurons rich in ACh, AChE and ChAc may not respond to iontophoretic application of ACh (Karczmar and Nishi, unpublished). Furthermore, the cholinoceptive neurons are not always cholinergic, that is, a cholinoceptive neuron may not be always activated in the process of natural transmission by ACh released by the presynaptic cell.

The proof of cholinergicity of a central synapse is not easy, as it depends primarily on the demonstration at a particular site that the pharmacology of the cholinoceptivity on the one hand and of the transmission on the other is similar; indeed, for technical reasons the demonstration of the release of ACh from a neuron upon its presynaptic stimulation is well-nigh impossible in the case of the central nervous system. Thus, besides the spinal synapse between the motor collateral and the Renshaw cell (Eccles et al., 1954; Eccles, 1969), relatively few sites of proven cholinergicity can be adduced today. In this symposium, J.W. Phillis presented such an evidence for the cerebral cortex, and in the past he described similar evidence for several thalamic nuclei (Phillis, 1971). Controversies on this point rage with regard to cerebellar Purkinjě cells, hippocampal pyramidal neurons, certain synapses of the reticular formation, hypothalamus, etc. Yet, in view of the wide spread of the components of the cholinergic system and of the cholinoceptive neurons in the brain, as well as because so many activities of the brain are sensitive to the cholinergic agonists and antagonists, bona fide cholinergic synapses should be widely spread and well-nigh ubiquitous in the central nervous system.

CENTRAL MUSCARINIC AND NICOTINIC RESPONSES

The Renshaw cell exhibits a primarily nicotinic response and delayed, weak muscarinic response[2]. Generally however, the central cholinoceptive and/or cholinergic responses are muscarinic in nature; rather few sites - including certain loci in the thalamus (Tebecis, 1970a and b; Phillis, 1971) and in the reticular formation (cf. for review Pradhan and Dutta, 1971) - respond nicotinically. The problem is that, on the basis of the more readily studied peripheral cholinergic synapses, frequently endowed with both muscarinic and nicotinic sites (Koketsu, 1969; Nishi, 1970), the former are only facilitatory in nature, i.e. they facilitate the cholinergic potential but do not initiate it (cf. Table 4 in Karczmar et al., 1972). Furthermore, both in the case of the ganglion and in that of the few central responses that could be adequately studied the muscarinic response is slow, delayed, and does not operate in terms of the opening of the ionic gates (Krnjević, 1969; Krnjević et al., 1971; Nishi, 1970; Karczmar et al., 1970, 1971), this latter characteristic being commensurate with an effective excitatory potential.

2) Muscarinic response is evoked by ACh and such muscarinic agents as mecholyl, and is blocked by atropinics; the nicotinic response is evoked by ACh, nicotine and nicotinic substances, and blocked by d-tubocurarine and by curaremimetics.

Yet, as stated at this Symposium by J.R. Phillis, the muscarinic responses may be transmittive in the case of the CNS; furthermore, certain functional and behavioral phenomena such as the EEG patterns, thermo-control, thirst, etc., are exquisitely sensitive to muscarinic and atropinic drugs. In this presentation I will stress the participation of ACh in multisynaptic circuitry of a net inhibitory character; from this viewpoint it does not matter whether ACh activates such circuitry as a transmitter, or increases its excitability as a muscarinic facilitator. It should be emphasized in this context that ACh may act directly at certain central cholinoceptive neurons as an inhibitory transmitter; this was described at this Symposium by J.R. Phillis, and elsewhere by York (1967), Anderson and Curtis (1964) and others (cf. Phillis, 1971, and Curtis and Crawford, 1969). While the methods used do not lend themselves to the unequivocal demonstration that the ACh responses in question are indeed directly inhibitory rather than dependent on activation of an inhibitory interneuron, the phenomenon itself is consistent with the main speculation of this paper.

INVOLVEMENT OF THE ACH SYSTEM WITH OTHER TRANSMITTERS

Besides throwing a light on the meaning of the nicotinic versus muscarinic responses, the investigations of the peripheral sites offer another interesting lead for the central studies. For years now, the peripheral studies indicate that the cholinergic system is closely coupled with other transmission systems, and, furthermore, that this coupling results frequently in net inhibitions. The term "cholinergic link" was coined by Burn (1966) to explain phenomena which appeared to lead, via cholinergic activations, to the release and physiological activity of catecholamines (CA's) (cf. also Campbell, 1970, and Karczmar, 1967). It is of particular interest that in some cases the cholinergic link may result in the release of an inhibitory substance (cf. also Koelle, 1969a and b); both pre- and post-synaptic sites may be involved in such inhibitions (Nishi, 1970; Christ and Nishi, 1971). There is evidence that a CA may be involved in the inhibitory postsynaptic potential, activated by ACh, of the sympathetic ganglion as well as in the block of the release of the excitatory transmitter (Eccles, 1955; Nishi, 1970; Libet and Kobayashi, 1969), and that ACh may activate serotonin - dependent inhibitions involved in the intestinal reflexes (Kottegoda, 1970). It is of further interest that these ACh-dependent activations of these inhibitions are muscarinic in character.

For several years we and others (Varagić, 1966; Martin and Eades, 1967) studied the central coupling between the ACh and other systems; I will summarize our own findings, illustrating this coupling in terms of neurochemical, behavioral and neuropharmacological data.

Changes in central catecholamines dependent on ACh levels. In rabbits, the administration of the anticholinesterase (antiChE) diisopropyl phosphofluoridate (DFP) led to significant increases in brain dopamine (DA) and decreases in brain norepinephrine (NE; Glisson and Karczmar, 1971, 1972). These effects were related quantitatively to the inhibition of AChE and to the increase in the central concentrations of ACh. The effects were particularly marked in animals in which levels of CA's were increased following pretreatment with L-DOPA and monoamine oxidase inhibitors; the quantitative and sometimes qualitative differences between the effects of increased levels of ACh in non-treated and MAOI-DOPA pretreated animals cannot be explained at present.

Interestingly enough, the elevation of brain DA appeared to be at least in part mediated by peripheral actions of DFP: quaternary atropinics were even more effective than atropine in blocking this effect of DFP, while intraventricular injection of DFP proved less effective than its systemic administration in elevating DA levels. On the other hand, the analysis of the NE depletion by DFP characterized this effect as strictly central in nature.

To explain these phenomena, it may be speculated that cholinergic or cholinoceptive neurons may activate via interneurons DA synthesis as well as NE depletion; the former effect may also depend on the peripheral input. This speculative explanation raises the question of the central location of the cholinergic or cholinoceptive neurons in question. Indeed, a "balance" and/or agonist - antagonist relationship between ACh and CA's were recently postulated for a variety of pharmacological, behavioral and clinical reasons in the case of the striate (cf. Krnjević, 1970), hypothalamus (cf. for instance Lomax, 1969; Lomax et al., 1969), or limbic system (Stein, 1969). While these speculations were devised to explain specific phenomena arising from the ACh-CA interplay at specific sites, it is of interest that the effects described at present were wide-spread; while they were very marked in the case of, for instance, the hypothalamus and several other brain parts, they were less marked in the striate. Rather than specific, area-limited phenomenon, our data seem to describe a general, ubiquitous phenomenon.

Another point may be raised. Our speculation (cf. Karczmar, 1971) implies that the hypothetical cholinergic link is a one-way affair, and some of our other data support his notion. Pharmacological maneuvers which induced increases and decreases in brain DA and NE levels did not affect brain AChE or ACh levels (Barnes et al., 1973a and b; cf. also Sethy and Van Woert, 1973). On the other hand, changes in brain levels of serotonin may induce marked changes in those of ACh.

Behavioral data. If a coupling between the neurotransmitters is a general phenomenon, it may be expected that behavioral phenomena cause simultaneous changes in several neurotransmitter and related substances. This may be a trite statement or a truism: as the brain function is a multitransmitter phenomenon, one may expect that, say, learning, aggression, or "stress" will affect more than one transmitter. In fact, the relative lack of the pertinent data is simply due to the fact that frequently the investigators engaged in this type of research study only one type of substance. That this trend is over is illustrated at this Symposium by R.D. Myers' description of the thermocontrol and appetitive phenomena, P. Morgane's of the control of sleep (cf. also Jouvet, 1972), Brust-Carmona's of learning, and D.J. Reiss' of aggression.

Let me illustrate by means of a particularly clear-cut example the multitransmitter effect of a specific behavioral paradigm. In this case the latter could be quantitated: we dealt here with mice stressed by means of an inescapable foot shock, and the multitransmitter effect - the increase in serotonin and the decrease in ACh and NE - could be related to the "amount" of stress, i.e. to the number of shocks (Fig. 1; cf. also Karczmar et al., 1973). Furthermore, it could be shown at least in the case of serotonin that the increase of its levels was accompanied with an increase in its synthesis (turnover) which was dose-effect wise related to the number of shocks to which the mice were exposed. It is of further interest that the effect was subject to conditioning; when, after the stress effect subsided the mice were returned to the stress chamber but not shocked, they exhibited neurochemical changes similar to those shown by the mice which were actually shocked.

This example serves to emphasize an additional point; when the animals were exposed to avoidance conditioning procedures and received in its course a number of shocks similar to that which they were exposed to in the "stress" paradigm, the effect on serotonin, NE and ACh was minimal. It may be suggested that learning and conditioning constitute, contrary to "stress" and "frustration", a homeostatic and adaptive process (Scudder, 1971; Hanigan and Scudder, 1973; Karczmar et al., 1973; Karczmar, 1973), and thus must result in minimal changes of the brain levels of brain amines, although both types of processes may affect the turnover values.

Neuropharmacological data. The neuropharmacological phenomena to be described at present are particularly relevant in the context of this presentation. They not only illustrate in neuropharmacological terms the coupling between ACh and other, particularly CA, systems, but also they demonstrate the unique capacity of ACh as a desynchronizing substance; this characteristic of ACh may be related in turn to my speculation on its participation in the inhibitory circuitry.

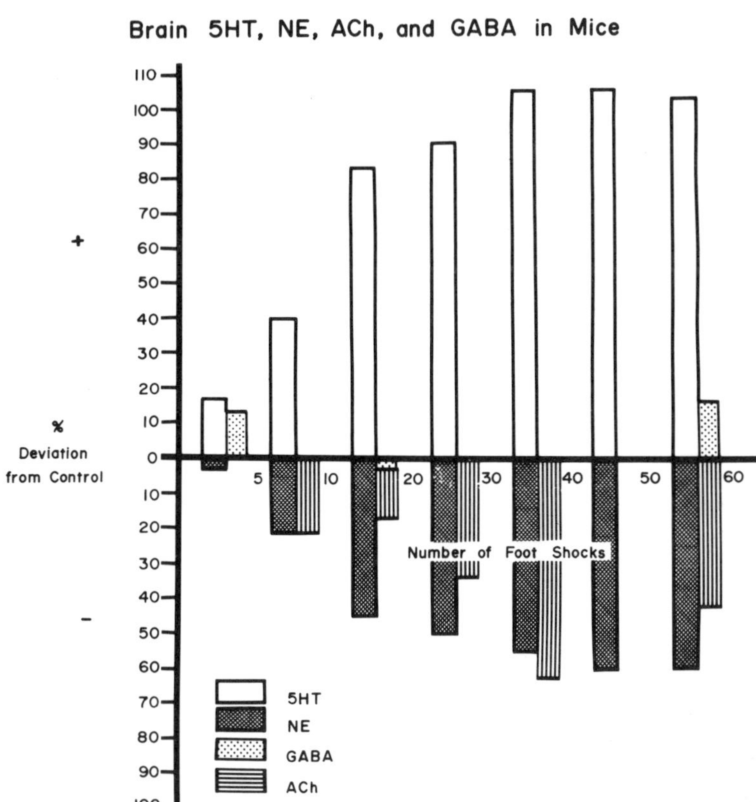

Fig. 1. The effect of stress on brain serotonin (5HT), norepinephrine (NE), acetylcholine (ACh) and γ-aminobutyric acid (GABA). Inescapable footshock was delivered to caged SCI mice at 1 shock/min. The seven groups of mice, 10 mice/group, were given from 5 to 60 shocks (abscissa). Brain values for NE, 5HT, ACh and GABA were expressed in percent of controls (ordinates); GABA values are preliminary. NE and 5HT were measured fluorospectrophotometrically, ACh - in terms of P^{32} phosphorylcholine after high voltage paper electrophoretic separation (cf. Reid et al., 1971), and GABA - photometrically in terms of NADPH produced during the conversion of GABA to succinic acid. From Cosgrove et al. (1973), and Kaplan and Scudder (1973); cf. also Richardson et al. (1972).

Let us state at the outset that central accumulation of ACh concomitant with the systemic administration of antiChE's produces EEG desynchronization and flattening (EEG "arousal"). This desynchronizing effect of antiChE's is very specific for these compounds; indeed, they are mg per mg more potent in this respect than even amphetamines, the classical EEG alerting substances. It is of additional interest that, as pointed out already in 1952 by Wikler, this EEG "arousal" was not accompanied by a corresponding behavioral arousal, and Wikler (1952) referred to this paradoxical phenomenon as that of the "divorce" between EEG and behavioral indices.

That the cholinergic stimulation of the diffused reticulo-thalamico-cortical system is involved in this phenomenon was proposed originally by Himwich (1962; Rinaldi and Himwich, 1955). Certain other mechanisms may underlie the desynchronizing action of antiChE's, namely their antirecruitment effect. Recruitment is a slow, waxing and waning cortical EEG pattern induced by slow electrical stimulation of non-specific, midline thalamic nuclei; it is related to α waves, to sleep spindles and to synchronization phenomena. As shown originally by Longo and Silvestrini (1957), antiChE's and central accummulation of ACh exert a potent and specific antirecruitment effect. While the stimulus threshold for recruitment may be elevated when the EEG is desynchronized behaviorally or by amphetamines, yet at threshold the recruitment is essentially unchanged under these circumstances; on the other hand, antiChE's readily attenuate recruitment under all conditions (Longo, 1962; Van Meter and Karczmar, 1971; Van Meter, 1970). Now, my first neuropharmacological example of ACh-CA coupling deals with the antirecruitment action of antiChE's. Indeed, we could show that this action depends on the presence of NE, as it was prevented by alpha blockers and specific depleters of NE and restored by the administration of DOPA; additional pharmacological and neurochemical data also supported this conclusion (Van Meter and Karczmar, 1967, 1971).

The point to be emphasized is that the desynchronizing action of antiChE's does not depend solely upon their antirecruitment action: even after depletion of NE or in the presence of α blockers, antiChE's while not capable anymore of blocking the recruitment are still effectively desynchronizing the EEG.

In fact, under these circumstances the EEG desynchronization by antiChE's acquires a special character, that of paradoxical sleep. The investigations of V.G. Longo and myself (Karczmar et al., 1970) have shown that following CA depleters the antiChE's produce an exact pharmacological model of paradoxical sleep, with all its EEG and EMG characteristics. Thus, ACh-CA coupling underlies the desynchronizing antirecruitment action, while in the absence of this coupling ACh may induce another type of desynchronization related to the paradoxical sleep. Parenthetically, this line of evidence may explain Wikler's (1952) "divorce" phenomenon (cf. above). The antiChE-induced EEG alerting may be behaviorally related to the paradoxical sleep rather than to wakefulness; normally, the consequences of the ACh-CA coupling prevent to an extent the expression of this relationship and shift the behavior towards wakefulness, while the uncoupling of the ACh-CA system induces the full-fledged paradoxical sleep (Karczmar, 1971; George et al., 1964)[3].

ACh AND CONVULSIONS

It is pertinent to consider the subject of convulsions in the present context. Gastaut and Fischer-Williams (1959) described EEG seizures as examples of hypersynchrony; in view of the potent and specific desynchronizing actions of antiChE's it is of interest to describe the interplay between antiChE's, ACh and seizures.

It was already pointed out that, at small doses, centrally effective antiChE's produce desynchrony. However, at higher doses of antiChE's, particularly of the long-acting, organophosphorus type, EEG seizure results. Our data indicate that, following antiChE's, desynchrony is quantitatively related to moderate central accummulation of ACh, while seizures occur at higher - about 4 to 6 times normal - levels of ACh (Karczmar, 1973). It can be speculated that the excessive levels of ACh produce blockade at sites at which normally ACh would activate desynchronization, and that ACh desynchronization is antagonistic to EEG hypersynchrony (Karczmar, 1973). In fact, as soon as even very minor regeneration of AChE has a chance to occur following the administration of DFP, the convulsions cease (Karczmar and Koppanyi, 1953; Karczmar, 1973).

Further, indirect evidence for this notion arises from the effects of electroconvulsive seizure upon ACh metabolism. The post-seizure effect is that of ACh depletion (Essman, 1972; Karczmar, 1973); furthermore, this depletion is accompanied by markedly increased ACh synthesis (Karczmar, 1973)[4]. These events may,

3) Limitations of space prevent me from discussing in this context the slow sleep effects of ACh applied locally to certain brain structures reported by Hernandez-Peón (1965; Hernandez-Peón and Chavez-Ibarra, 1963); these results could not always be duplicated by others (cf. Babb et al., 1971 and Marczynski, 1967 for further references).

4) The ACh synthesis was measured in terms of ACh accummulation after the complete inhibition of AChE. This method may be not as reliable as that employed recently by Reid et al. (1971) and Hanin et al. (1972).

speculatively, suggest that an active desynchrony-hypersynchrony antagonism arises in the course of the ECS, resulting in maximal mobilization of ACh; ACh synthesis is raised without being able to compensate for its actual need. Indirect evidence pointing in the same direction was obtained in our laboratory as the brain levels of ACh of several rodent strains and genera were found to be inversely related to ECS threshold (Scudder et al., 1966; Karczmar et al., 1973; Karczmar, 1973). It is of interest that in this case ACh levels paralleled those of CA's, well known to be inversely proportional to the effectiveness of the ECS (Toman, 1963). Finally, on different grounds, Vazquez and Krip (1973) also suggested an anticonvulsive role for the central cholinergic system.

This conceptualization constitutes a simplification necessitated by space limitations. One must account for instance for the well-known fact that ACh induces seizures or neuronal spikes when applied at many central structures (cf. for instance Hubbard and Rayport, 1968; Babb et al., 1971; Machne and Unna, 1963). This direct neuronal convulsive action of ACh may constitute an action of ACh independent of that discussed here. Furthermore, the data of Babb et al. (o.c.) may also suggest that the ACh-induced seizure arises only after a delay and after a period of desynchronization; thus, it may result, as I believe is the case, with the experiments adduced here, from the block of desynchronizing sites due to excessive concentration of ACh.

ACh-ACTIVATED INHIBITORY CIRCUITRY AND DESYNCHRONIZATIONS

It is tempting to ascribe the desynchronizing actions of ACh to its participation in inhibitions. This phenomenon may well depend on coupling between ACh and inhibitory substances; the evidence for this concept for the peripheral synapses and for certain central phenomena was described above.

Furthermore, central synaptic correlates for the notion of the participation of the cholinergic synapses in inhibitory circuitry may be adduced at present. Indeed, the classical central cholinergic synapse between the motor collateral and the Renshaw cell constitutes an excellent example of cholinergic activation, via a glycine-releasing interneuron, of postsynaptic inhibition (Eccles et al., 1954; Eccles, 1969). Several other instances of cholinergic activation of postsynaptic hyperpolarizing inhibitions were suggested elsewhere (Karczmar, 1969, 1970, 1971; Phillis, 1971; Iwata et al., 1971; Kidokoro et al., 1968).

More recent data suggest that ACh may be similarly involved in activating the other, important type of inhibition, the presynaptic depolarizing inhibition (Eccles, 1969; Eccles et al.,

1963). It was pointed out by several investigators (Kiraly and Phillis, 1961; Koketsu, Karczmar and Kitamura, 1969), that pharmacological analysis of the dorsal root potential (DRP) resulting from the presynaptic inhibition of the dorsal, afferent impulses indicates the cholinergic nature of the DRP. Further analysis of this phenomena by means of sucrose gap technique led us (Koketsu et al., 1969) to conclude that actually the cholinergic characteristics of the DRP depend on the cholinergic activation of a cholinoceptive interneuron involved in the DR-DRP and VR-DRP reflexes, rather than on direct depolarizing action of ACh upon the afferent nerve terminal. Our most recent investigations indicate that the inhibitory transmitter involved directly in the depolarization of the terminal is γ-aminobutyric acid (GABA) (Nishi et al., 1973a and b; cf. also Barker and Nicoll, 1973); our current work may demonstrate that we deal here with a polysynaptic chain which includes a cholinoceptive interneuron capable of activating a gabaminergic cell. It is more than of passing interest that the same transmitter, GABA, is involved in <u>inhibitory postsynaptic hyperpolarizations</u> (as in the case of the cortical neurons, Krnjević and Schwartz, 1967) as well as in <u>inhibitory presynaptic depolarizations</u>; we endeavored recently to explain this phenomenon on the basis of the differences between the ionic characteristics of the two types of neurons involved (Nishi et al., o.c.).

Finally, several instances of interplay between ACh and both NE and DA, two CA's exhibiting inhibitory actions centrally (cf. Curtis and Crawford, 1969) and peripherally (at the ganglion for instance; cf. Nishi, 1970), were adduced in this presentation. It was suggested that DA acts as an inhibitor of cholinergic neurons in the nigrostriatal pathway; this proposed link seems to be directionally opposite to that ascribed for ACh-CA's in this presentation. However, the concept of the CA-ACh link in the nigrostriate was criticized recently (Krnjevic, 1970), and future investigations may well demonstrate that ACh-DA links constitute additional examples of ACh-activated inhibitions, to be added to the ACh-activated pre- and post-synaptic inhibitions discussed above. Furthermore, additional types of linkages between ACh and inhibitory substances were suggested by Koelle (1969).

How may we speculatively relate the ACh-dependent inhibitory circuitry to EEG desynchronization? Central neurons may exhibit synchronizing tendencies due to their own firing characteristics (Morell, 1967), to the activities of the circuitry of the thalamic nuclei (Andersen and Andersson, 1968), and to those of other pacemakers. When inhibitory influences are made to bear, synchronously acting neuronal populations may be subdivided, and EEG-wise the subpopulations may become out of phase with respect to each other. Related mechanisms may be also proposed - the ACh-sensitive thalamus is a good example of such mechanisms, as its synchronizing activities should be readily blocked via ACh-activated inhibitions.

Some behavioral implications of this speculation were described above with regard to Wikler's "divorce" phenomenon and to the paradoxical sleep. It must be also pointed out that at this symposium R.D. Myers and D.J. Reis while describing specific effects of ACh pointed out that it is misleading to speak of cholinergic centers in the context of the control of thirst, aggression or body temperature. It may be more pertinent to consider the desynchronizing and indirectly inhibitory actions of ACh as illustrating a generalized organizational effect of the latter on the brain function; this mechanism may render the brain, previously synchronized and thus not set for a specific behavior, capable of a specific function, which would depend then on a cybernetic interplay - expressed EEG-wise as desynchronization - between neuronal subpopulations.

REFERENCES

Andersen, P. and Andersson, S.A., 1968, "Physiological Basis of Alpha Rhythm," Appleton-Century-Crafts, New York.

Anderson, P. and Curtis, D.R., 1964, The excitation of thalamic neuronesby acetylcholine, Acta Physiol. Scand. 61: 85-99.

Aprison, M.H., 1965, Research approaches to problems in mental illness: Brain neurohumor-enzyme systems, and behavior, in "Progress in Brain Research," (W.A. Himwich and J.E. Schade, eds.), pp. 48-80, 16.

Babb, T.L., Babb, M., Mahnke, J.H. and Verzeano, M., 1971, The action of cholinergic agents on the electrical activity of the non-specific nuclei of the thalamus, Int. J. Neurol. 8: 198-210.

Barker, J.L. and Nicoll, R.A., 1973, The pharmacology and ionic dependency of amino acid responses in the frog spinal cord, J. Physiol. 228: 259-277.

Barnes, L., Cann, F., Karczmar, A.G. and Longo, V.G., 1973a, 5-HTP and DOPA responses in 6-OH dopamine and 5,6-dihydrotryptamine treated mice, Fed. Proc. 32: 276.

Barnes, L., Cann, F., Karczmar, A.G., Kindel, G. and Longo, V.G., 1973b, Effects of L-DOPA on behavior and on brain amines in mice trated with 6-hydroxydopamine, Pharmacol. and Behav. 1: 35-40.

Bovet-Nitti, F., 1965, Action of nicotine on conditioned behavior in naive and pretrained rats. II. Complex forms of acquired behavior, in "Symposium on Tobacco Alkaloids and Related Compounds," (U.S.V. Euler, ed.), pp. 137-143, Pergamon Press, Oxford.

Burn, J.H., 1966, Introductory remarks. Section V. Adrenergic transmission, Pharmac. Rev. 18: 459-470.

Campbell, G., 1970, Autonomic nervous supply to effector cells, in "Smooth Muscle," (E. Bülbring, A.F. Brading, A.W. Jones and T. Tomita, eds.), pp. 451-495, The Williams and Wilkins Co., Baltimore.

Christ, D.D. and Nishi, S., 1971, Site of adrenalin blockade in the superior cervical ganglion of the rabbit, J. Physiol. 213: 107-117.

Cosgrove, K.A., Scudder, C.L. and Karczmar, A.G., 1973, Some aspects of acute quantitative shock on mouse whole brain levels of acetylcholine and choline, The Pharmacologist, 15: 255.

Curtis, D.R. and Crawford, J.M., 1969, Central synaptic transmission -microelectrophoretic studies, Ann. Rev. Pharmacol. 9: 209-250.

Domino, E.F., 1968, Cholinergic mechanisms and the EEG, EEG Clin. Neurophysiol. 24: 292-293.

Domino, E.F., Yamamoto, K. and Dren, A.T., 1968, Role of cholinergic mechanisms in states of wakefulness and sleep, Brain Res. 28: 113-133.

Eccles, R.M., 1955, Intracellular potentials recorded from a mammalian sympathetic ganglion, J. Physiol. 130: 572-584.

Eccles, J.C., 1969, Historical Introduction, in "Central Cholinergic Transmission and its Behavioral Aspects," (A.G. Karczmar, ed.), pp. 90-94, Fed. Proc. 28.

Eccles, J.C., 1969, "The Inhibitory Pathways of the Central Nervous System," Charles C. Thomas (Publ.), Springfield, Illinois.

Eccles, J.C., Fatt, P. and Koketsu, K., 1954, Cholinergic and inhibitory synapses in a pathway from motor axon collaterals to motoneurons, J. Physiol. (Lond.), 216: 524-562.

Eccles, J.C., Schmidt, R.F. and Willis, W.D., 1963, Pharmacological studies of synaptic inhibition, J. Physiol. 168: 500-530.

Essman, W.B., 1972, Neurochemical changes in ECS and ECT, Seminars in Psychiat. 4: 67-79.

Gastaut, H. and Fischer-Williams, M., 1959, The physiopathology of epileptic seizures, in "Handbook of Physiology, Section I: Neurophysiology," (J. Field, ed.), pp. 329-363, American Physiol. Soc., Washington, D.C.

George, R., Haslett, W.L. and Jenden, D.J., 1964, A cholinergic mechanism in the brain stem reticular formation: induction of paradoxical sleep, Internatl. J. Neuropharmacol. 3: 541-552.

Glisson, S.N. and Karczmar, A.G., 1971, Atropine methyl nitrate block of diisopropyl fluorophosphate effect on dopamine in rabbit brain, Fed. Proc. 30: 382.

Glisson, S.N., Karczmar, A.G. and Barnes, L., 1972, Cholinergic effects on adrenergic transmitters in rabbit brain parts, Neuropharmacol. 11: 465-477.

Hanigan, W. and Scudder, C.L., 1973, A systems view of seizure patterning, Gen. Syst. Bull. 4: 3-13.

Hanin, I., Massarelli, R. and Costa, E., 1972, An approach to the study of the biochemical pharmacology of cholinergic function, in "Studies of Neurotransmitters at the Synaptic Level," (E. Costa, L.L. Iversen and R. Paoletti, eds.), pp. 181-202, Adv. Biochem. Psychopharmacol. 6.

Hernández-Peón, R., 1965, Central neurohumoral transmission in sleep and wakefulness, in "Progress in Brain Research, Sleep Mechanisms," (K. Akert, C. Bally and J.P. Schadé, eds.), pp. 96-116, Elsevier, Amsterdam.

Hernández-Peón, R. and Chavez Ibarra, G., 1963, Sleep induced by electrical or chemical stimulation of the forebrain, Electroenceph. Clin. Neurophysiol. Suppl. 24: 188-198.

Himwich, H.E., 1962, Reticular activating system - current concepts of function, Chapter 28, in "Psychosomatic Medicine," (J.H. Nodine and J.H. Meyer, eds.),pp. 211-220, Lea and Fabiger, Philadelphia.

Hubbard, J.H. and Rayport, M., 1968, Cholinergic induction of thalamo-cortical seizures in cats, EEG Clin. Neurophysiol. 24: 189.

Iwata, N., Sakai, Y. and Deguchi, T., 1971, Effects of physostigmine on the inhibition of trigeminal motoneurons by cutaneous impulses in the cat, Exp. Brain Res. 13: 519-522.

Jouvet, M., 1972, Some monoaminergic mechanisms controlling sleep and waking, in "Brain and Human Behavior," (A.G. Karczmar and J.C. Eccles, eds.), pp. 131-160, Springer-Verlag, Berlin.

Kaplan, R.L. and Scudder, C.L., 1973, The effect of unavoidable footshock stress on γ-aminobutyric acid levels in mouse brain, The Pharmacologist, 15: 259.

Karczmar, A.G., 1967, Pharmacologic, toxicologic and therapeutic properties of anticholinesterase agents, in "Physiological Pharmacology," (W.S. Root and F.G. Hofman, eds.), pp. 163-322, Vol. 3, Academic Press, New York.

Karczmar, A.G., 1969, Is the central cholinergic system overexploited? in "Central Cholinergic Transmission and its Behavioral Aspects," (A.G. Karczmar, ed.), Fed. Proc. 28: 47-157.

Karczmar, A.G., 1970, Central cholinergic pathways and their behavioral implications, in "Principles of Psychopharmacology," (W.G. Clark and J. del Giudice, eds.), pp. 57-86, Academic Press, New York.

Karczmar, A.G., 1971, Possible mechanisms underlying the so-called "Divorce" phenomena of EEG desynchronizing actions of anticholinesterases, Presented at the Regional Midwest EEG Meetings, April 1971, Hines V.A. Hospital.

Karczmar, A.G., 1971, Neurophysiological behavioral and neurochemical correlates of the central cholinergic synapses, in "Advances in Neuropsychopharmacology," (O. Vinar, Z. Votava and P.B. Bradley, eds.), pp. 455-480, North-Holland Publ. Co., Amsterdam.

Karczmar, A.G., 1973, Brain acetylcholine and seizures, in "Psychobiology of Electroconvulsive Therapy," (M. Fink, ed.), Winston and Sons, Publ., New York (In Press).

Karczmar, A.G. and Koppanyi, T., 1953, Central effects of diisopropyl fluorophosphonate in urodele larvae, Schmiedebergs Arch. f. expt. Path. 219: 261-270.

Karczmar, A.G. and Nishi, S., 1971, The types and sites of cholinergic receptors, in "Advances in Cytopharmacology," (F. Clementi and B. Ceccarelli, eds.), 1: 301-318, Raven Press, New York.

Karczmar, A.G., Longo, V.G. and Scotti de Carolis, A., 1970, A pharmacological model of paradoxical sleep: the role of cholinergic and monoamine systems, Physiology and Behavior 5: 175-182.

Karczmar, A.G., Nishi, S. and Blaber, L.C., 1970, Investigations, particularly by means of the anticholinesterase agents, of the multiple peripheral and central cholinergic mechanisms and of their behavioral implications, Acta Vitaminologica et Enzymologica 24: 131-189.

Karczmar, A.G., Nishi, S. and Blaber, L.C., 1972, Synaptic modulations, in "Brain and Human Behavior," (A.G. Karczmar and J.C. Eccles, eds.), pp. 63-92, Springer-Verlag, Berlin.

Karczmar, A.G., Scudder, C.L. and Richardson, D., 1973, Interdisciplinary approach to the study of behavior in related mice types, in "Neurosciences Research," (I. Kopin, ed.), 5: 159-244, Academic Press, New York.

Kidokoro, Y., Kubota, K., Shuto, S. and Sumino, R., 1968, Possible interneurons responsible for reflex inhibition of motoneurons of jaw-closing muscles from inferior dental nerve, J. Neurophysiol. 31: 709-716.

Kiraly, M.K. and Phillis, J.W., 1961, Action of some drugs on the dorsal root potentials of the isolated toad spinal cord, Brit. J. Pharmacol. 17: 224-231.

Koelle, G.B., 1963, Cytological distribution and physiological functions of cholinesterases, in "Cholinesterases and Anticholinesterase Agents," (G.B. Koelle, ed.), Handbch. d. Exper. Pharmakol., Ergänzungswk., 15: 187-298, Springer-Verlag, Berlin.

Koelle, G.B., 1969, Significance of acetylcholinesterase in central synaptic transmission, in "Symposium on Central Cholinergic Transmission and its Behavioral Aspects," (A.G. Karczmar, ed.), Fed. Proc., 28: 147-157.

Koelle, G.B., 1969b, Pharmacology of synaptic transmission, in "Basic Mechanisms of the Epilepsies," (H.H. Jasper, A.A. Ward, Jr., and A. Pope, eds.), pp. 195-211.

Koketsu, K., 1969, Cholinergic synaptic potentials and the underlying ionic mechanisms, in "Central Cholinergic Transmission and its Behavioral Aspects," (A.G. Karczmar, ed.), Fed. Proc., 28: 101-112.

Koketsu, K., Karczmar, A.G. and Kitamura, R., 1969, Acetylcholine depolarization of the dorsal root nerve terminals in the amphibian spinal cord, Int. J. Neuropharmacol., 8: 329-336.

Kottegoda, S.R., 1970, Peristalsis of the small intestine, in "Smooth Muscle," (E. Bülbring, A.F. Brading, A.W. Jones and T. Tomita, eds.), pp. 525-541, The Williams and Wilkins Co., Baltimore.

Krnjević, K., 1969, Central cholinergic pathways, in "Central Cholinergic Transmission and its Behavioral Aspects," (A.G. Karczmar, ed.), Fed. Proc., 28: 113-120.

Krnjević, K., 1970, Dopamine, acetylcholine and excitatory amino acids in nigrostriatal transmission, in "L-DOPA and Parkinsonism," (A. Barbeau and F.H. McDowell, eds.), pp. 189-190, F.A. Davis Co., Philadelphia.

Krnjević, K. and Schwartz, S., 1967, The action of γ-aminobutyric acid on cortical neurones, Exptl. Brain Res. 3: 320-336.

Krnjević, K., Pumain, R. and Renaud, L., 1971, The mechanism of excitation by acetylcholine in the cerebral cortex, J. Physiol. 215: 247-268.

Lewis, P.R. and Shute, C.C.D., 1966, The distribution of cholinesterase in cholinergic neurons demonstrated with electrone microscope, J. Cell Science, 1: 381-390.

Libet, B. and Kobayashi, H., 1969, Generation of adrenergic and cholinergic potentials in sympathetic ganglion cells, Science, 164: 1530-1532.

Lomax, P., 1969, Drugs and body temperature, Int. Rev. Neurobiol. 12: 1-43.

Lomax, P., Foster, R.S. and Kirkpatrick, W.E., 1969, Cholinergic and adrenergic interaction in the flermoregulatory centers of the rat, Brain Res. 15: 431-438.

Longo, V.G., 1962, Electroencephalographic atlas for pharmacological research, Elsevier Publ. Co., Amsterdam.

Longo, V.G. and Silvestrini, G., 1957, Action of eserine and amphetamine on the electrical activity of rabbit brain, J. Pharmacol. Exptl. Therap. 120: 160-170.

Machne, X. and Unna, K.R.W., 1963, Actions of the central nervous system, in "Cholinesterases and Anticholinesterase Agents," (G.B. Koelle, ed.), Hndbch, d. exper. Pharmakol. Ergänzungswk., 15: 679-700, Springer-Verlag, Berlin.

Marczynski, T.J., 1967, Topical application of drugs to subcortical brain structures and related aspects of electrical stimulation, Ergebn. d. Physiol. Biol. Chem. Exp. Pharmakol. 59: 86-159.

Martin, W.R. and Eades, C.G., 1967, Pharmacological studies of spinal cord adrenergic and cholinergic mechanisms and their relation to physical dependence on morphine, Psychopharmacologia 11: 195-223.

Morrell, F., 1967, Electrical signs of sensory coding, in "The Neurosciences," (J.C. Quarton, T. Melnechuk and F.O. Schmitt, eds.), pp. 452-468, The Rockefeller Press, New York.

Nishi, S., 1970, Cholinergic and adrenergic receptors at sympathetic preganglionic nerve terminals, Fed. Proc. 29: 1457-1465.

Nishi, S., Minota, S. and Karczmar, A.G., 1973b, The GABA-mediated depolarization of Primary afferent neurons, The Physiologist, (In Press).

Nishi, S., Minota, S. and Karczmar, A.G., 1973c, Primary afferent neurons: The ionic mechanism of GABA-mediate depolarization, Science (In Press).

Phillis, J.W., 1971, The pharmacology of thalamic and geniculate neurons, Int. Rev. Neurobiol. 14: 1-48.

Pradhan, S.N. and Dutta, S.N., 1971, Central cholinergic mechanism and behavior, Int. Rev. Neurobiol. 14: 173-231.

Reid, W., Haubrich, D. and Krishna, G., 1971, Enzymatic radioassay for estimating brain levels of acetylcholine and choline, Anal. Biochem., 42: 392-396.

Richardson, D., Karczmar, A.G., and C.L. Scudder, 1972, Behavioral neurochemical correlates of stress, Proc. Fifth Int. Cong. Pharmacol., p. 192.

Rinaldi, F. and Himwich, H., 1955, Cholinergic mechanisms involved in function of mesodiencephalic activating system, Arch. Neurol. Psychiat. 73: 396-402.

Scudder, C.L., 1971, The brain: A neurohumorally regulated ultrahomeostat. Gen. Syst. Bull. 3: 2-11.

Scudder, C.L., Karczmar, A.G., Everett, G.M., Gibson, E. and Rifkin, M., 1966, Brain catechol and serotonin levels in various strains and genera of mice and a possible interpretation for the correlations of amine levels with electroshock latency and behavior, Int. J. Neuropharmacol., 5: 343-351.

Sethy, V.H. and Van Woert, M.H., 1973, Effect of L-DOPA on brain acetylcholine and choline in rats, Neuropharmacol. 12: 27-31.

Stein, L., 1968, Chemistry of reward and punishment, in "Psychopharmacology. A Review of Progress 1957 to 1967," (D.H. Efron, ed.), pp. 105-124, U.S. Govt. Printing Office, PHS Publ. No. 836, Washington, D.C.

Tebécis, A.K., 1970a, Properties of cholinoceptive neurons in the medial geniculate nucleus, Brit. J. Pharmacol. 38: 117-137.

Tebécis, A.K., 1970b, Studies on cholinergic transmission in the medial geniculate nucleus, Brit. J. Pharmacol. 38: 138-147.

Toman, J., 1963, Some aspects of central nervous pharmacology, Ann. Rev. Pharmacol. 3: 153-184.

Van Meter, W., 1970, "Responses to anticholinesterases," Ph.D. Thesis, Loyola University, Chicago, Illinois.

Van Meter, W. and Karczmar, A.G., 1967, Effects of catecholamine depletion on anticholinesterase activity in the central nervous system, Fed. Proc. 26: 651.

Van Meter, W.G. and Karczmar, A.G., 1971, An effect of physostigmine on the central nervous system of rabbits, related to brain levels of norepinephrine, Neuropharmacoi. 10: 379-390.

Varagić, V. and Kristić, M., 1966, Adrenergic activation by anticholinesterases, Pharmac. Rev. 18: 796-800.

Vazquez, A.J. and Krip, G., 1973, Evidence for an inhibitory role for acetylcholine, catecholamines, and serotonin on the cerebral cortex, in "Chemical Modulation of Brain Function - A Tribute to J.E.P. Toman," (H.C. Sabelli, ed.), pp. 137-159, Raven Press, Publs., New York.

Wikler, A., 1952, Pharmacologic dissociation of behavior and EEG sleep patterns in dogs: Morphine, n-allyl normorphine and atropine, Proc. Soc. Exptl. Biol. Med. 79: 261-265.

York, D.H., 1967, The inhibitory action of dopamine on the neurons of the caudate nucleus, Brain Res. 5: 263-266.

ABSTRACT:

The presence of central cholinergic synapses is frequently evaluated in terms of the distribution of acetylcholine, acetylcholinesterase and choline acetylase. The interesting fact is that neurons containing these entities do not have to be cholinoceptive; for instance, early in ontogeny the spinal ganglia cells contain the components of the cholinergic system, but do not respond to acetylcholine. Even cholinoceptivity does not prove the presence of cholinergic transmission, and appropriate pharmacological analysis of the presynaptic response as compared to that of the cholinoceptive response is needed for such a proof. Altogether, there are relatively few proven central cholinergic synapses, although cholinoceptivity is widely spread in the CNS.

The better known peripheral cholinergic synapses exhibit several synaptic mechanisms and sites which contribute to numerous feedbacks and modulations; altogether, they may be referred to as little brains. Some of these mechanisms can be demonstrated in the case of the central cholinergic synapses. Furthermore, both intra- and inter-synaptic involvements may be demonstrated, and the coupling of central cholinergic synapses with other transmitter systems will be stressed. Frequently, these couplings indicate that the cholinergic synapses link "one-way" with other systems; the increment in brain acetylcholine induces changes in brain dopamine and norepinephrine, but the changes in catecholamines do not seem to affect brain acetylcholine. The couplings between the various neurotransmitter systems are illustrated EEG- and behavior-wise with regard to paradoxical sleep, desynchronization and arousal phenomena, stress, and tremor. Altogether, the concept of the interaction between cholinergic and other systems is consistent with our data which indicate that changes in several neurotransmitters and related substances result from behaviorally or drug-induced changes in brain acetylcholine.

In terms of brain organization, these couplings amount to cholinergic participation in inhibitory circuitry. Such circuitry frequently results in either post- or pre-synaptic inhibition. Our data on the cholinergic contribution to post- and pre-synaptic inhibition due to GABA will be described; the meaning of this contribution for the brain excitability and seizures will be discussed.

Altogether, the cholinergic contribution to inhibitory circuitry results in desynchronization of the activities of neuronal populations. Very likely, the older models of "cholinergic centers" for aggression and thirst have to be redrawn in terms of cybernetic systems and of the capacity of the cholinergic synapses to subdivide and desynchronize neuronal populations.

A MODEL OF THE VERTEBRATE NERVOUS SYSTEM BASED LARGELY ON DISINHIBITION: A KEY ROLE OF THE GABA SYSTEM

Eugene Roberts

Division of Neurosciences, City of Hope National Medical Center, Duarte, California 91010

I would like to be able to utilize neurochemical information already currently available in such a way as to lead to a better understanding of problems of nervous system function and malfunction. In order to begin to achieve this it is necessary to have some model of normal nervous system function. The present essay follows several such efforts on my part (Roberts, 1966a; Roberts, 1966b; Roberts _et al._, 1964; Roberts and Matthysse, 1970; and Roberts, 1972), and is an attempt to look at the role of the γ-aminobutyric (GABA) system in somewhat larger dimensions than usual.

A GENERAL DESCRIPTION OF THE ORGANISM IN ITS ENVIRONMENT

The overall behavioral model I wish to employ is a "homing" or hedonic model (Roberts and Matthysse, 1970). The primary assumption of this model is that the behavior of a healthy waking organism is aimed at attaining a state in which maximal internal homeostatic adjustment is achieved. In the waking organism the equilibrium condition may be described as the state of well-being, and can be associated with terms such as comfort, pleasure, ease, absence of anxiety, satisfaction, etc. Displacement from an equilibrium state, which may differ in set point from one time to another, may result in feeling of anxiety, discomfort, or pain. Continual changes taking place in the external environment and in the internal metabolism of the organism generally act counter to the maintenance or achievement of such a state as that described above and tend to displace it away from equilibrium. At all times an awake organism finds itself in a multisensory environment that it scans internally and externally for physical and chemical changes with specialized receptors, and it responds to the patterning of the relative values of the effective sensory cues, an abstraction of the environmental realities. The changing pattern in the perceived environment, external and internal, is the stimulus for the organism.

Experience of the external and internal environment can occur only through the sensory receptors, which are acted upon by physical and chemical changes. The proprioceptors sense what goes on in the muscles, tendons, joints, mesenteries, and the blood vessel walls of the organism itself. The exteroceptors and interoceptors deal with the occurrences outside of the organism, the former monitoring the outside world from the surfaces of the organism and the latter giving information about the events largely concerned with the processing of food in the mouth and in the digestive cavity. The sensory information undergoes neural transformation (partially innate and partially learned) at various levels in the nervous system in such a way that the organism experiences an abstraction of reality, not reality itself. Thus, in the visual system, similarity and contrast, onset and termination of stimuli, motion and geometrical shape are singled out for perceptual attention. A still higher level of perceptual integration takes place before the organism can respond adaptively to the pattern of sensory input. The stimulus pattern is brought into relation with memories of past stimuli and their consequences (associative integration) and with the drive state of the organism (emotional integration). Some information processing system must call up from memory storage stimulus patterns similar to the one at hand, and decide whether or not the stimulus presents opportunities for need satisfaction or potential threats to survival (or symbolic equivalents thereof). A growing body of evidence suggests that at least some of this higher-order perceptual integration takes place in the hippocampus which, through an abundant stream of fibers, brings abstractions of integrated stimulus patterns, invested with their full associative and emotional significance, immediately into interplay with the central visceromotor system (CVS) (hypothalamus, septum, and paramedian region of the mesencephalon) (Nauta, 1972). Thus, the environment, acting through the receptors on neural circuits in the CVS, has effects, which in turn have important influences on behavior, particularly at times when a new environmental setting requires the organism to develop new ways of adjustment. The key to the CVS system appears to be in the hypothalamus, which plays a central role in the coordination of physiological processes within the organism so that a relative constancy of steady states (homeostasis) within the organism tends to be maintained at all times. The hypothalamus is strategically situated to exert a major influence upon practically all those effector systems that are qualified to adjust and preserve the constancy of the internal environment and has both a short and long range function in restoring the state of well-being and enhancing future adaptive capacity by facilitating plastic change throughout the CNS (see Roberts and Matthysse, 1970, for detailed discussion).

The result of the impingement and processing of information is behavior, some reaction of the organism to the environment and/or action upon the environment as a result of activity of the effector systems, the neuronal systems that give signals to muscles, glands, etc. Not only is the processing of information within an organism the result of the coordinated activity of a concatenation of cybernetic neural units, but also the organism and its physical and social environment form a cybernetic unit. Consideration either of the isolated organism or of the environment alone becomes meaningless in this context. The organism acts on the external environment, the consequences of this action are fed back to the organism through the receptors, the results are monitored and analyzed, and corrective feedback signals occur within and between neuronal and neuroendocrine subsystems in such a way as to result in a modification of the action pattern of the organism. In a healthy organism this process continues in such a way that the behavioral options chosen tend to bring the organism back to the equilibrium or hedonic state while subserving its survival and reproductive needs.

There are a number of ways in which homeostatic behavior can be upset. Thus, the intensity and complexity of the environmental input may be increased beyond the capacity of the organism to handle it; the numbers of response options available could become too few as a result of physical or social restrictions; nonlethal lesions arising from numerous causes could occur in many regions of the CNS resulting in malfunctioning neural circuitry and upsets in neuroendocrine balances; many neuroeffective drugs could be taken or administered which could cause local or global disturbances in neural function, etc.

WHAT ARE NEURONAL CIRCUITS LIKE IN PRINCIPLE?

I would like to think of all neurons as possessing an innate capacity for firing spontaneously.[1] I believe that if a particular neuron were isolated from its biological context and maintained under suitable environmental circumstances, it would exhibit a characteristic firing pattern occurring close to its maximal potential rate. For each neuron this pattern would be a unique resultant of the interaction of its genetic potential with the environmental influences that had acted upon it up to the moment of observation. Each neuron would speak with its own voice, since multiple environmental gradients exist from the time of earliest development, and no two very similar neurons in an organism could be identical in every respect. Differences probably could be detected with sufficiently subtle methods even in adjacent neurons subserving similar functions and originally arising from the same embryonic area.

Most neurons in their normal environments are members of neural circuits in intact organisms and largely have ceded their autonomy, while becoming citizens of an integrated neuronal community. Many neurons do not fire spontaneously at their maximal rates; and some do not fire spontaneously at all. Glial cells might serve as one major restraining influence on spontaneous activity of neurons by removing substances from the extraneuronal environment in the regions of synapses, by adding substances to it, or by preventing diffusion of substances liberated from neurons in such a way as to shunt depolarizing ionic currents, thereby decreasing intrinsic excitatory levels of neuronal membranes below their spontaneous firing levels. Another type of inhibitory influence could be exerted by the effects of transmitters liberated upon neurons from inhibitory neurons as some GABA neurons may be. Such neurons could be tonically active, spontaneously firing cells or could be releasing inhibitory transmitter constantly without an action potential, the rate of release being determined by the degree of polarization. Inhibitory transmitters could act as chemical voltage clamps. If a neuron were held in check both by glia and a tonically active inhibitory neuron, release from inhibition could be achieved by direct depolarization of the inhibited cell, by inhibition of the inhibitory interneurons, or by a combination of both.

If I were to design a neuronal circuit that would fire reliably on demand, I would insure the fail-safe regulation of the circuit as a whole by putting the pacemaker neuron of the circuit under the control both of tonically active inhibitory neurons and glial cells, while only wrapping the intracircuit neurons in a glial restraining blanket. The firing of the neural circuit as a whole from pacemaker neuron to effector could be of major behavioral consequence to the organism, while the occasional escape from inhibition of an individual neuron within the circuit would not necessarily be of great importance, as long as the circuit as a whole were inoperative.

Most of us had been conditioned by our past training automatically to consider neural circuits as excitatory. We tended to think of the first neurons in the circuits as being excited by some input and passing on excitatory or depolarizing messages via excitatory transmitters in such a way that there results a progressive excitation passed from neuron to neuron until the final neuron in the circuit may depolarize an effector cell, muscle or gland. Now, recent findings have uncovered neural circuitry that operates to a considerable extent with neurons that liberate inhibitory transmitters.

Let us suppose that in a particular instance a neuron controls, through a single interneuron, the activity of a motor neuron, and that the motor neuron, if left alone, could discharge spontaneously at a rapid rate causing a muscle fiber to contract. Let us also suppose that the interneuron is a tonically active inhibitory neuron from which, in the absence of input from the first neuron in the series, an inhibitory chemical transmitter substance is liberated at such a rate that the membrane potential of the motor neuron is clamped at a level below the firing level, so that the muscle fiber is not caused to contract. If the first neuron in the series is inhibitory, an increase in its rate of discharge will decrease the inhibition from the interneuron on the motor neuron. As a result the rate of firing of the motor neuron will increase, and this will increase the rate of contraction of the muscle fiber. To achieve the same result as above if the first neuron in the series liberates an excitatory transmitter, one need only interpose an additional inhibitory interneuron in the circuit so that the signal would go from excitatory neuron to inhibitory neuron to inhibitory neuron to motor neuron.

Indeed, recent findings have uncovered neural circuitry that operates to a considerable extent with inhibitory neurons that liberate inhibitory transmitter. Almost all of the activity <u>within</u> the stomatogastric ganglion in the lobster is integrated by mechanisms involving inhibition and disinhibition rather than by direct excitation, while the final motor messages to the stomach muscles are excitatory (Maynard, 1972). The cerebellar cortex has two known excitatory inputs, the climbing fibers to the Purkinje cells and mossy fiber endings on the granule cells, and an inhibitory input from the locus coeruleus. All of the known interneurons in the cerebellar cortex (stellate, basket, and Golgi type II cells) are inhibitory, as are the only output neurons, the Purkinje cells.

HOW A NERVOUS SYSTEM MAY BE PUT TOGETHER

All normal or adaptive activity in nervous systems is a result of the coordinated dynamic interplay of excitation and inhibition, within and between neuronal subsystems. A basic inadequacy in this interplay could lead to an abnormal (disease) process such as is found in Parkinson's disease, Huntington's chorea, schizophrenia, epilepsy, depression, manic-depressive disorder, hyperkinetic behavior in children, psychosomatic disorders, etc.

Wherever one looks at nervous systems as a whole or at functional subunits, one is impressed by the absence of permissivity and democracy. All of the neurons are not firing independently of each other, nor is the activity of the systems the result of a statistical consensus. Instead, it seems that a nervous system and its subunits consist largely of poised genetically preprogrammed circuits which are released for action by neurons (command neurons) that are strategically located at junctions in neuronal hierarchies dealing with both sensory input and effector output. In some instances it has been possible to identify a hierarchy of controls within a neuronal population. A command neuron seems to be in control of the whole unit, a ganglion, for example; and its activity is the key to the activity of the ganglion.

In a hierarchical segmental system, such as has been found in crustacea and analyzed in the case of the system involved in control of swimmeret movement in the crayfish (Ikeda and Wiersma, 1964; Wiersma and Ikeda, 1964), the activity of a command neuron of a particular segment is controlled to a considerable extent by the neuronal activities of the segments above it in the hierarchy, with the head ganglion or brain exerting the highest level of control. If the communication between the segments is interrupted by cutting the connectives, then command neurons within each particular segment assume control of the activity of the segment. From considerable experimental evidence it appears that segmental pacemaker neurons, like the circuits they control, largely are inhibited and that the degree of inhibition is controlled from above. A decrease in inhibition allows them to fire, thereby releasing the preprogrammed circuits over whose activity they preside. A striking instance of this is the release of the highly coordinated stereotypical sexual behavior of the male praying mantis when his head is bitten off by the female (Roeder *et al.*, 1960). At successive stages of development of the guppy *embryo*, strychnine enhances motility of progressively more complex and coordinated nature, suggesting that active inhibitory processes are responsible for the early relative immotility rather than lack of development of excitatory synaptic functions (Pollack and Crain, 1972). Almost everyone is familiar with the sight of a headless chicken running. Although similar patterns of relationships are discernible in the nervous systems of higher vertebrate forms, the order of complexity is greatly increased. An example of this in humans may be the stereotypical postural effects and paranoia produced by overdoses of amphetamines (Randrup and Munkvad, 1970). Paranoid thinking may be a complicated, but stereotypical, genetically preprogrammed process can be evoked by skillful demagogues as well as by drugs.

In behavioral sequences, innate or learned, many genetically preprogrammed circuits may be released to function at varying rates and in various combinations by inhibition of neurons which are tonically holding in check pacemaker cells with capacity for spontaneous activity. If the pacemaker neurons are triggering a circuit related to the regulation of a vital function such as heart action or respiration, the inhibitory neurons might act in such a way as to vary the rate of the discharge of the pacemaker neuron. On the other hand, if a behavioral sequence involves the voluntary movement of a limb muscle, the pacemaker neuron might be held in complete check by the tonic action of inhibitory neurons and might be allowed to discharge in a graded manner only related to demand. In this view excitatory input to pacemaker neurons would have largely a modulatory role. Thus, disinhibition, acting in conjunction with intrinsic pacemaker activity and often with modulatory excitatory input, appears to be one of the major organizing principles in nervous system function. Disinhibition may act as a switch, turning on a specific coherent neural pattern which is otherwise actively and continuously inhibited, as well as play a role in the organization of sequential and alternating discharges among separate groups of elements (Maynard, 1972).

When one deals with whole neuronal systems in intact organisms, the situation becomes quite complex. For instance, the overall effect of inhibition at the synaptic level may be either activation or inhibition of the system. Thus, inhibition of inhibitory neurons may lead to disinhibition or excitation of the system as a whole; inhibition of excitatory neurons may lead to inhibition of the system; inhibition of excitatory neurons that act on inhibitory neurons may lead to disinhibition or excitation, etc. It is important to realize that the overall physiological or behavioral effects of an excitatory or inhibitory synapse depend on the circuit of which it is a part. The Purkinje cells of the cerebellar cortex, which convey the output of that structure, make only inhibitory connections with neurons in a number of deeper cerebellar structures. The activity of the Purkinje cells is modified by powerful inhibitory synapses formed by presynaptic endings of basket cells on their somas. Therefore, basket cell activity has a disinhibiting or activating effect on the activity of those cells that receive an input from the Purkinje cells. GABA probably is the transmitter substance employed by both the Purkinje and basket cells. Motor neurons of the spinal cord which employ acetylcholine as transmitter, in addition to their primary excitatory effect on muscles, send excitatory collateral fibers to inhibitory interneurons that lie in ventral regions of the cord. These neurons, in turn, may inhibit the same motor neurons that excite them as well as other motor neurons in the vicinity. Thus, release of an excitatory transmitter may result in physiological inhibition. Similarly, behavioral inhibition or depression must not be equated in a one-to-one way with synaptic inhibition.

THE SYSTEM OF NEURONS UTILIZING GABA IS A MAJOR INHIBITORY SYSTEM IN THE CNS

I would like to summarize briefly some current knowledge about the GABA system, a subject with which my colleagues and I have been concerned for over 20 years. The first report of the presence of GABA in uniquely large concentrations in the vertebrate CNS was made in 1950 (Roberts and Frankel, 1950).

A consideration of the biochemical, pharmacological, and physiological data subsequently developed about the function of GABA and currently available about nervous system function, in general, suggested that the major neural system exerting tonic inhibition on pacemaker neurons might be the system of inhibitory neurons utilizing GABA as transmitter (Roberts and Kuriyama, 1968; Roberts and Hammerschlag, 1972; and Roberts, 1972). GABA neurons are present ubiquitously in the CNS of vertebrate species, and on a quantitative basis GABA is much more extensively and relatively more evenly distributed throughout the various brain regions than the neuronal systems that employ other known neural transmitters such as acetylcholine, the catecholamines, or serotonin (Roberts, 1972).

An outline of the chief known reactions of GABA is shown in Fig. 1. GABA is formed in the CNS of vertebrate organisms to a large extent, if not entirely, from L-glutamic acid. The reaction is catalyzed by an L-glutamic acid decarboxylase (GAD I), an enzyme found in mammalian organisms only in the CNS, largely in gray matter. For a number of years it was assumed that there was only one GAD in the vertebrate organism and that it was located entirely in neurons in the CNS. With more sensitive methods it was found that GAD activity can be detected in glial cells, kidney, adrenal and pituitary glands, and in blood vessels. The GAD in the latter tissues (GAD II) shows different properties from the neuronal enzyme (Wu and Roberts, 1973) and does not cross-react with antibodies to it.

We now have succeeded in purifying GAD I to homogeneity (Wu, Matsuda, and Roberts, 1973). It has a molecular weight of 85,000, a sharp pH optimum at 7.0, and catalyzes the rapid α-decarboxylation only of L-glutamic acid of the naturally-occurring amino acids and to a very slight extent that of L-aspartic acid. Further studies of the properties of this enzyme are in progress. Henceforth, when mention is made of GAD, it will be assumed that reference is being made to GAD I. Further work also is in progress in our laboratories on the purification and properties of GAD II. The reversible transamination of GABA with α-ketoglutarate is catalyzed by an aminotransferase, GABA-T, which in the CNS is found chiefly in the gray matter, but also is found in other tissues. The products of the transaminase reaction are succinic semialdehyde and glutamic acid.

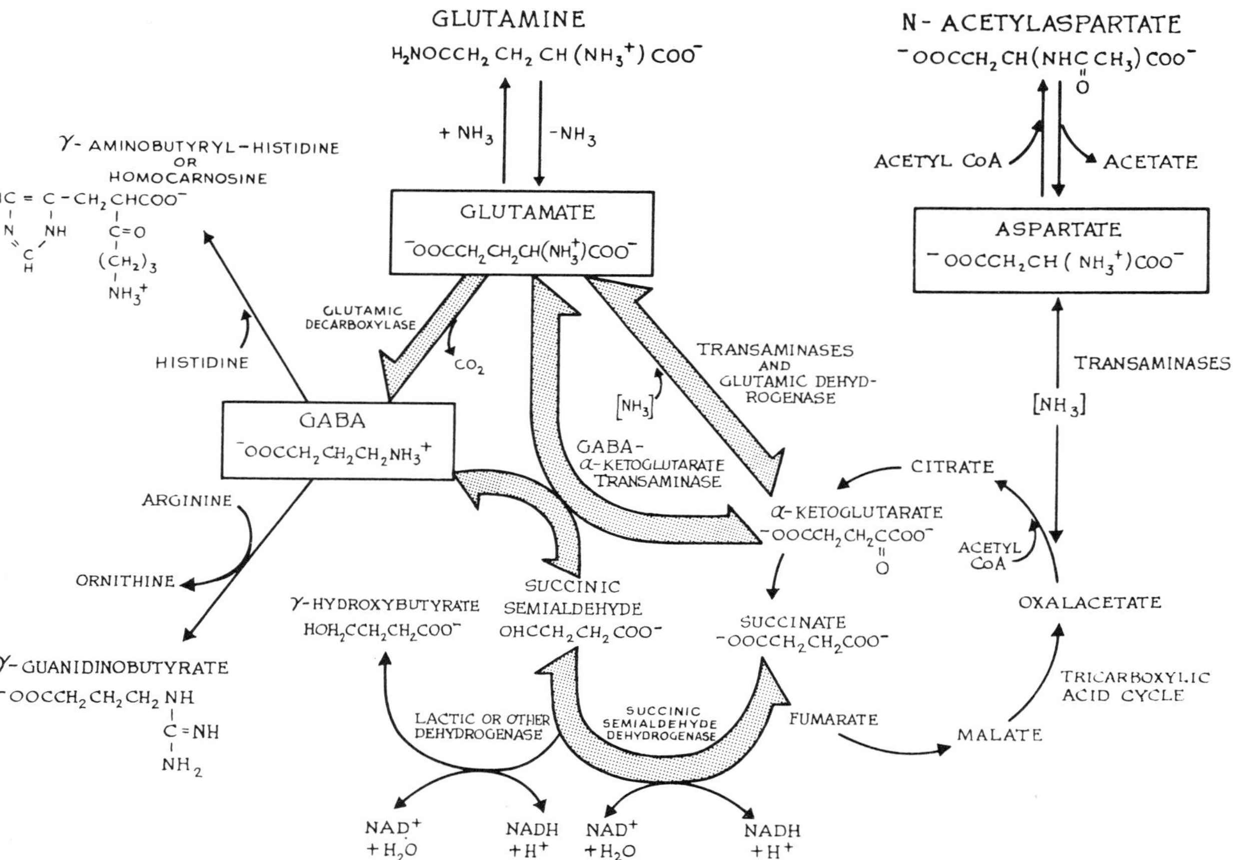

Figure 1. Some metabolic relations of 2-aminobutyric acid.

A dehydrogenase is present which catalyzes the oxidation of succinic semialdehyde to succinic acid, which in turn can be oxidized via the reactions of the tricarboxylic acid cycle. Recently we have succeeded in purifying GABA-T to homogeneity, and its properties are being studied in detail (Schousboe et al, 1973). The enzyme has a molecular weight of 109,000, a pH optimum of 8.05, and can be split into two unequal subunits. Of the keto acids tested only α-ketoglutarate was an amino group acceptor. However, of a series of amino acids tested β-alanine, δ-aminovaleric acid, and β-aminoisobutyric acid were effective amino donors in the reaction catalyzed by the enzyme. In lobsters and in other invertebrates similar enzymes are present in both peripheral and central nervous systems (Otsuka et al, 1966).

The steady state concentrations of GABA in various brain areas normally are governed by the GAD activity and not by the GABA-T. In many inhibitory nerves, both GAD and GABA are present and are distributed throughout the neuron, the GAD being somewhat more highly concentrated in the presynaptic endings than elsewhere. The GABA-T is contained in mitochondria of all neuronal regions, but it seems to be richer in the mitochondria of those neuronal sites onto which GABA might be liberated. Such regions would be expected to exist in perikarya and dendrites that receive inhibitory inputs and possibly in the glial and endothelial cells that are in the vicinity of inhibitory synapses (Roberts and Kuriyama, 1968; Roberts and Hammerschlag, 1972).

There is evidence for presynaptic release of GABA. Stimulation of axons of several nerves inhibiting different lobster muscles was shown to result in the release of GABA in amounts related to the extent of stimulation, while stimulation of the excitatory nerve did not produce GABA release (Otsuka et al, 1966). Data showing the liberation of GABA on stimulation of specific inhibitory neurons in the vertebrate nervous system are extremely difficult to obtain, but there are many experiments that indicate that this does take place (Jasper et al., 1965; Mitchell and Srinivasan, 1969; Srinivasan et al., 1969; Obata and Takeda, 1969; and Bradford, 1970).

The ionic basis of the inhibitory effect of GABA on the postsynaptic regions of vertebrate and invertebrate neurons is known (see Roberts and Kuriyama, 1968, for review). Applied GABA alters the membrane conductance to chloride ions with the membrane potential staying near the resting level. GABA also has a presynaptic inhibitory action at the crayfish neuromuscular junction, imitating the action of the natural inhibitory transmitter by increasing permeability to chloride, thus decreasing the probability of release of quanta of excitatory transmitter (Takeuchi and Takeuchi, 1966). There is as yet no hint about the manner in which chemical or physical interaction of GABA with membranes produces increases in the chloride ion conductance of the membranes, and to date all of our attempts to isolate the GABA receptor have failed. What is lacking is a high affinity-label for the active site. Recent evidence suggests that GABA may be the transmitter mediating presynaptic inhibition in a number of regions in the vertebrate CNS (Davidson and Southwick, 1971; Nicoll, 1971; Barker and Nicoll, 1972), and that it may act by producing depolarization of primary afferent terminals. Although it was suggested that sodium may be the chief ion involved in the mechanism (Barker and Nicoll, 1972), recent unpublished data in several laboratories suggest the participation of the chloride ion, as in the invertebrate nervous system.

The cessation of action of a synaptically active substance could be brought about by the removal of the substance from the sensitive sites by destruction, by transport, or by diffusion. In the case of GABA it is likely that active transport out of the synaptic gap is the major inactivating mechanism (Roberts and Kuriyama, 1968).

WHERE GABA FUNCTIONS

Biochemical analytical data have shown the presence of GAD and GABA in many regions of the vertebrate CNS. However, in no instance have the functional relationships been worked out to the same extent as they have in the cerebellum (Roberts and Kuriyama, 1968). GABA probably mediates the inhibitory actions of stellate and basket cells upon the Purkinje cells and of the Golgi type 2 cells at the mossy fiber endings. It also transmits the inhibitory messages that go from the Purkinje cells to each other and to the cells in the intracerebellar nuclei. GABA also takes part in information processing beyond the Purkinje cell synapses. For example, rabbit oculomotor neurons may be inhibited by GABA neurons found in the vestibular nuclear complex (Obata and Highstein, 1970).

In the retina the cellular layers all contained GABA and GAD activity, higher contents of GABA being accompanied by higher levels of GAD activity (Kuriyama *et al*., 1968). The highest values for GABA and GAD were found in the third layer, the ganglion cell layer, and the next highest in the receptor-containing layer. The region of the optic fibers showed no GABA and only traces of GAD. Therefore, the GABA neurons probably are not the retinal ganglion cells, whose axons furnish the bulk of the fibers of the optic nerve. The GABA neurons of the retina, therefore, probably are indigenous to the retina. Although it appears likely that at least some of the amacrine cells are GABA neurons, further work is required to determine which of them are the GABA neurons.

Recent data suggest strongly that GABA plays an important regulatory role in the activity of the hippocampus, possibly mediating the inhibition of the pyramidal cells by the basket cells (Fonnum and Storm-Mathisen, 1969; Curtis *et al.*, 1970).

There is convincing physiological evidence that inhibition in the mammalian cortex may be mediated by GABA neurons (Dreifuss *et al.*, 1969). In the spinal cord the chief enzymes of GABA metabolism, GAD and GABA-T, are higher in the dorsal than in ventral gray matter (Albers and Brady, 1959; Salvador and Albers, 1959) and recent data suggest that there is an extensive system of indigenous GABA neurons in the dorsal part of the cord and that these may be involved in the inhibitory mechanisms that produce the dorsal root potential (Miyata and Otsuka, 1972) and presynaptic inhibition in the cord (Davidoff, 1972). Thus, GABA neurons may control the input to the first synapse in the CNS.

Recently we have achieved the immunohistochemical localization of GAD in rat cerebellum, hippocampus, and spinal cord employing antibody prepared against the purified mouse brain enzyme. After the first descriptive phase of the work, numerous opportunities will be open to determine whether or not deficiencies might exist in the relative numbers of GABA neurons or defects may occur in their relationships to other types of neurons in pathological states in humans, such as Huntington's chorea, schizophrenia, epilepsy, etc. This will be possible because the antibodies produced in rabbits to GAD purified from mouse brain were found to cross-react with the GAD from human brain, microcomplement fixation tests showing the mouse and human enzymes to be quite similar (Saito *et al.*, 1973).

THE RELATIONS OF THE GABA SYSTEM TO OTHER TRANSMITTER SYSTEMS

Inhibitory neurons interact with other inhibitory neurons regardless of the transmitter which they employ. For example, the Purkinje cells, which probably are GABA-releasing neurons (Obata *et al.*, 1967), are inhibited by basket and stellate cells, which also probably are GABA-releasing neurons (Curtis *et al.*, 1970; Woodward *et al.*, 1971); and, likewise, Purkinje cells are inhibited through noradrenergic endings of nerve fibers (Bloom *et al.*, 1971), which have their origin in the locus coeruleus (Olson and Fuxe, 1971). Thus, inhibitory neurons utilizing GABA as a transmitter can be inhibited by other GABA neurons and by noradrenergic neurons. Serotonin-releasing neurons from the Raphé nuclei inhibit noradrenergic neurons in the brain stem (Jouvet, 1969). It is likely that in the complex arrangements of various regions of the CNS a variety of combinations of inhibitory neurons can act upon each other. In most instances in which they have been studied by iontophoretic application, the biogenic amines have been found to exert an inhibitory action on neurons (Bradley, 1968; Curtis and Crawford, 1969).

Excitatory information coming in from receptors can go to neurons in the spinal cord or in the brain stem increasing their probability of firing. Sensory input comes in via excitatory endings that may impinge directly on pacesetter neurons and also on inhibitory neurons (GABA-releasing, catecholaminergic, serotonergic) lying entirely in the immediate neighborhood, which would disinhibit pacesetter neurons by liberating them from the tonic inhibition of GABA neurons. The identity of the excitatory transmitter (or transmitters) is unknown with certainty, but it may be a polypeptide, possibly hypothalamic Substance P (Otsuka _et al._, 1972). In addition, some of the GABA neurons could inhibit adjacent sensory endings by presynaptic inhibition via axo-axonic connections. If a stimulus is sufficiently great, it may affect cholinergic spinal motor neurons so that their discharge would elicit a spinal reflex prior to detailed processing of the incoming signals in CNS regions beyond the spinal cord segment involved. The latter situation must be clearly distinguished from that in which motor units would be used as part of a coordinated behavioral sequence resulting from a processing of the incoming stimulus pattern in the higher centers of the CNS. Contrary to much previous belief, it is obvious that the simple spinal reflex cannot be used as a model for CNS function.

As mentioned previously, indigenous GABA neurons probably are richest in the substantia gelatinosa and the upper laminae of the dorsal horn. Inhibitory interneurons employing glycine as transmitter seem to be richer in the ventral portions of the spinal cord (Aprison _et al._, 1970). Neurons in the substantia gelatinosa appear to modulate the input from primary cutaneous afferents in the spinal cord. Fluorescent microscopic studies in mice and rats have shown there to be a system of fine longitudinally oriented fibers in the substantia gelantinosa which appear to be noradrenergic; while in the cat morphologically similar terminals probably contain serotonin (Fuxe _et al._, 1970). The noradrenergic fibers appear to be in synaptic contact with each other as well as with dendrites of other neurons in the substantia gelantinosa. The latter probably are descending fibers from the pons. From the above discussion it is reasonable to surmise that there may be synapses between the noradrenergic or serotonergic fibers with the GABA neurons in the substantia gelantinosa.

Directly, and/or after processing in the spinal cord, the sensory input can enter the brain stem, where it may act upon monoaminergic interneurons lying entirely within the brain stem that, when activated, can inhibit the tonically active GABA neurons that are holding in check intersystem monoaminergic neurons. This action could result in the disinhibition and, therefore, activation, of the input of the brain stem neurons to the higher brain centers. Ascending pathways from the reticular formation of the brain stem are very important in increasing the background activity of neurons in higher brain centers. When the connections between the brain stem region and the cortex are cut, the activity of cortical neurons either stops completely or the frequency of discharge becomes much lower than in the brain of an intact animal, while stimulation of this region in an intact animal increases the activity of cortical neurons. In a study of neuronal units in rabbit visual cortex it was shown that destruction of the mesencephalic reticular formation or its blockade by drugs reduced the frequency of appearance of groups of spikes from a given cortical neuron but did not change the number of spikes appearing in a group when a neuron did fire (Velikaya and Sycheva, 1970). This decrement in discharge frequency was not observed when the lateral geniculate body or the superior colliculi, specific relay stations of the visual system, were destroyed. This suggests that impulses from the brain stem serve to release cortical units for firing in their own characteristic fashion, and that the pathways from the brain stem are much more important for maintaining cortical activity than those from the specific sensory nuclei. There is abundant evidence that the excitatory information of the brain stem to the cortex largely is mediated by a neuronal system employing norepinephrine (NE) as a transmitter (Fuxe et al., 1970). The NE innervation comes along a dorsal NE bundle that originates from NE neurons in the locus coeruleus area of the pons, and recent data suggest that a single NE neuron from this area can innervate both cerebral and cerebellar cortices and also in its course can give off collaterals to the colliculi, the geniculate bodies and part of the thalamus (Olson and Fuxe, 1971). This system of neurons may be, at least to some extent, involved in EEG arousal. Since most of the direct cellular experiments with NE have shown that this substance exerts an inhibitory effect on the neurons being studied, the only logical way in which it can be envisioned that a neuronal system employing NE as a transmitter may be an excitatory one is if it were to be acting through inhibition of inhibitory interneurons, or by disinhibition. If we postulate that the primary inhibition within the cortex is exerted by a system of tonically active inhibitory GABA interneurons, then the well known system of NE endings in the cortex might be acting in such a way as to inhibit the GABA neurons which are themselves holding cortical activity in check. Although there exists a diffuse network of very fine, varicose NE nerve terminals in practically all parts of the cerebral cortex, the highest density of terminals appeared in the four outer cortical layers (Fuxe et al., 1968). GABA forming neurons probably have by far the highest density in the four outer cortical layers of the occipital or visual cortex in the monkey, the highest being in layers 3-4A (Albers and Brady, 1959). Since this is also a region of high density of NE nerve endings, it is not unreasonable to presume that the NE neurons, activated by incoming sensory input, send signals from the brain stem region to the cortex releasing the activity of cortical pyramidal neurons from inhibition by GABA neurons. The input of brain stem NE neurons occurs both to the inhibitory neurons which are restraining the activity of cortical pacesetter neurons and to centers involved in perceptual integration, such as the hippocampus, cerebellum, thalamus, basal ganglia, etc. The input from brain stem neurons is not sufficient in itself to release a preprogrammed effector circuit. Instead, I would like to suggest that perceptual integration must take place within the various analyzing regions of the nervous system (see Roberts and Matthysse, 1970, for pertinent discussion) and that from the central visceromotor system there must come either a GO or a NO GO signal. Within the structures of this system the emotionally significant aspects of the total stimulus pattern, external and internal, are summated, and a signal is generated which causes the organism to act by releasing or withholding behavior options already available to it. If a decision is reached that a particular behavioral option is _not_ to be employed, then an inhibitory signal may go out to the cortical pacesetter neurons related to the control of that behavioral option, so that they will be sure not to discharge at all or, at least, not to increase their basal rate of discharge. On the other hand, if a decision is reached that a particular behavioral option should be employed, then there might be inhibition of the same tonically active inhibitory GABA neurons that were acted on by the inhibitory brain stem (NE) neurons, so that the additive effect would result in the disinhibition of appropriate pacesetter neurons. From the above, it is suggested that the disinhibitory signal from brain stem neurons alone is a necessary, but not sufficient condition for the release of the activity of pacesetter cortical neurons in an intact conscious organism. As a result of the disinhibition of pacesetter neurons in a higher brain center, excitatory signals to monoaminergic interneurons located in the brain stem disinhibit descending intersystem monoaminergic neurons which, together with other disinhibitory and excitatory influences communicated by the same pacesetter neurons through the pyramidal tract, release neurons in the spinal cord from inhibition by tonically active GABA neurons. This finally allows motor neurons to fire and cause those effectors to respond which are involved in the particular behavioral option that has been selected. Thus, spinal circuits would be released for action at least in part by the disinhibitory action of descending NE fibers.

Recently, striking experimental support has appeared for some aspects of the above scheme (Forssberg and Grillner, 1973). Acute spinal cats ordinarily show neither postural nor locomotor activity. However, after intravenous administration of Clonidine [2-(2,6-dichlorophenylamine)-2-imidazoline hydrochloride], a specific stimulator of central noradrenergic receptors that passes the blood brain barrier, it was possible to elicit walking behavior on a moving treadmill with a speed that can be adjusted by the speed of the treadmill, and in the best preparations "this locomotion looks normal to the eye with smooth alternating movements in all joints." Thus, stimulation of noradrenergic receptors in the cord combined with stimulation by a treadmill can release the expression of neural programs for coordinated postural control and locomotion that are located entirely in the cord! Can this knowledge be applied to problems of paraplegia, etc?

The type of interaction discussed above illustrates the general principle by which I believe the nervous system as a whole, as well as many neuronal subsystems, might operate so that if one were to go into detailed consideration of the relations of the neuronal subsystems involved in mechanisms of perceptual integration, in hypothalamic relations, etc., one would come out with the same general type of scheme, of course varying in detail and complexity from region to region.

APPLICATION OF THE ABOVE CONSIDERATIONS TO PROBLEMS OF DISEASE

The successful operation of a nervous system such as discussed above requires a coordination of neural activity which can determine from birth, or even before, the ability of an individual to prevent the too-frequent firing of preprogrammed circuits of behavioral options spontaneously or maladaptively and to maintain within physiological limits the rates of operation of continuously needed neural circuits, such as those required for heart function, respiration, maintenance of blood pressure, etc., under a variety of environmental circumstances. When gross malfunctions of the coordination of inhibitory and disinhibitory neuronal systems with which we are concerned are found at birth there may result lethal effects either through generalized seizures or cessation of operation of some vital function or, alternatively, some obviously severe neurological dysfunctions may occur.

In the model presented above GABA neurons are envisioned to play a key role at all levels, from setting the gain on the sensitivity of sensory receptors to coordinating the function of the systems involved in perceptual integration and in reaching the decisions with regard to which neural circuits should be released for use at a particular time. For example, if in an individual there should be a paucity or defective function of horizontal GABA neurons in layer IV of the motor cortex, this individual might be expected to be more susceptible than normal to occurrence of grand mal seizures. If such a problem should exist in the region of the globus pallidus, postural control would be expected to be defective. If the GABA system were inadequate in those regions of the hypothalamus dealing with food intake, hyperphagia or anorexia nervosa might result. If GABA neurons in the dorsal horn of the spinal cord were inadequately functional, there might be an inordinately great sensitivity to tactile and thermal stimulation and inadequate spatial and temporal discrimination of the stimuli. If there were a defect in GABA function in the retina, visual perception and integration might be faulty.

Let us examine the situation in which there is a relatively small, but continuous, degree of incoordination between the GABA system and other transmitter systems because for some reason the inhibitory GABA neurons have a considerably lower than normal effectiveness upon their recipient neurons. The defect may be restricted to a local brain region, may include several regions, or may be global throughout the CNS, depending on a variety of hereditary and/or developmental factors. Under relatively simple environmental conditions the nervous system in such an individual could function in an apparently adequately adaptive manner, which to an outside observer might appear to be in the normal range. As the complexity and intensity of environmental inputs is increased, there would be a correlated increase of disinhibitory influences acting upon the GABA neurons, largely via the monoaminergic systems. If the capacities of the monoaminergic systems to deliver their characteristic transmitters are normal, then under conditions of environmental stress, which are considered for most to be within the normal adaptive range, imbalances often might arise. Thus, if special hypothalamic regions are affected, greater than normal degrees of changes in responses might be observed in emotional reactivity, heart and respiratory functions, blood pressure, galvanic skin response, insulin secretion, liberation of gastric acid, motility of colon, etc. Herein may lie one of the bases for psychosomatic medicine. Those systems in the nervous system that are most poorly controlled will tend to break down under stress and to produce peripheral symptoms that are consequent to such a breakdown. Almost all of us have an Achilles heel. I believe that to a considerable extent this reflection of our physiological breaking point starts from a relatively poorly compensated region in the CNS. In this regard it is interesting that there are families in which there is a common maladaptive response to stress. There are families in which members tend to respond to stress with gastrointestinal symptoms, cardiac problems, skin outbreaks, respiratory ailments, etc. This suggests that there may be strong hereditary factors involved, although learning factors cannot be entirely eliminated in most instances.

THE GABA SYSTEM IN SCHIZOPHRENIA

The current data (Kety and Matthysse, 1972) seem to be in general agreement with the idea that the maladaptive function characteristic of susceptibility to schizophrenic disorder could arise in an organism in which there is a defect, genetic in origin, in some primary coordinative components found throughout the whole CNS. Since all normal or adaptive activity in nervous systems is a result of the coordinated dynamic interplay of excitation and inhibition, a basic inadequacy in this interplay could lead to a multi-stage development of abnormal interactional patterns with the environment which eventually results in the constellation of symptoms typical of schizophrenia (Meehl, 1962; Meehl, 1964).

Let us suppose that in an individual who is born with a susceptibility to schizophrenia, for some reason, the inhibitory GABA neurons have a considerably lower than normal effectiveness upon their recipient neurons. Under relatively simple environmental conditions the nervous system in such an individual could function in a barely adequate manner, which to an outside observer might appear to be in the normal range. As the complexity and intensity of environmental inputs is increased, there would be a correlated increase of inhibitory influences acting upon the inhibitory GABA neurons, largely via the monoaminergic systems. If the capacities of the monoaminergic systems to deliver their characteristic transmitters are normal, then, under what are considered to be normal circumstances for most, imbalances often would arise, more than normal numbers of behavioral options would be released, and greater than normal degrees of changes in responses would be observed in heart and respiratory functions, blood pressure, galvanic skin response, etc. It would be expected that such imbalances might have critical and lasting effects early in infancy; unusual birth trauma, the continual pain of colic, inappropriate handling by the mother, etc., might play an important role in increasing the probability of a poor prognosis for such an individual. However, if the environmental stresses to which a susceptible individual is exposed are graded in intensity with time of development in a manner attuned to his needs so that a number of adaptive behavioral options gradually are developed through learning, such an individual may not become schizophrenic.

Since many environmental stresses faced by individuals in society result from human contacts, and schizotypic individuals seem particularly sensitive to this kind of stress, one result of learning and self-knowledge in such individuals might be the development of skills and activities in which they deal with abstract ideas or material objects, or both, but minimally with people. Laboratory research, writing, composing music, painting, working in individual crafts, and even such activities as exploration and mountain climbing may give the requisite degree of relatively long periods of social isolation to maintain a sufficiently simple input for an individual to be a "compensated schizotype." Many inherited aspects of an individual's physiological makeup not related directly to the schizotaxic defect may make it more or less possible even for a good rearing regime to protect him from becoming schizophrenic. If at any time compensatory mechanisms and activities become inadequate, then initially inappropriate responses to environmental patterns may result in an increased stressful feedback from the environment with a further incapacity to respond appropriately (positive feedback) resulting in concomitant discomfort and anxiety. Any or all information-processing systems within such an individual may become rate-limiting in function. There may be varying degrees of failure of control of motor and autonomic functions or of patterns of expression of rage, aggression, eating, drinking, sexual activities, etc. Chaotic states would be expected to exist in those centers involved in perceptual integration; and, insofar as physical appearance, verbal expression, and pictorial representation through drawing give an insight into the state of an individual's adequacy of perceptual integration or reality orientation, "cognitive slippage" and aberrations of body image would be observed.

Once given the situation that the deviations from homeostatic equilibrium are so severe and prolonged that a cessation of adaptive behavior occurs, how could one attempt to restore transmitter balance? According to the present hypothesis one way would be to decrease the extent of impingement of the effective disinhibitory transmitters, chiefly the monoamines. This would be the equivalent of adding negative feedback to the system, and would allow the less than normally effective GABA neurons to maintain their tenuous hold on the pacesetter neurons requiring their restraints for participation as cooperative citizens in the various cybernetic information-processing systems. One obvious maneuver would be to decrease the sensory input by simplifying the environment, particularly in terms of decreasing the complexity of social interaction. This actually should decrease the monoaminergic input to the GABA neurons. A pharmacological way in which to achieve such a result would be to decrease the effectiveness of the monoamines within the nervous system by blocking, at least to some extent, the number of receptive sites for these substances on the membranes of the GABA neurons. It is currently believed that the major antipsychotic tranquilizers, such as chlorpromazine and haloperidol, actually might be exerting their action by blocking postsynaptic receptor sites for dopamine and NE. Eventually an understanding of the mechanism of action of drugs like those above on neuronal membranes at the molecular level, whose clinical utility originally was arrived at empirically, may give us insights into the nature of the basic disturbance in schizophrenia. Analysis of drug action in treatment of schizophrenic patients and in animal experiments and of the effects of amphetamines led to the suggestion that there may be abnormalities in some neuronal systems in schizophrenia in which dopamine neurons participate (Matthysse, 1973; Klawans, 1972). Current data suggest that the defect may not lie in the function of the dopamine neurons, themselves, but rather in their activity relative to other neuronal systems, such as the GABA system, for example.

In addition to the use of major tranquilizers, an additional or synergistic approach to the correction of an imbalance between the above systems, in which the GABA system may be the weak component, might be to attempt to devise some methods by which the potencies or firing rates of GABA neurons could be enhanced. That this might, indeed, be achieved has been suggested by the finding that administration of diphenylhydantoin (Dilantin), a widely used antiepileptic drug, results in a remarkable and prolonged increase in cats in the firing rates of cerebellar Purkinje cells, known GABA neurons (Julien and Halpern, 1971).

THE GABA SYSTEM IN THE BASAL GANGLIA

Although the nature of all of the connections among the basal ganglia and between them and other brain structures is not known and all of their functions have not been delineated, it appears that they largely are concerned with processing information related to proprioceptive, vestibular, and visual stimuli in the service of coordinating postural mechanisms. As far as it is possible for me to understand current neuroanatomical opinion, it seems that the caudate nucleus, putamen, and substantia nigra all exchange fibers with each other and that efferent outputs from the caudate and putamen go to the globus pallidus, which also may receive some fibers from the substantia nigra. The globus pallidus has two-way communication with the subthalamic nucleus. There are thalamic and cortical inputs to the caudate and putamen. The final results of the computations in the basal ganglia are sent out via a fiber system from the globus pallidus to the ventral lateral nuclei of the thalamus.

The globus pallidus and substantia nigra have the highest contents of GABA and highest activities of GAD in the brain (Fahn and Coté, 1969; Perry *et al.*, 1971a; Perry *et al.*, 1971b; Okada *et al.*, 1971). A recent detailed regional analysis for GABA in human substantia nigra showed there to be an uneven distribution, the content being higher in the pars reticulata than in the pars compacta and highest in the middle portion at the border between the two regions (Kanazawa *et al.*, 1973). Stimulation of the head of the caudate produced inhibition in nigral neurons and it was concluded that caudato-nigral fibers inhibit nigral cells monosynaptically (Yoshida and Precht, 1971). This inhibition was blocked by picrotoxin (Precht and Yoshida, 1971). A microiontophoretic study of nigral neurons that were inhibited by caudate stimulation showed them also to be strongly inhibited by GABA, but not by glycine, acetylcholine, or dopamine (Feltz, 1971). GABA levels in the substantia nigra were considerably reduced after destruction of the striatum by suction (Kim _et al._, 1971). The above results all are in keeping with the interpretation that there are GABA inhibitory synapses on dendrites of nigral neurons in the pars reticulata formed by axons that are striatal in origin. Since there is monosynaptic inhibition in pallidal neurons mediated by axon collaterals of the caudato-nigral fibers, it would appear possible that also here GABA plays an inhibitory role. On the other hand, destruction of the substantia nigra did not show significant reduction of GAD activity in the caudate nucleus, even when dopamine was reduced to undetectable levels (Hockman _et al._, 1971). The latter shows that there probably are no nigro-caudatal GABA fibers. Pallidal lesions were shown to reduce nigral GAD activity, but not that of the caudate (McGeer _et al._, 1971). The latter results suggest that the globus pallidus also may exert inhibitory effects in the substantia nigra via a GABA input and that the known pallidal inhibitory effects in the ventral lateral thalamic nuclei also might be mediated by fibers from pallidal GABA neurons. The inhibitory dopamine tracts from the nigral neurons to the striatum are, of course, well known.

Normal relations in the basal ganglia must involve _minimally_ a coordinated functioning of different groups of inter- and intrasystem neurons whose transmitters are GABA, dopamine, acetylcholine (McGeer _et al._, 1971), and an excitatory transmitter whose action externally applied glutamic acid can mimic (Krnjevic, 1970).

Let us suppose that the basal ganglia contain preprogrammed neural circuits for patterned postural control that are held in tonic inhibition by indigenous, closely lying GABA neurons. The chief switching mechanisms for turning on the patterned activities within the non-nigral regions may be the dopamine fibers emanating from the substantia nigra. Afferent inputs to the nigral neurons may release patterns of firing. In analogy to the activating effects of noradrenergic input to the cortex, the nigro-fugal fibers release dopamine in the caudate and putamen, inhibiting indigenous tonically inhibitory GABA neurons and, acting together with excitatory and/or disinhibitory inputs from the thalamus and cortex, release specific coded neural patterns in a sequential manner for which the pacesetters may be acetylcholine neurons. The results of this activity are communicated to the pallidum and thence to regions in the thalamus where integration with other incoming information takes place. The final postural instructions are then sent to the appropriate regions of the motor cortex, where after further refinement the activity of appropriate pyramidal neurons is released to signal the effectors. The circuits that are fired in the basal ganglia inform the other units about their activity via intersystem inhibitory GABA fibers, particularly to the appropriate nigral neurons, thus preventing their own further activation until the need arises again.

How does the above model fit in with what is known about the basal ganglia? If postural-regulating circuits can only be released for firing when there is sufficient inhibitory action of dopaminergic neurons on GABA neurons, a relative deficiency or excess in monoamine input should result in the failure of coordination of postural behavior. Hypofunction or nonfunction of the nigral neurons does lead to the failure to release appropriate circuits necessary for maintenance of postural control. Indeed, a "striatal dopamine deficiency" produced by naturally induced or experimental destruction of the nigral neurons or by blocking of their function by decreasing the content of dopamine (α-methyl tyrosine or reserpine) or by blocking the action of dopamine on its postsynaptic sites (chlorpromazine or haloperidol) leads to akinesia and rigidity, the classic symptoms of Parkinson's disease (Hornykiewicz, 1973)--the programs for postural controls cannot be released. The ameliorative effects of treatment with high doses of L-Dopa in parkinsonism probably largely can be attributable to an increased transmission by the remaining striatal dopaminergic neurons. Relative hyperactivity of the nigral dopamine neurons should lead to inappropriate release of action patterns; the reversible amphetamine-induced stereotyped behavioral patterns in animals and humans may be attributable to the extra release of monoamines that this drug produced in the basal ganglia (Randrup and Munkvad, 1970). If the nigral dopamine neurons act directly on the GABA neurons, their destruction typically should lead to atrophic changes in the GABA neurons; in untreated Parkinson's disease the GAD activity in the caudate was found to be reduced to less than half the normal value (Hornykiewicz, 1973). Does treatment with L-Dopa bring striatal GAD levels back to normal?

In contrast to what one sees with dopamine deficiency, specific degeneration or functional blockade of GABA neurons in the basal ganglia should lead to the expression of inappropriate and dissociated movement patterns. Because of its ubiquitous occurrence and function throughout the CNS, it has not been possible to observe effects of pharmacological blockade of GABA that are specific to the basal ganglia; usually running fits followed by tonic-clonic convulsions are induced. However, recently relatively specific and remarkable decreases in GABA levels (Perry _et al_., 1973) and GAD activities (Bird _et al_., 1973) have been found in the substantia nigra, putamen-globus pallidus, and caudate nucleus of patients with Huntington's chorea, a hereditary disease in which slight inconstant irregular choreiform movements eventually progress into constant twitching, stretching, gesturing, facial grimaces, etc. The occasionally helpful effects of haloperidol, a blocker of dopamine receptors, in patients with Huntington's chorea might be attributable to a decreased extent of inhibition of GABA neurons, allowing the less-than-normally effective GABA neurons to maintain their tenuous hold on the neurons requiring their restraints. Since orally or parenterally administered GABA usually does not pass the blood-brain barrier and often causes undesirable peripheral effects, the search is on for a suitable GABA-mimetic substance as a possible treatment in Huntington's chorea! Imidazoleacetic acid is a GABA-mimetic substance with very interesting pharmacological properties that passes the blood-brain barrier (Roberts and Simonsen, 1966; Roberts and Simonsen, 1970; Marcus _et al._, 1971; van Balgooy _et al._, 1972; Tunnicliff _et al._, 1972; Krnjević and Phillis, 1963; McGeer _et al._, 1961). It will shortly be in clinical testing in Huntington's chorea. Perhaps imidazoleacetic acid and related substances, alone or together with haloperidol, might be effective in alleviating some of the symptoms of this dread disorder.

WHERE DO WE GO FROM HERE?

The model of the nervous system outlined in this paper drew on information derived from anatomical, biochemical, physiological, pharmacological, and behavioral observations. Tests of the various hypotheses related to this model can be performed using the previously employed techniques, or refinements thereof. However, proofs of deduced relationships will only be possible against a background of an understanding of the precise relationships of neurons utilizing the various neurotransmitters studied at histological and ultra-structural levels in relatively well-known regions of the CNS, such as the cerebellum, hippocampus, and retina. Immunohistochemical methods employing antibodies to the rate-limiting enzymes in bio-synthesis of the known transmitters appear to be the tools of choice.

FOOTNOTES

[1]This spontaneous activity could occur either continuously or in an oscillatory fashion in the absence of input from other neurons or of changes in the ambient environment. When oscillations occur, their rhythms may be set by the rate of progressive depolarization of the membranes which may occur during successive firings, until a potential is reached at which an ion pump is activated. The ion pump might repolarize the membrane to its resting level, or even may hyperpolarize it, and stop the cell firing. Over a period of time, ion leakage or ion-linked membrane transport processes could again result in depolarization to the firing level, causing repetition of the cycle. In spontaneously firing neurons which show a continuous regular rhythm of discharge, the depolarizing influences and ion pumps may be more closely synchronized than in the oscillatory ones. The activity of such neurons also may be greatly modified by liberation of transmitters from neighboring neurons, affecting subsynaptic membranes in such a way as to result in

specific changes in the membrane conductances of ions. As a consequence the activities of the restorative and/or electrogenic ion pumps could be modified. Furthermore, membrane properties, and therefore activity, of such neurons may be altered by many other influences in the immediate extracellular environment which might affect their physical state, such as pH, concentrations of small charged molecules (organic and inorganic), hormones, availability of water, exogenously administered drugs, and temperature, for example.

REFERENCES

Albers, R. W., and Brady, R. O., 1959, The distribution of glutamic decarboxylase in the nervous system of the rhesus monkey, J. Biol. Chem. 234:926.

Aprison, M. H., Davidoff, R. A., and Werman, R., 1970, Glycine: its metabolic and possible transmitter roles in nervous tissue, Handbook of Neurochem. 3:381.

Barker, J. L., and Nicoll, R. A., 1972, Gamma-aminobutyric acid: role in primary afferent depolarization, Science 176:1043.

Bird, E. D., Mackay, A. V. P., Rayner, C. N., and Iversen, L. L., 1973, Reduced glutamic-acid-decarboxylase activity of post-mortem brain in Huntington's chorea, Lancet, May 19:1090,

Bloom, F. E., Hoffer, B. J., and Siggins, G. R., 1971, Studies on norepinephrine-containing afferents to Purkinje cells of rat cerebellum. II. Sensitivity of Purkinje cells to norepinephrine and related substances administered by microiontophoresis, Brain Research 25:523.

Bradford, H. F., 1970, Metabolic response of synaptosomes to electrical stimulation: release of amino acids, Brain Research 19:239.

Bradley, P. B., 1968, Synaptic transmission in the central nervous system and its relevance for drug action, Internat. Rev. Neurobiol. 11:1.

Curtis, D. R., and Crawford, J. M., 1969, Central synaptic transmission--microelectrophoretic studies, Ann. Rev. Pharmacol. 9:209.

Curtis, D. R., Duggan, A. W., Felix, D., and Johnston, G. A. R., 1970, GABA, bicuculline and central inhibition, Nature 226:1222.

Curtis, D. R., Felix, D., and McLennan, H., 1970, GABA and hippocampal inhibition, Br. J. Pharmacol. 40:881.

Davidoff, R. A., 1972, Penicillin and presynaptic inhibition in the amphibian spinal cord, Brain Research 36:218.

Davidson, N., and Southwick, C. A. P., 1971, Amino acids and presynaptic inhibition in the rat cuneate nucleus, J. Physiol. (Lond.) 219:689.

Dreifuss, J. J., Kelly, J. S., and Krnjevic, K., 1969, Cortical inhibition and γ-aminobutyric acid, Exp. Brain Research 9:137.

Fahn, S., and Coté, L. J., 1969, Regional distribution of γ-aminobutyric acid (GABA) in brain of the rhesus monkey, J. Neurochem. 15:209.

Feltz, P., 1971, γ-Aminobutyric acid and a caudato-nigral inhibition, Can. J. Physiol. Pharmacol. 49:1113.

Fonnum, F., and Storm-Mathisen, J., 1969, GABA synthesis in rat hippocampus correlated to the distribution of inhibitory neurones, Acta Physiol. Scand. 76:35A.

Forssberg, H., and Grillner, S., 1973, The locomotion of the acute spinal cat injected with clonidine i.v., Brain Research 50:184.

Fuxe, K., Hamberger, B., and Hökfelt, T., 1968, Distribution of noradrenaline nerve terminals in cortical areas of the rat, Brain Research 8:125.

Fuxe, K., Hökfelt, T., and Ungerstedt, U., 1970, Morphological and functional aspects of central monoamine neurons, Int. Rev. Neurobiol. 13:93.

Hockman, C. H., Lloyd, K. G., Farley, I. J., and Hornykiewicz, O., 1971, Experimental midbrain lesions: neurochemical comparison between the animal model and Parkinson's disease, Brain Research 35:613.

Hornykiewicz, O., 1973, Parkinson's disease: from brain homogenate to treatment, Fed. Proc. 32:183.

Ikeda, K., and Wiersma, C. A. G., 1964, Antogenic rhythmicity in the abdominal ganglia of the crayfish: the control of swimmeret movements, Comp. Biochem. Physiol. 12:107.

Jasper, H. H., Khan, R. T., and Elliott, K. A. C., 1965, Amino acids released from the cerebral cortex in relation to its state of activation, Science 147:1448.

Jouvet, M., 1969, Biogenic amines and states of sleep, Science 163:32.

Julien, R. M., and Halpern, L. M., 1971, Diphenylhydantoin: evidence for a central action, Life Sci. 10(Pt. 1):575.

Kanazawa, I., Miyata, Y., Toyokura, Y., and Otsuka, M., 1973, The distribution of γ-aminobutyric acid (GABA) in the human substantia nigra, Brain Research 51:363.

Kety, S. S., and Matthysse, S. (eds.), 1972, Prospects for research on schizophrenia. A report based on an NRP work session. Neurosci. Res. Prog. Bull. 10: pp.370-507.

Kim, J. S., Bak, I. J., Hassler, R., and Okada, Y., 1971, Role of γ-aminobutyric acid (GABA) in the extrapyramidal motor system. 2. Some evidence for the existence of a type of GABA-rich strio-nigral neurons, Exp. Brain Res. 14:95.

Klawans, H. L., 1972, Pathophysiology of schizophrenia and the striatum, Diseases of the Nervous System 33:711.

Krnjevic, J., 1970, Dopamine, acetylcholine and excitatory amino acids in nigrostriatal transmission, in "L-Dopa and Parkinsonism" (A. Barbeau and F. H. McDowell, eds.), pp. 189-191, F. A. Davis Co., Philadelphia.

Krnjevic, K., and Phillis, J. W., 1963, Actions of certain amines on cerebral cortical neurones, Br. J. Pharmac. Chemother. 20:471.

Kuriyama, K., Sisken, B., Haber, B., and Roberts, E., 1968, The gamma-aminobutyric system in rabbit retina, Brain Research 9:165.

Marcus, R. J., Winters, W. D., Roberts, E., and Simonsen, D. G., 1971, Neuropharmacological studies of imidazole-4-acetic acid actions in the mouse and rat, Neuropharmacology 10:203.

Matthysse, S., 1973, Antipsychotic drug actions: a clue to the neuropathology of schizophrenia? Fed. Proc. 32:200.

Maynard, D. M., 1972, Simpler networks, Ann. N. Y. Acad. Sci. 193:59.

McGeer, E. G., McGeer, P. L., and McLennan, H., 1961, The inhibitory action of 3-hydroxytyramine, gamma-aminobutyrate acid (GABA) and some other compounds towards the crayfish stretch receptor neuron, J. Neurochem. 8:36.

McGeer, P. L., McGeer, E. G., Fibiger, H. C., and Wickson, V., 1971, Neostriatal choline acetylase and cholinesterase following selective brain lesions, Brain Research 35:308.

McGeer, P. L., McGeer, E. G., Wada, J. A., and Jung, E., 1971, Effects of globus pallidus lesions and Parkinson's disease on brain glutamic acid decarboxylase, Brain Research 32:425.

Meehl, P. E., 1962, Schizotaxia, schizotypy, schizophrenia, Am. Psychol. 17:827.

Meehl, P. E., 1964, Manual for Use with Checklist of Schizotypic Signs, Minneapolis: University of Minnesota Medical School, p. 29.

Mitchell, J. F., and Srinivasan, V., 1969, Release of ^{3}H-γ-amino-butyric acid from the brain during synaptic inhibition, Nature 224:663.

Miyata, Y., and Otsuka, M., 1972, Distribution of γ-aminobutyric acid in cat spinal cord and the alteration produced by local ischaemia, J. Neurochem. 19:1833.

Nauta, W. J. H., 1972, The central visceromotor system: a general survey, in "Limbic System Mechanisms and Autonomic Function," (C. H. Hockman, ed.) pp. 21-38, Charles C. Thomas, Springfield, Illinois.

Nicoll, R. A., 1971, Pharmacological evidence for GABA as the transmitter in granule cell inhibition in the olfactory bulb, Brain Research 35:137.

Obata, K., and Highstein, S. M., 1970, Blocking by picrotoxin of both vestibular inhibition and GABA action on rabbit oculomotor neurones, Brain Research 18:538.

Obata, K., and Takeda, K., 1969, Release of γ-aminobutyric acid into the fourth ventricle induced by stimulation of the cat's cerebellum, J. Neurochem. 16:1043.

Obata, K., Ito, M., Ochi, R., and Sato, N., 1967, Pharmacological properties of the postsynaptic inhibition by Purkinje cell axons and the action of γ-aminobutyric acid on Deiters' neurones, Exp. Brain Res. 4:43.

Okada, Y., Nitsch-Hassler, C., Kim, J. S., Bak, I. J., and Hassler, R., 1971, Role of γ-aminobutyric acid (GABA) in the extrapyramidal motor system, Exp. Brain Res. 13:514.

Olson, L., and Fuxe, K., 1971, On the projections from the locus coeruleus noradrenaline neurons: the cerebellar innervation, Brain Research 28:165.

Otsuka, M., Iversen, L. L., Hall, Z. W., and Kravitz, E. A., 1966, Release of gamma-aminobutyric acid from inhibitory nerves of lobster, Proc. Nat. Acad. Sci. (U.S.A.) 56:1110.

Otsuka, M., Konishi, S., and Takahashi, T., 1972, A further study of the motoneuron-depolarizing peptide extracted from dorsal roots of bovine spinal nerves, Proc. Jap. Acad. 48:747.

Perry, T. L., Berry, K., Hansen, S., Diamond, S., and Mok, C., 1971a, Regional distribution of amino acids in human brain obtained at autopsy, J. Neurochem. 18:513.

Perry, T. L., Hansen, S., Berry, K., Mok, C., and Lesk, D., 1971b, Free amino acids and related compounds in biopsies of human brain, J. Neurochem. 18:521.

Perry, T. L., Hansen, S., and Kloster, M., 1973, Huntington's chorea. Deficiency of γ-aminobutyric acid in brain, New England J. Med. 288:337.

Pollack, E. D., and Crain, S. M., 1972, Development of motility in fish embryos in relation to release from early CNS inhibition, J. Neurobiol. 3:381.

Precht, W., and Yoshida, M., 1971, Blockage of caudate-evoked inhibition of neurons in the substantia nigra by picrotoxin, Brain Research 32:229.

Randrup, A., and Munkvad, I., 1970, Biochemical, anatomical and psychological investigations of stereotyped behavior induced by amphetamines, in "International Symposium on Amphetamines and Related Compounds" (E. Costa and S. Garattini, eds.) pp. 279-332, Academic Press, New York.

Roberts, E., 1966a, The synapse as a biochemical self-organizing microcybernetic unit, Brain Research 1:117.

Roberts, E., 1966b, Models for correlative thinking about brain, behavior, and biochemistry, Brain Research 2:109.

Roberts, E., 1972, An hypothesis suggesting that there is a defect in the GABA system in schizophrenia, Neurosci. Res. Pro. Bull. 10:468.

Roberts, E., and Frankel, S., 1950, γ-Aminobutyric acid in brain, Fed. Proc. 9:219.

Roberts, E., and Hammerschlag, R., 1972, Amino acid transmitters, in "Basic Neurochemistry" (R. W. Albers, G. J. Siegel, R. Katzman, and B. W. Agranoff, eds.) pp. 131-165, Little, Brown and Company, Boston.

Roberts, E., and Kuriyama, K., 1968, Biochemical-physiological correlations in studies of the γ-aminobutyric acid system, Brain Research 8:1.

Roberts, E., and Matthysse, S., 1970, Neurochemistry at the crossroads of neurobiology, Ann. Rev. Biochem. 39:777.

Roberts, E., and Simonsen, D. G., 1966, A hypnotic and possible analgesic effect of imidazoleacetic acid in mice. Biochem. Pharmacol. 15:1875.

Roberts, E., and Simonsen, D. G., 1970, Some properties of cyclic 3',5'-nucleotide phosphodiesterase of mouse brain: effects of imidazole-4-acetic acid, chlorpromazine, cyclic 3',5'-GMP, and other substances, Brain Research 24:91.

Roberts, E., Wein, J., and Simonsen, D. G., 1964, γ-Aminobutyric acid (GABA), vitamin B_6, and neuronal function--a speculative synthesis, Vitamins and Hormones 22:503.

Roeder, K. D., Tozian, L., and Weiant, E. A., 1960, Endogenous nerve activity and behavior in the mantis and cockroach. Insect Physiol. 4:45.

Saito, K., Wu, J.-Y., and Roberts, E., 1973, Immunochemical comparisons of vertebrate glutamic acid decarboxylase, Brain Research, In Press.

Salvador, R. A., and Albers, R. W., 1959, The distribution of glutamic-γ-aminobutyrate transaminase in the nervous system of the rhesus monkey, J. Biol. Chem. 234:922.

Schousboe, A., Wu, J.-Y., and Roberts, E., 1973, Purification and characterization of the 4-aminobutyrate-2-ketoglutarate transaminase from mouse brain, Biochemistry, In Press.

Srinivasan, V., Neal, M. J., and Mitchell, J. F., 1969, The effect of electrical stimulation and high potassium concentrations on the efflux of [^{3}H]γ-aminobutyric acid from brain slices, J. Neurochem. 16:1235.

Takeuchi, A., and Takeuchi, N., 1966, On the permeability of the presynaptic terminal of the crayfish neuromuscular junction during synaptic inhibition and the action of γ-aminobutyric acid, J. Physiol. (Lond.) 183:433.

Tunnicliff, G., Wein, J., and Roberts, E., 1972, Effects of imidazole-acetic acid on brain amino acids and body temperature in mice, J. Neurochem. 19:2017.

van Balgooy, J. N. A., Marshall, F. D., and Roberts, E., 1972, Metabolism of intracerebrally administered histidine, histamine, and imidazoleacetic acid in mice and frogs, J. Neurochem. 19:2341.

Velikaya, R. R., and Sycheva, T. M., 1970, Role of the reticular formation of the brainstem in background activity of cortical neurons, Neirofiziologiya 2:43. In English in Neuroscience Transaltions 15:11, 1970-71.

Wiersma, C. A. G., and Ikeda, K., 1964, Interneurons commanding swimmeret movements in the crayfish, Procambarns clarki (Girard), Comp. Biochem. Physiol. 12:509.

Woodward, D. J., Rushmer, D., Hoffer, B. J., Siggins, G. R., Oliver, A. P., and Armstrong, C., 1971, Evidence for the presence of stellate cell inhibition in frog cerebellum and for mediation of this inhibition by gamma-aminobutyric acid, Fed. Proc. 30:318.

Wu, J.-Y., and Roberts, E., 1973, Comparative studies of L-glutamate decarboxylases from mouse brain and kidney. Trans. Am. Soc. Neurochem. 4:70.

Wu, J.-Y., Matsuda, T., and Roberts, E., 1973, Purification and characterization of glutamate decarboxylase from mouse brain, J. Biol. Chem. 248:3029.

Yoshida, M., and Precht, W., 1971, Monosynaptic inhibition of neurons of the substantia nigra by caudato-nigral fibers, Brain Research 32:225.

ABSTRACT:

All normal or adaptive activity in nervous systems is a result of the coordinated dynamic interplay of excitation and inhibition, within and between neuronal subsystems. A basic inadequacy in this interplay could lead to an abnormal (disease) process such as is found in Parkinson's disease, Huntington's chorea, schizophrenia, epilepsy, depression, manic-depressive disorder, hyperkinetic behavior in children, psychosomatic disorders, etc.

In behavioral sequences, innate or learned, genetically preprogrammed circuits may be released to function at varying rates and in various combinations by inhibition of neurons which are tonically holding in check pacemaker cells with capacity for spontaneous activity. If the pacemaker neurons are triggering a circuit related to the regulation of a vital function such as heart action or respiration, the inhibitory neurons might act in such a way as to vary the rate of the discharge of the pacemaker neurons. On the other hand, if a behavioral sequence involves the voluntary movement of a limb muscle, the pacemaker neuron might be held in complete check by the tonic action of inhibitory neurons and might be allowed to discharge in a graded manner only related to demand. In this view, excitatory input to pacemaker neurons would have largely a modulatory role. Thus, disinhibition appears to be one of the major organizing principles in nervous system function.

Biochemical, pharmacological, and physiological data currently available about nervous system function suggested that a major neural system exerting tonic inhibition on pacesetter neurons could be the GABA system. GABA neurons are present ubiquitously in the CNS of vertebrate species, and on a quantitative basis GABA is much more extensively and relatively more evenly distributed throughout the various brain regions than the neuronal systems that employ other known neural transmitters such as acetylcholine, the catecholamines, or serotonin. The possible relations of the transmitter systems utilizing the latter transmitters to the GABA system will be discussed from the point of view of normal and abnormal nervous system function.

LIST OF CONTRIBUTORS

ALEMAN, V., Sección de Neurobiología, Centro de Investigación y de Estudios Avanzados del IPN, México, D.F., Mexico.

ALVAREZ-LEEFMANS, J., Physiology Department, Faculty of Medicine, National University of Mexico, México 20, D.F., Mexico.

ANTÓN-TAY, F., Investigador, Instituto de Investigaciones Biomédicas, Universidad Nacional Autónoma de México, Ciudad Universitaria, México, D.F., Mexico. (Session Chairman)

BAYON, A., Sección de Neurobiología, Centro de Investigación y de Estudios Avanzados del IPN, México, D.F., Mexico.

BLOOM, F. E., Laboratory of Neuropharmacology, DSMHR, NIMH, St. Elizabeths Hospital, Washington, D. C. 20032.

BROUGHTON, R., Faculties of Medicine and Psychology, University of Ottawa, Ottawa, Canada.

BRUST-CARMONA, H., Physiology Department, Faculty of Medicine, National University of Mexico, México 20, D.F., Mexico.

DRUCKER-COLÍN, R. R., Instituto Miles de Terapéutica Experimental Apdo. Postal 22026, México 22, D.F., Mexico.

FERIA V., A., División de Investigación, Instituto Mexicano del Seguro Social, México, D.F., Mexico. (Session Chairman)

FERNÁNDEZ-GUARDIOLA, A., Instituto de Investigaciones Biomédicas, Ciudad Universitaria, UNAM, México 20, D.F., Mexico.

FIBIGER, H. C., Kinsmen Laboratory of Neurological Research, Department of Psychiatry, University of British Columbia, Vancouver, Canada.

GOLD, P. E., Department of Psychobiology, University of California, Irvine, California.

GÓMEZ-PUYOU, A., Profesor de Bioquímica, Facultad de Medicina, Universidad Nacional Autónoma de México, Ciudad Universitaria, México, D.F., Mexico. (Session Chairman)

GRINBERG-ZYLBERBAUM, J., Physiology Department, Faculty of Medicine, National University of Mexico, México 20, D.F., Mexico.

GRINSTEIN, S., Department of Physiology, Escuela Nacional de Ciencias Biológicas, Instituto Politécnico Nacional, México D.F., Mexico.

HALL, G., Tobacco Research Council, Yorkshire, England. (Session Chairman)

HATTORI, T., Kinsmen Laboratory of Neurological Research, Department of Psychiatry, University of British Columbia, Vancouver, Canada.

HIMWICH, H. E., Thudichum Psychiatric Research Laboratory, Galesburg State Research Hospital, Galesburg, Illinois 61401.

IZQUIERDO, I., Departamento de Fisiologia, Farmacologia e Biofisica, Instituto de Biociencias, Universidade Federal de Rio Grande do Sul (UFRGS), Porto Alegre, RS, Brazil.

KARCZMAR, A. G., Department of Pharmacology and Institute for Mind, Drugs and Behavior, Loyola University Medical Center, Maywood, Illinois 60153.

MALER, L., Kinsmen Laboratory of Neurological Research, Department of Psychiatry, University of British Columbia, Vancouver, Canada.

MCGAUGH, J. L., Department of Psychobiology, University of California, Irvine, California.

MCGEER, E. G., Kinsmen Laboratory of Neurological Research, Department of Psychiatry, University of British Columbia, Vancouver, Canada.

MCGEER, P. L., Kinsmen Laboratory of Neurological Research, Department of Psychiatry, University of British Columbia, Vancouver, Canada.

MOLINA, J., Sección de Neurobiología, Centro de Investigación y de Estudios Avanzados del IPN, México D.F., Mexico.

MONNIER, M., Physiological Institute, University of Basel, Vesalgasse 1, CH-4051 Basel, Switzerland.

MORGANE, P. J., Laboratory of Neurophysiology, The Worcester Foundation for Experimental Biology, Shrewsbury, Massachusetts.

MOSNAIM, A. D., University of Health Sciences, The Chicago Medical Schook, 2020 West Ogden Avenue, Chicago, Illinois 60612.

MYERS, R. D., Laboratory of Neuropsychology, Purdue University, Lafayette, Indiana 47907.

NARASIMHACHARI, N., Thudichum Psychiatric Research Laboratory, Galesburg State Research Hospital, Galesburg, Illinois 61401.

PHILLIS, J. W., Department of Physiology, College of Medicine, University of Saskatchewan, Saskatoon, Canada.

PRADO-ALCALÁ, R., Physiology Department, Faculty of Medicine, National University of Mexico, México 20, D.F., Mexico.

RADULOVAČKI, M., Department of Pharmacology, College of Medicine, University of Illinois, Chicago, Illinois.

REIS, D. J., Laboratory of Neurobiology, Department of Neurology, Cornell University Medical College, 1300 York Avenue, New York, New York 10021.

ROBERTS, E., Division of Neurosciences, City of Hope National Medical Center, Duarte, California 91010.

RUSSEK, M., Department of Physiology, Escuela Nacional de Ciencias Biológicas, Instituto Politécnico Nacional, México D.F., Mexico.

SABELLI, H. C., University of Health Sciences, The Chicago Medical School, 2020 West Ogden Avenue, Chicago, Illinois 60612.

SCHOENENBERGER, G. A., Biol. chem. Res. Div., Department of Surgery, University of Basel, Basel, Switzerland.

SINGH, V. K., Kinsmen Laboratory of Neurological Research, Department of Psychiatry, University of British Columbia, Vancouver, Canada.

STERN, W. C., Laboratory of Neurophysiology, The Worcester Foundation for Experimental Biology, Shrewsbury, Massachusetts.

TAPIA, R., Instituto de Biología, Universidad Nacional Autónoma de México, México 20, D.F., Mexico.

VAZQUEZ, A. J., University of Health Sciences, The Chicago Medical School, 2020 West Ogden Avenue, Chicago, Illinois 60612.

ZARCO CORONADO, I., Physiology Department, Faculty of Medicine, National University of Mexico, México 20, D.F., Mexico.

AUTHOR INDEX

SUBJECT INDEX